Markus Sauer, Johan Hofkens,
and Jörg Enderlein

Handbook of Fluorescence Spectroscopy and Imaging

Related Titles

Krämer, R. / Krämer, C.

Fluorescent Molecular Labels and Probes

2011

ISBN: 978-3-527-32720-1

Geddes, Chris D.

Metal-Enhanced Fluorescence

2010

ISBN-13: 978-0-470-22838-8

Goldys, Ewa M.

Fluorescence Applications in Biotechnology and Life Sciences

2009

ISBN-13: 978-0-470-08370-3 -

Wolfbeis, Otto S. (ed.)

Fluorescence Methods and Applications
Spectroscopy, Imaging, and Probes
Annals of the New York Academy of Sciences

2008

ISBN-13: 978-1-57331-716-0

Albani, Jihad Rene

Principles and Applications of Fluorescence Spectroscopy

2007

ISBN-13: 978-1-4051-3891-8

Vogel, W., Welsch, D.-G.

Quantum Optics

2006

ISBN: 978-3-527-40507-7

Valeur, B.

Molecular Fluorescence
Principles and Applications

2002

ISBN: 978-3-527-29919-5

Markus Sauer, Johan Hofkens, and Jörg Enderlein

Handbook of Fluorescence Spectroscopy and Imaging

From Single Molecules to Ensembles

WILEY-VCH Verlag GmbH & Co. KGaA

The Authors

Prof. Dr. Markus Sauer
Julius-Maximilians- University
Biotechnology & Biophysics
Am Hubland
97074 Würzburg
Germany

Prof. Dr. Johan Hofkens
Katholieke Univ. Leuven
Department of Chemistry
Celestijnenlaan 200F
3001 Heverlee
Belgium

Dr. Jörg Enderlein
Georg-August University
Institute of Physics
Friedrich-Hund-Platz 1
37077 Göttingen
Germany

Library of Congress Card No.: applied for

British Library Cataloguing-in-Publication Data
A catalogue record for this book is available from the British Library.

Bibliographic information published by the Deutsche Nationalbibliothek
The Deutsche Nationalbibliothek lists this publication in the Deutsche Nationalbibliografie; detailed bibliographic data are available on the Internet at http://dnb.d-nb.de.

Composition Thomson Digital, Noida, India
Printing and Binding Fabulous Printers Pte. Ltd., Singapore
Cover Design Adam Design, Weinheim

Printed in Singapore
Printed on acid-free paper

ISBN: 978-3-527-31669-4

Contents

Handbook of Fluorescence Spectroscopy and Imaging. M. Sauer, J. Hofkens, and J. Enderlein

ISBN: 978-3-527-31669-4

Preface

Fluorescence spectroscopy and imaging have become powerful and widely used methods in almost any laboratory around the globe for the non-invasive study of polymers, inorganic materials, cells, and tissues. With the development of elaborate single-molecule fluorescence techniques, energy and electron transfer methods have experienced a renaissance and today they are used successfully to study protein folding, molecular interactions, and the motion and interaction of individual molecules, even in living cells with high temporal resolution. New methods that have been introduced to increase the optical resolution of microscopy beyond the diffraction barrier enable researchers to study cellular structures in fixed and living cells with unprecedented resolution, which seemed impossible to achieve only a few years ago. With the ongoing success of fluorescence spectroscopy and microscopy, fundamental knowledge about the chemical and photophysical properties of fluorophores lie at the heart of single-molecule sensitive experiments.

This book is intended to give interested readers the fundamental knowledge necessary for planning and designing successful fluorescence spectroscopy and imaging experiments with high spatiotemporal resolution. After a basic introduction to fluorescence spectroscopy, different fluorophores and their properties, we describe the details of specific fluorescence labelling of target molecules. A chapter that explains our current understanding of fluorophore photophysics and photobleaching pathways in single-molecule fluorescence experiments is followed by chapters giving introductions to fluorescence correlation spectroscopy and also to energy and electron transfer. Finally, the book introduces various super-resolution imaging methods and gives examples of how single-molecule spectroscopy can be used for the successful study of enzyme kinetics, conformational dynamics of biopolymers, protein folding, and for diagnostic applications.

The book is mainly aimed at graduate students and researchers who want to begin using the advanced fluorescence techniques of single-molecule fluorescence spectroscopy and imaging. However, we hope that even experts will find it a useful resource to consult for the planning and discussion of their experiments.

Würzburg, Leuven, Göttingen, November 2010

Markus Sauer
Johan Hofkens
Jörg Enderlein

Handbook of Fluorescence Spectroscopy and Imaging. M. Sauer, J. Hofkens, and J. Enderlein

ISBN: 978-3-527-31669-4

1
Basic Principles of Fluorescence Spectroscopy

1.1
Absorption and Emission of Light

As fluorophores play the central role in fluorescence spectroscopy and imaging we will start with an investigation of their manifold interactions with light. A fluorophore is a component that causes a molecule to absorb energy of a specific wavelength and then re-remit energy at a different but equally specific wavelength. The amount and wavelength of the emitted energy depend on both the fluorophore and the chemical environment of the fluorophore. Fluorophores are also denoted as chromophores, historically speaking the part or moiety of a molecule responsible for its color. In addition, the denotation chromophore implies that the molecule absorbs light while fluorophore means that the molecule, likewise,emits light. The umbrella term used in light emission is luminescence, whereas fluorescence denotes allowed transitions with a lifetime in the nanosecond range from higher to lower excited singlet states of molecules.

In the following we will try to understand why some compounds are colored and others are not. Therefore, we will take a closer look at the relationship of conjugation to color with fluorescence emission, and investigate the absorption of light at different wavelengths in and near the visible part of the spectrum of various compounds. For example, organic compounds (i.e., hydrocarbons and derivatives) without double or triple bonds absorb light at wavelengths below 160 nm, corresponding to a photon energy of >180 kcal mol^{-1} (1 cal = 4.184 J), or >7.8 eV (Figure 1.1), that is, significantly higher than the dissociation energy of common carbon-to-carbon single bonds.

Below a wavelength of 200 nm the energy of a single photon is sufficient to ionize molecules. Therefore, photochemical decomposition is most likely to occur when unsaturated compounds, where all bonds are formed by σ-electrons, are irradiated with photon energies >6.2 eV. Double and triple bonds also use π-electrons in addition to a σ-bond for bonding. In contrast to σ-electrons, which are characterized by the rotational symmetry of their wavefunction with respect to the bond direction, π-electrons are characterized by a wavefunction having a node at the nucleus and rotational symmetry along a line through the nucleus. π-bonds

Handbook of Fluorescence Spectroscopy and Imaging. M. Sauer, J. Hofkens, and J. Enderlein

ISBN: 978-3-527-31669-4

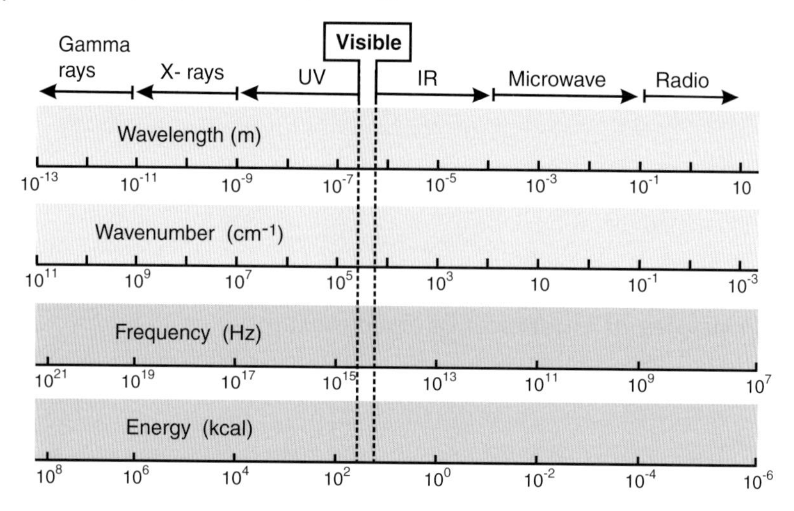

Figure 1.1 The electromagnetic spectrum.

are usually weaker than σ-bonds because their (negatively charged) electron density is further from the positive charge of the nucleus, which requires more energy. From the perspective of quantum mechanics, this bond weakness is explained by significantly less overlap between the component π-orbitals due to their parallel orientation. These less strongly bound electrons can be excited by photons with lower energy. If two double bonds are separated by a single bond, the double bonds are termed conjugated. Conjugation of double bonds further induces a red-shift in the absorption (a so-called bathochromic shift). All fluorophores that have a high absorption in the visible part of the spectrum possess several conjugated double bonds.

Above 200 nm only the two lowest energy transitions, that is, $n \rightarrow \pi^*$ and $\pi \rightarrow \pi^*$, are achieved as a result of the energy available from the photons. When sample molecules are exposed to light having an energy that matches a possible electronic transition within the molecule, some of the light energy will be absorbed as the electron is promoted to a higher energy orbital. As a simple rule, energetically favored electron promotion will be from the highest occupied molecular orbital (HOMO), usually the singlet ground state, S_0, to the lowest unoccupied molecular orbital (LUMO), and the resulting species is called the singlet excited stateS_1. Absorption bands in the visible region of the spectrum correspond to transitions from the ground state of a molecule to an excited state that is 40–80 kcal mol^{-1} above the ground state. As mentioned previously, in saturated hydrocarbons in particular, the lowest electronic states are more than 80 kcal mol^{-1} above the ground state, and therefore they do not absorb light in the visible region of spectrum. Such substances are not colored. Compounds that absorb in the visible region of the spectrum (these compounds have color) generally have some weakly bound or delocalized electrons. In these systems, the energy difference between the lowest LUMO and the HOMO corresponds to the energies of quanta in the visible region.

On the other side of the electromagnetic spectrum, there is a natural limit to long-wavelength absorption and emission of fluorophores, which is in the region of 1 μm [1]. A dye absorbing in the near-infrared (>700 nm) has a low-lying excited singlet state and even slightly lower than that, a metastable triplet state, that is, a state with two unpaired electrons that exhibits biradical character. Even though no generally valid rule can be formulated predicting the thermal and photochemical stability of fluorophores, the occupation of low-lying excited singlet and triplet states potentially increases the reactivity of fluorophores. Therefore, it is likely that fluorophores with long-wavelength absorption and emission will show less thermal and photochemical stability, due to reactions with solvent molecules such as dissolved oxygen, impurities, and other fluorophores. In addition, with increasing absorption, that is, with a decreasing energy difference between S_1 and S_0, the fluorescence intensity of fluorophores decreases owing to increased internal conversion. That is, with a decreasing energy difference between the excited and ground state, the number of options to get rid of the excited-state energy by radiationless deactivation increases. Hence, most known stable and bright fluorophores absorb and emit in the wavelength range between 300 and 700 nm.

Fluorophores with conjugated doubled bonds (polymethine dyes) are essentially planar, with all atoms of the conjugated chain lying in a common plane linked by σ-bonds. π-electrons, on the other hand, have a node in the plane of the molecule and form a charge cloud above and below this plane along the conjugated chain (Figure 1.2). The visible bands for polymethine dyes arise from electronic transitions involving the π-electrons along the polymethine chain. The wavelength of these bands depends on the spacing of the electronic levels. The absorption of light by fluorophores such as polymethine dyes can be understood semiquantitatively by applying the free-electron model proposed by Kuhn [2, 3]. The arrangement of alternating single–double bonds in an organic molecule usually implies that the π-electrons are delocalized over the framework of the "conjugated" system. As these π-electrons are mobile throughout the carbon atom skeleton containing the alternating double bonds, a very simple theoretical model can be applied to such a system in order to account for the energy of these electrons in the molecule. If one makes the

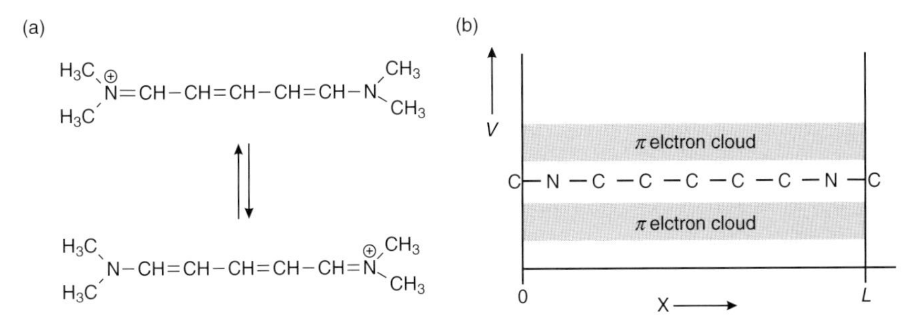

Figure 1.2 (a) Limiting structures of a resonance hybrid of a simple positively charged cyanine dye. (b) The π-electron cloud of the cyanine dye as seen from the side in a simplified potential energy (V) trough of length L.

seemingly drastic assumption that the several π-electrons that comprise the system are non-interacting (presumably, if the π-electrons are delocalized over the $-C{=}C{-}C{=}C{-}C{=}C-$ framework, they spread out, minimizing repulsion between them), then one can view the energetics of this system as arising from the simple quantum mechanical assembly of one-electron energy levels appropriate to the *particle in the box model.* In this case, one considers the potential energy of the electron as being constant throughout the length of the molecular box and then rising to infinity at each end of the conjugated portion of the molecule. As an example, consider a positively charged simple cyanine dye. The cation can "resonate" between the two limiting structures shown in Figure 1.2a, that is, the wavefunction for the ion has equal contributions from both states. Thus, all the bonds along this chain can be considered equivalent, with a bond order of 1.5, similar to the C–C bonds in benzene.

Assuming that the conjugated chain extends approximately one bond length to the left and right beyond the terminal nitrogen atoms, application of the Schroedinger equation to this problem results in the well known expressions for the wavefunctions and energies, namely:

$$\psi_n = \sqrt{\frac{2}{L}}\sin\left(\frac{n\pi x}{L}\right)$$

and

$$E_n = \frac{n^2\,h^2}{8\,mL^2}$$

where

n is the quantum number ($n = 1, 2, 3, \ldots$) giving the number of antinodes of the eigenfunction along the chain
L is the "length" of the (one dimensional) molecular box
m is the mass of the particle (electron)
h is Planck's constant
x is the spatial variable, which is the displacement along the molecular backbone.

Each wavefunction can be referred to as a molecular orbital, and its respective energy is the orbital energy. If the spin properties of the electron are taken into account along with the *ad hoc* invocation of Pauli's exclusion principle, the model is then refined to include spin quantum numbers for the electron (½) along with the restriction that no more than two electrons can occupy a given wavefunction or level, and the spin quantum numbers of the two electrons occupying a given energy level are opposite (spin up and spin down). Thus, if we have N electrons, the lower states are filled with two electrons each, while all higher states are empty provided that N is an even number (which is usually the case in stable molecules as only highly reactive radicals posses an unpaired electron). This allows the electronic structure for the π-electrons in a conjugated dye molecule to be constructed. For example, for the conjugated molecule $CH_2{=}CH{-}CH{=}CH{-}CH{=}CH_2$ 6 π-electrons have to be considered. The lowest energy configuration, termed the electronic ground state, corresponds to the six electrons being in the lowest three orbitals. Higher energy configurations are constructed by promoting an electron from the HOMO with quantum number $n=3$ to the LUMO with $n=4$. This higher energy arrangement is called the electronically excited singlet state. The longest wavelength absorption band corresponds to the energy difference between these two states, which is then given by the following expression:

$$\Delta E = E_{\mathrm{LUMO}} - E_{\mathrm{HOMO}} = \frac{h^2}{8\,mL^2}\left(n_{\mathrm{LUMO}}^2 - n_{\mathrm{HOMO}}^2\right)$$

The energy required for this electronic transition can be supplied by a photon of the appropriate frequency, given by the Planck relationship:

$$E = h\,\nu = hc/\lambda$$

where

h is Planck's constant
ν is the frequency
c is the speed of light
λ is the wavelength.

Because the ground state of a molecule with N π-electrons will have $N/2$ lowest levels filled and all higher levels empty, we can write $n_{\mathrm{LUMO}} = N/2 + 1$ and $n_{\mathrm{HOMO}} = N/2$:

$$\Delta E = \frac{h^2}{8\,mL^2}(N+1) \quad \text{or} \quad \lambda = \frac{8\,mc}{h}\frac{L^2}{N+1}$$

This indicates that to a first approximation the position of the absorption band is determined only by the chain length and the number of delocalized π-electrons. Good examples for this relationship are symmetrical cyanine dyes of the general formula shown in Figure 1.2a.

1.2
Spectroscopic Transition Strengths

The probability of a molecule changing its state by absorption or emission of a photon depends on the nature of the wavefunctions of the initial and final states, how strongly light can interact with them, and on the intensity of any incident light. The probability of a transition occurring is commonly described by the transition strength. To a first approximation, transition strengths are governed by selection rules that determine whether a transition is allowed or disallowed. In the classical theory of light absorption, matter consists of an array of charges that can be set into motion by the oscillating electromagnetic field of the light. Here, the electric dipole oscillators set in motion by the light field have specific natural characteristics, that is, frequencies, ν_i, that depend on the material. When the frequency of the radiation is near the oscillator frequency, absorption occurs, and the intensity of the radiation decreases on passing through the substance. The intensity of the interaction is known as the oscillator strength, f_i, and it can be thought of as characterizing the number of electrons per molecule that oscillate with the characteristic frequency, ν_i. Therefore, practical measurements of the transition strength are usually described in terms of f_i. The oscillator strength of a transition is a dimensionless number that is useful for comparing different transitions. For example, a transition that is fully allowed quantum mechanically is said to have an oscillator strength of 1.0. Experimentally, the oscillator strength, f, is related to the intensity of absorption, that is, to the area under an absorption band plotted versus the frequency:

$$f = \frac{2303\ mc}{\pi\ N_A\ e^2\ n} \int \varepsilon(\nu) d\nu$$

where

ε is the molar absorptivity
c is the velocity of light
m is the mass of an electron
e is the electron charge
n is the refractive index of the medium,
N_A is Avogadro's number.

The integration is carried out over the frequency range associated with the absorption band.

The quantum mechanical description, which is the most satisfactory and complete description of the absorption of radiation by matter, is based on time-dependent wave mechanics. Here, a transition from one state to another occurs when the radiation field connects the two states. In wave mechanics, the connection is described by the transition dipole moment, μ_{GE}:

$$\mu_{GE} = \int \psi_G \mu \psi_E dv$$

where

ψ_G and ψ_E are the wavefunctions for the ground and excited state, respectively
dv represents the volume element.

The transition dipole moment will be nonzero whenever the symmetry of the ground and excited states differ. For example, ethylene ($CH_2{=}CH_2$) has no permanent dipole moment, but if ψ_G is a π-molecular orbital and ψ_E is a π^*-molecular orbital, then $\mu_{\pi\pi^*}$ is not zero. The direction of the transition moment is characterized by the vector components: $\langle\mu_x\rangle_{GE}$, $\left\langle\mu_y\right\rangle_{GE}$, and $\langle\mu_z\rangle_{GE}$. It has to be pointed out that for most transitions the three vectors are not all equal, that is, the electronic transition is polarized. The ethylene $\pi \rightarrow \pi^*$ transition, for example, is polarized along the C=C double bond. The magnitude of the transition is characterized by its absolute-value squared, which is called the dipole strength, D_{GE}:

$$D_{GE} = |\mu_{GE}|^2$$

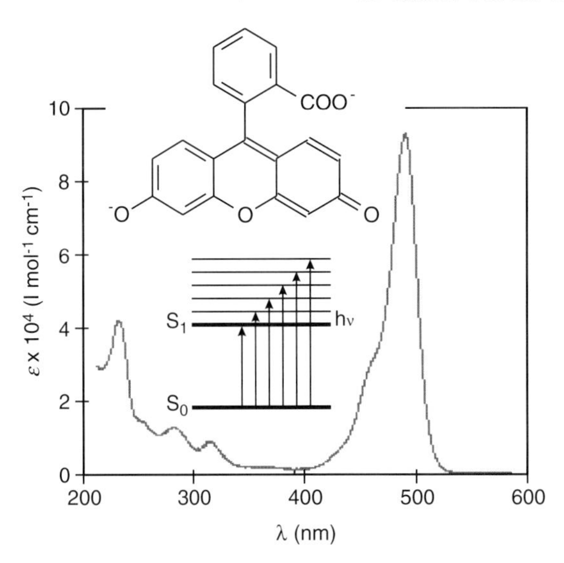

Figure 1.3 Molecular structure and absorption spectrum of the dianion fluorescein in ethanol, pH 9.0. The inset also shows the electronic ground and first excited singlet state level and possible absorptive transitions involving different vibronic states.

A peculiarity of the absorption spectra of organic dyes as opposed to atomic spectra is the width of the absorption band, which usually covers several tens of nanometers. This is easy to understand recalling that a typical dye molecule is composed of several tens of atoms, giving rise to manifold vibrations of the skeleton. These vibrations together with their overtones densely cover the spectrum between a few wave numbers and $3000\,cm^{-1}$. Furthermore, most of these vibrations are coupled to the electronic transitions through the change in electron densities over the bonds constituting the conjugated chain. That is, after electronic excitation the electron density changes, which is associated with a change in bond length. Quantum mechanically this means that transitions have occurred from the electronic and vibrational ground state S_0 of the molecule to an electronically and vibrationally excited state S_1. This results in broad absorption spectra like that shown for the fluorescein dianion in Figure 1.3 and depends on how many of the vibrational sublevels spaced at $h\nu\ (n + ½)$, with $n = 0, 1, 2, 3, \ldots$, are reached and what the transitions moments of these sublevels are.

1.3
Lambert–Beer Law and Absorption Spectroscopy

Lambert–Beer Law is a mathematical means of expressing how light is absorbed by matter (liquid solution, solid, or gas). The law states that the amount of light emerging from a sample is diminished by three physical phenomena: (i) the amount of absorbing material (concentration c), (ii) the optical path length l, that is, the distance the light must travel through the sample, and (iii) the probability that the photon of that particular energy will be absorbed by the sample (the extinction coefficient ε of the substance). Considering a sample of an absorbing substance placed between two parallel windows that transmit the light and supposing that the light of intensity I_0 is incident from the left, propagates along the x direction, then the intensity I decreases smoothly from left to right and exits with an intensity I_t. If the sample is homogeneous, the fractional decrease in light intensity is the same across a small interval dx, regardless of the value of x. As the fractional decrease for a solution depends linearly on the concentration of the absorbing molecule, the fractional change in light intensity dI/I can be written as:

$$-\frac{dI}{I} = \alpha c\, dx$$

where

α is a constant of proportionality.

Because neither α nor c depends on x, integration between limits I_0 at $x = 0$ and I_t at $x = l$, provides

$$\ln \frac{I_0}{I_t} = \alpha\, cl \quad \text{or} \quad I_t = I_0\, e^{-\alpha cl}$$

For measurements made with cuvettes of different path lengths, the transmitted intensity, I_t, decreases exponentially with increasing path length. Alternatively, the transmitted intensity decreases exponentially with increasing concentration of an absorbing solute. The absorbance or optical density, A, is defined as base 10 rather than natural logarithms,

$$A = \log \frac{I_0}{I_t} = \varepsilon\, c\, d$$

where

$\varepsilon = \alpha/2.303$ is the molar extinction coefficient (or molar absorptivity) with units M^{-1} cm^{-1}, when the concentration, c, and the path length, d, are given in molarity, M, and cm, respectively.

The Lambert–Beer Law shows that the absorbance is proportional to the concentration of the solute and path length with ε as the proportionality constant. The relationship between absorbance and transmission, $T = I_t/I_0$ is given by

$$A = -\log T$$

Because the absorption intensity depends strongly on wavelength, the wavelength at which the measurement was performed always has to be specified. The wavelength dependence of ε or of A is known as the absorption spectrum of the compound (Figure 1.3).

When measuring absorption spectra, several error sources have to be considered. Firstly, it should be known that a small but significant portion of light is lost by reflection at the cuvette windows. Corrections for this effect, as well as for absorption by the solvent (in addition to absorption by the solute), are usually made by performing a parallel measurement (double-beam method) using a cuvette containing only solvent. The transmitted intensity of this second measurement is then used as I_0 in the Lambert–Beer expression. Secondly, deviations can arise from inhomogeneous samples, light scattering by the sample, dimerization or aggregation at higher concentrations, or changes in equilibrium. The most common consequence is that the measured absorbance does not increase linearly with increasing concentration or path length. Finally, excitation and subsequent fluorescence emission can significantly deteriorate the shape of the absorption spectrum of solutes with high fluorescence quantum yield at higher concentrations.

One of the most widely used applications of absorption spectroscopy is the determination of the concentration of substances in solution. Through the knowledge of the extinction coefficient, ε, absolute concentrations can be easily calculated using the Lambert–Beer relationship ($c = A/\varepsilon d$). However, it has to be pointed out here that the extinction coefficients of common fluorophores, which are generally in the range of 10^4–$10^5\,\mathrm{l\,mol^{-1}\,cm^{-1}}$, do not represent inherently constant parameters. Conjugation of fluorophores to or interactions with other molecules can change the extinction coefficient by influencing, for example, the planarity of the conjugated π-electron system, thereby affecting the transitions strength. Likewise, the extinction coefficient can vary with the solvent. Furthermore, the dimerization of fluorophores and formation of higher order aggregates in solution and the solid state can induce dramatic color changes, that is, changes of the extinction coefficients.

1.4 Fluorophore Dimerization and Isosbestic Points

Deviations of Lambert–Beer Law behavior for the solution spectra of organic dyes is generally attributed to aggregation of dye molecules. Aggregation of dye molecules, as measured by deviations from ideality, has been found in several classes of solvents for concentration ranges typically in the order of from 10^{-6} to 10^{-4} M. Molecular aggregates are macroscopic clusters of molecules with sizes intermediate between crystals and isolated molecules. In the mid-1930s Scheibe [4] and, independently Jelley [5] discovered that when increasing the concentration of the dye pseudoisocyanine (PIC) in water, a narrow absorption band arises, red shifted to the monomer band. The narrow absorption band was ascribed to the optical excitation of the aggregates formed. To form the simplest aggregate, a dimer, the dye–dye interaction must be strong enough to overcome any other forces which would favor solvation of the monomer.

Usually, dye aggregates are classified on the basis of the observed spectral shift of the absorption maximum relative to the respective absorption maximum of the monomer. For the majority of possible dimer geometries, two absorption bands arise, one at higher energy relative to the monomer band, termed H-type aggregates (absorption band shifted hypsochromic), and at lower energy relative to the monomer

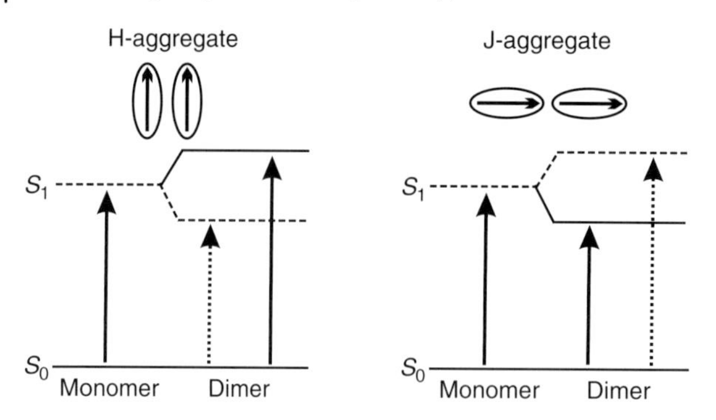

Figure 1.4 Simplified schematic of exciton theory to explain the different absorption and fluorescence behaviors of H- and J-aggregtes.

band, termed J-type or Scheibe-type aggregates (absorption band shifted bathochromic) [6–12]. J-type aggregates exhibit a bent or head-to-tail structure and usually show fluorescence with an intensity that fairly often surpasses that of the monomeric dyes. In contrast, it is known that the fluorescence of face-to-face-stacked H-type dimer aggregates (sandwich-type dimers) is strongly quenched. In fact, with the exception of a few examples [13], the non-emissive character of the excited state has become commonly accepted as a general feature of H-aggregates.

According to exciton theory of Kasha *et al.* [7], in J-aggregates, only transitions to the low energy states of the exciton band are allowed and, as a consequence, J-aggregates are characterized by a high fluorescence quantum yield. In contrast, H-aggregates are characterized by a large Stokes-shifted fluorescence that has a low quantum yield (Figure 1.4). After exciting the H-exciton band, a rapid downwards energy relaxation occurs to the lower exciton states that exhibit vanishingly small transition dipole moments. Therefore, their fluorescence is suppressed and a low fluorescence yield characterizes the H-aggregates. These two different types of aggregates, the J- and H-aggregates, are distinguished by the different angle α between the molecular transition dipole moments and the long aggregate axis.

When $a > 54.7°$, H-aggregates are formed and when $a < 54.7°$, J-aggregates are formed. In general, when there is interaction between two or more molecules in the unit cell of the aggregate, two or more excitonic transitions with high transition moment are observed and the original absorption band is split into two or more components. This splitting depends on the distance between the molecules, the angle of their transition dipole moments with the aggregate axis, the angle of the transition dipole moments between neighboring molecules and the number of interacting molecules. The appearance of isosbestic points in the absorption spectrum with increasing dye concentration provides good evidence for an equilibrium between monomeric and dimeric species and enables the association constant to be calculated, in addition to the spectra of the pure monomer and dimer.

A wavelength at which two or more components have the same extinction coefficient is known as an isosbestic wavelength or isosbestic point. When only two absorbing compounds are present in solution, one or more isosbestic points are

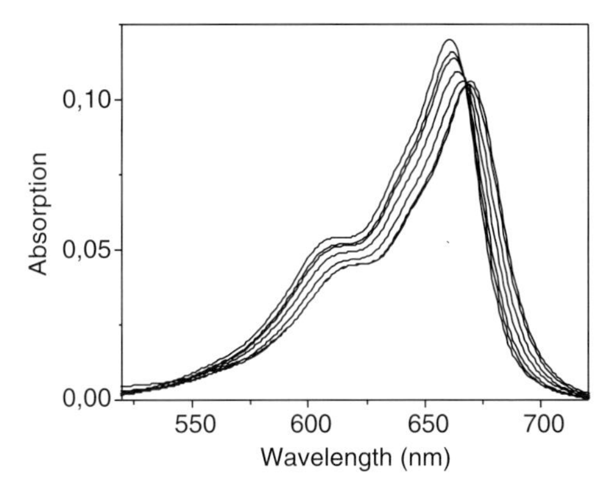

Figure 1.5 Absorption spectrum of the fluorophore MR121 measured in PBS (phosphate-bufferd saline), pH 7.4, with increasing tryptophan concentration (0–50 mM). The appearance of isosbestic points at 531 and 667 nm indicates the formation of 1 : 1 complexes for tryptophan concentrations less than 20 mM [12].

frequently encountered in the absorption spectrum. In multicomponent solutions isosbestic points almost never occur because the probability that three or more compounds have identical molar absorbances at any wavelength is negligibly small. Owing to this low probability, the occurrence of two or more isosbestic points demonstrates the presence of two and only two components absorbing in the observed spectral region. However, this rule does not apply if two chemically distinct components have identical absorption spectra (e.g., adenosine triphosphate, ATP, and adenosine diphosphate, ADP). In this case the entire spectrum represents a set of isosbestic points for these two components alone. Isosbestic points are especially useful for the study of equilibrium reactions involving absorbing reactants and products. Here, the presence of isosbestic points can be used as evidence that there are no intermediate species of significant concentration between the reactants and products. For example, besides dimer formation, several fluorophores (e.g., rhodamine and oxazine dyes) show isosbestic points in the absorption spectra upon addition of the amino acid tryptophan or the DNA nucleotide guanosine monosphosphate, dGMP, in aqueous solvents (Figure 1.5) [14–16]. Usually, the extinction of the fluorophores decreases slightly upon addition of dGMP or tryptophan and the absorption maxima shift bathochromically (that is, towards longer wavelengths) by up to ~10 nm. The appearance of isosbestic points suggests that an equilibrium is established between two species and reflect the formation of 1: 1 complexes between free fluorophores (F) and tryptohan (Trp). Both types absorb in the same region. At higher tryptophan concentrations (above 20 mM) slight deviations can be observed indicating a low probability of the formation of higher aggregated complexes.

The equilibrium state between free fluorophores and fluorophore/Trp-complexes can be described by [17]

$$F + \mathrm{Trp} \overset{K_g}{\longleftrightarrow} (\mathrm{FTr}p)$$

where the equilibrium constant, defined as

$$K_g = \frac{[\mathrm{FTr}p]}{[F][\mathrm{Trp}]}$$

can be calculated using the relationship given by Ketelaar *et al.* [18]

$$\frac{1}{\varepsilon - \varepsilon_F} = \frac{1}{K_g\left(\varepsilon_{\mathrm{FTr}p} - \varepsilon_F\right)}\frac{1}{[\mathrm{Trp}]} + \frac{1}{\varepsilon_{\mathrm{FTr}p} - \varepsilon_F}$$

Here ε, ε_F, and ε_{FTrp} are the apparent absorption coefficient, the absorption coefficient of the fluorophores and the complex under study, respectively. By the plotting of $(\varepsilon - \varepsilon_F)^{-1}$ versus $[\mathrm{Trp}]^{-1}$ the equilibrium constant, K_g, can be calculated from the intercept and the slope of the resulting straight line. For the addressed interactions between rhodamine and oxazine dyes with tryptophan or dGMP, equilibrium constants in the range of 50–200 M^{-1} have been measured [14–16, 19, 20].

1.5 Franck–Condon Principle

For relatively large fluorophores containing more than 30 atoms, such as the organic dye molecules generally used in fluorescence spectroscopy and imaging, many normal vibrations of differing frequencies are coupled to the electronic transition. In addition, collisions and electrostatic interactions with surrounding solvent molecules broaden the lines of vibrational transitions. Furthermore, every vibrational sublevel of every electronic state has superimposed on it a ladder of rotational states that are significantly broadened because of frequent collisions with solvent molecules, which seriously hinder rotation. This results in a quasicontinuum of states superimposed on every electronic level. The population of the levels in contact with the thermalized solvent molecules is determined by the Boltzmann distribution. In the quantum mechanical picture, vibrational levels and wavefunctions are those of quantum harmonic oscillators and rigid rotors (with some corrections for rotation–vibration coupling and centrifugal distortion) (Figure 1.6a).

The more realistic anharmonic potential (e.g., the Morse potential) describing the vibrational dynamics of a molecule is, in general, different for each electronic state. Excitation of a bound electron from the HOMO to the LUMO increases the spatial extent of the electron distribution, making the total electron density more diffuse, and often more polarizable. Thus, one finds in general that the bond length between two atoms in a diatomic molecule is larger and the bond strength is weaker in electronically excited states. On the other hand, a slightly weaker bond means that the force constant for vibrations will be lower, and the relationship between the force constant and the second derivative of the potential V ($k = d^2V/dx^2$) indicates that the weaker force constant in the excited state implies not only a lower vibrational frequency but simultaneously a broader spatial extent for the energy potential curve.

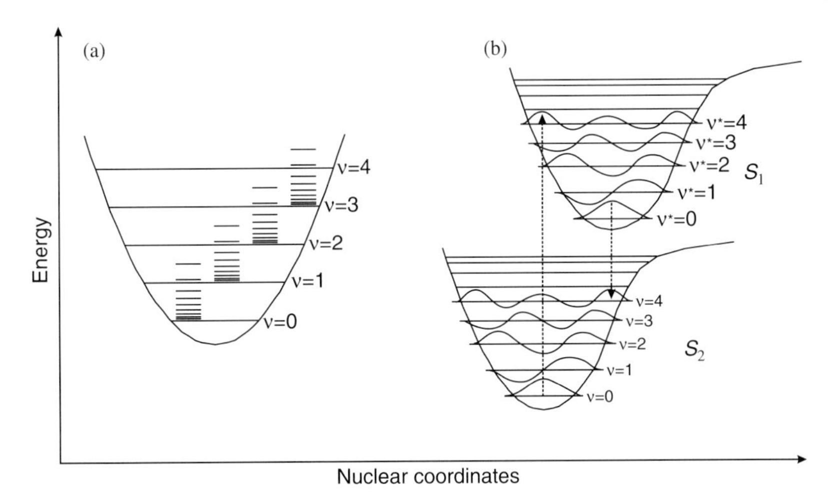

Figure 1.6 (a) Idealized potential energy curve for a diatomic molecule. In general, polyatomic molecules will have $3N-6$ vibrational modes ($\nu_1, \nu_2, \nu_3, \ldots$), where N is the number of atoms, and will be asymmetric rotors with three different inertial axes. For each of the $3N-6$ normal vibrations, a potential well exists with a rotational energy ladder. (b) Morse potentials for a molecule in the singlet ground state S_0 and first excited singlet state S_1 to demonstrate the Franck–Condon principle. As electronic transitions are very fast compared with nuclear motions, vibrational levels are favored when they correspond to a minimal change in the nuclear coordinates. The potential wells shown favor transitions between $\nu=0$ and $\nu=4$. In the simplest case of a diatomic molecule, the nuclear coordinate axis refers to the internuclear distance.

Figure 1.6b illustrates the principle for vibrational transitions in a molecule with Morse-like potential energy functions in both the ground and excited electronic states. At room temperature the molecule generally starts from the $\nu=0$ vibrational level of the ground electronic state as the vibrational energy will normally be 1000–3000 cm^{-1}, many times the thermal energy kT, which is about 200 cm^{-1} at room temperature. In contrast, pure rotational transitions require energies in the range of 100 cm^{-1}, that is, higher excited rotational levels are occupied at room temperature. Upon absorption of a photon of the necessary energy, the molecule makes a so-called vertical transition to the excited electronic state. The occurrence of vertical transitions on the potential energy curve is explained by the Franck–Condon principle and the Born–Oppenheimer approximation. The Born–Oppenheimer approximation is based on the fact that the proton or neutron mass is roughly 1870 times that of an electron and that electrons move much faster than nuclei. Thus, electronic motions when viewed from the perspective of the nuclear coordinates occur as if the nucleic were fixed in place. Applying the Born–Oppenheimer approximation to transitions between electronic energy levels, led Franck and Condon to formulate the Franck–Condon principle. Classically, the Franck–Condon principle is the approximation that an electronic transition is most likely to occur without changes to the position of the nuclei in the molecular entity and its environment. The quantum mechanical formulation of this principle is that the intensity of a vibrational transition is proportional to the square of the overlap integral between the vibrational wavefunctions of the two states that are involved in the transitions [21–24]. In other words, electronic transitions are essentially instantaneous compared with the time scale of nuclear motions. Therefore, if the molecule is to move to a new vibrational level during the electronic transition, this new vibrational level must be instantaneously compatible with the nuclear positions and momenta of the vibrational level of the molecule in the originating electronic state. In the semiclassical picture of vibrations of a simple harmonic oscillator, the necessary conditions can occur at the turning points, where the momentum is zero.

Because the electronic configuration of a molecule changes upon excitation (Figure 1.6b) the nuclei must move to reorganize to the new electronic configuration, which instantaneously sets up a molecular vibration. This is the reason why electronic transitions cannot occur without vibrational dynamics. After excitation has led to a transition to a nonequilibrium state (Franck–Condon state), the approach to thermal equilibrium is very fast in liquid solutions at room temperature, because a large molecule such as a fluorophore experiences at least 10^{12} collisions per second with solvent molecules, so that equilibrium is reached in a time of the order of one picosecond. Thus, the absorption is practically continuous all across the absorption band. In the electronic excited state, molecules quickly relax to the lowest vibrational level (Kasha's rule). Dependent on their fluorescence quantum yield they can then decay to the electronic ground state via photon emission. The Franck–Condon principle is applied equally to absorption and to fluorescence. Kasha's rule states that emission will always occur from the lowest lying electronically excited singlet state $S_{1,\nu=0}$. The applicability of the Franck–Condon principle in both absorption and emission, along with Kasha's rule, leads to the mirror symmetry of the absorption and the fluorescence spectrum of typical organic dye molecules (Figure 1.7a). Owing to the loss of vibrational excitation energy during the excitation/emission cycle fluorescence emission always occurs at lower energy, that is, spectrally red-shifted (the so-called Stokes shift). The Stokes shift, named after the Irish physicist George

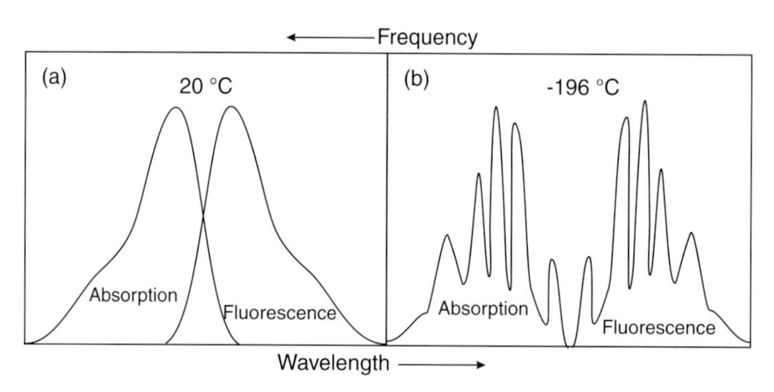

Figure 1.7 (a) Absorption and fluorescence spectrum of a typical organic dye molecule in solution (a) at room temperature (20 °C) and (b) at −196 °C.

G. Stokes, represents the basis for the high sensitivity of fluorescence spectroscopy compared with other spectroscopic methods because elastic scattering of the excitation light (Rayleigh-scattering) can be efficiently suppressed using appropriate filtering.

1.6 Temperature Effects on Absorption and Emission Spectra

On increasing the temperature, higher vibrational levels of the ground state are populated according to the Boltzmann-distribution and more and more transitions occur from these levels to higher vibrational levels of the first excited electronic state. As a result, the absorption spectrum becomes broader and the superposition of the different levels blurs most of the vibrational fine structure of the band. On the other hand, at lower temperatures, the spectral widths are usually reduced and the spectra exhibit enhanced vibrational information (Figure 1.7b). Therefore, dye solutions that form a clear organic glass when cooled down to 77 K show spectra comparable to theoretical calculations because of their well-resolved vibrational structure. Further cooling below the glass point, when the free movement of solvent molecules or parts thereof is inhibited, usually brings about no further sharpening of the spectral features. However, using a matrix of *n*-paraffins at temperatures below 20 K, very sharp, line-like spectra with a width of about only 1 cm^{-1} often appear, instead of the more diffuse-like band spectra of dye molecules. This so-called "Shpolski effect" [25] results because there are only a few different possibilities for solvation of the molecule in that specific matrix, and each of the different sites causes a series of spectral lines in absorption as well as in emission.

Generally speaking, the zero-phonon line and the phonon sideband jointly constitute the line shape of individual light absorbing and emitting molecules embedded into a transparent solid matrix. A phonon is a quantized mode of vibration occurring in a rigid crystal lattice, such as the atomic lattice of a solid. When the host matrix contains many chromophores, each will contribute a zero-phonon line and a phonon sideband to the absorption and emission spectra. The spectra originating from a collection of identical fluorophores in a matrix is said to be inhomogeneously broadened, because each fluorophore is surrounded by a somewhat different matrix environment that modifies the energy required for an electronic transition. In an inhomogeneous distribution of fluorophores, individual zero-phonon line and phonon sideband positions are therefore shifted and overlapping. The zero-phonon line is determined by the intrinsic difference in energy levels between the ground and excited states (corresponding to the transition between $S_{0,\nu=0}$ and $S_{1,\nu=0}$) and by the local environment. The phonon sideband is shifted to higher frequency in absorption and to lower frequency in fluorescence. The distribution of intensity between the zero-phonon line and the phonon sideband is strongly dependent on temperature. At room temperature the energy is high enough to excite many phonons and the probability of zero-phonon transitions is negligible. At lower temperatures, in particular at liquid helium temperatures, however, being dependent on the strength of coupling between the chromophore and the host lattice, the zero-phonon line of an individual molecule can become extremely narrow. On the other hand, the centers of frequencies of different molecules are still spread over a broad inhomogeneous band.

Inhomogeneous broadening results from defects in the solid matrix that randomly shift the lines of each individual molecule. Therefore, for each particular laser frequency, resonance is achieved for only a small fraction of molecules in the sample [26, 27]. In single-molecule spectroscopy (SMS) in solids at low temperature exactly this effect is used to isolate single fluorophores by their specific optical transition frequency (the zero-phonon line), which can be addressed selectively using a narrow line width laser source. The extreme sensitivity to their local environment of the sharp excitation lines of fluorophores at cryogenic temperatures enables the refined investigation of the coupling of chromophores with their surrounding matrix [27–30]. Usually, small changes in the absolute position of the purely electronic zero-phonon line absorption of single molecules were recorded by scanning the frequency of a narrow bandwidth laser used for excitation, and collecting the integral Stokes shifted fluorescence signal. The first optical detection of single molecules by Moerner and Kador in 1989 was performed in a solid at low temperature using a sensitive doubly modulated absorption method [28]. In 1990, Orrit and Bernard showed that fluorescence excitation spectra enhance the signal-to-noise ratio of single molecule lines dramatically [29]. The strong signals and the possibility to study an individual molecule for extended periods of time which is, for example, impossible in liquid solution because of rapid diffusion, paved the way for many other experiments, such as the measurement of external field effects by Wild *et al.* in 1992 [30].

The shape of the zero-phonon line is Lorentzian with a width determined by the excited-state lifetime T_{10}, according to the Heisenberg uncertainty principle. Without the influence of the lattice, the natural line width (full width at half maximum) of the fluorophore is $\gamma_0 = 1/T_{10}$. The lattice reduces the lifetime of the excited state by introducing radiationless decay mechanisms. At absolute zero, the lifetime of the excited state influenced by the lattice is T_1. Above absolute zero, thermal motions will introduce random perturbations to the fluorophores local environment. These perturbations shift the energy of the electronic transition, introducing a temperature dependent broadening of the line width. The measured width of a single fluorophore's zero-phonon line, the homogeneous line width, is then $\gamma_h(T) \geq 1/T_1$.

Considering an ensemble of molecules absorbing with very narrow homogenenous lines (ideally at zero temperature), that is, irradiated with a monochromatic excitation source, only the resonant molecules will absorb light significantly. Upon excitation, the molecule might undergo a number of possible photophysical and photochemical processes, such as intersystem crossing into long-lived triplet states that exhibit different absorption characteristics (generally lower) at the excitation wavelength. Thus, the sample will absorb less at the frequency of illumination during the lifetime of the triplet state. If an absorption spectrum of the sample is measured it will show a transient spectral hole at the laser frequency used to excite the sample (Figure 1.8). Some of the product states may have very long lifetimes, in particular if the molecule undergoes a chemical reaction. In that case, the spectral hole is

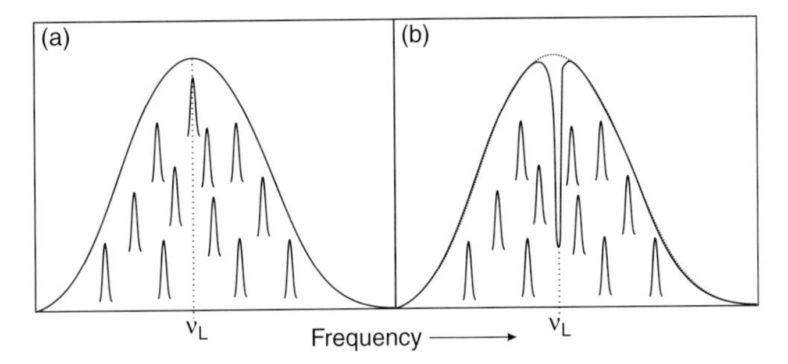

Figure 1.8 Hole-burning represents the modification of the optical properties of a material upon irradiation with light. Different environments in a disordered matrix shift the zero-phonon lines of single molecules at random. The resulting ensemble spectrum is inhomogeneously broadened as a result of the superposition of many narrow lines of individual molecules. The figure shows the inhomogeneous absorption spectrum of an ensemble of molecules before (a) and after (b) illumination at a specific laser frequency ν_L. The sharp spectral hole appears because the narrow lines of the excited molecule are shifted to new frequencies.

permanent. In photophysical processes, the conformation of neighboring atoms or groups of atoms can be modified by illumination, leading to a shift of the absorption line, much larger than the homogeneous width, but usually smaller than the inhomogeneous bandwidth. The resulting antihole is much broader than the hole and can only be seen after very deep and broad holes have been burned.

1.7
Fluorescence and Competing Processes

The electronic states of most organic molecules can be divided into singlet states and triplet states, where all electrons in the molecule are spin paired or one set of electron spins is unpaired, respectively. Upon excitation of fluorophores with light of suitable wavelength (which can be considered as an instantaneous process occurring at time scales of $\leq 10^{-15}$ s) the fluorophore generally resides in one of the many vibrational levels of an excited singlet state (see Jablonski diagram shown in Figure 1.9). The probability of finding the molecule in one of the possible excited singlet states, S_n, depends on the transition probabilities and the excitation wavelength. In other words, the occupation of singlet states is controlled by the interaction of the electron involved in the transition with the electric field of the excitation light. Upon excitation to higher excited singlet states, molecules relax through internal conversion to higher vibrational levels of the first excited singlet state, S_1, within 10^{-11}–10^{-14} s. Molecules residing in higher vibrational levels will quickly (10^{-10}–10^{-12} s) fall to the lowest vibrational level of this state via vibrational relaxation by losing energy to other molecules through collisions. From the lowest lying vibrational level of the first excited singlet state, the molecule can lose energy via radiationless internal conversion followed by vibrational relaxation.

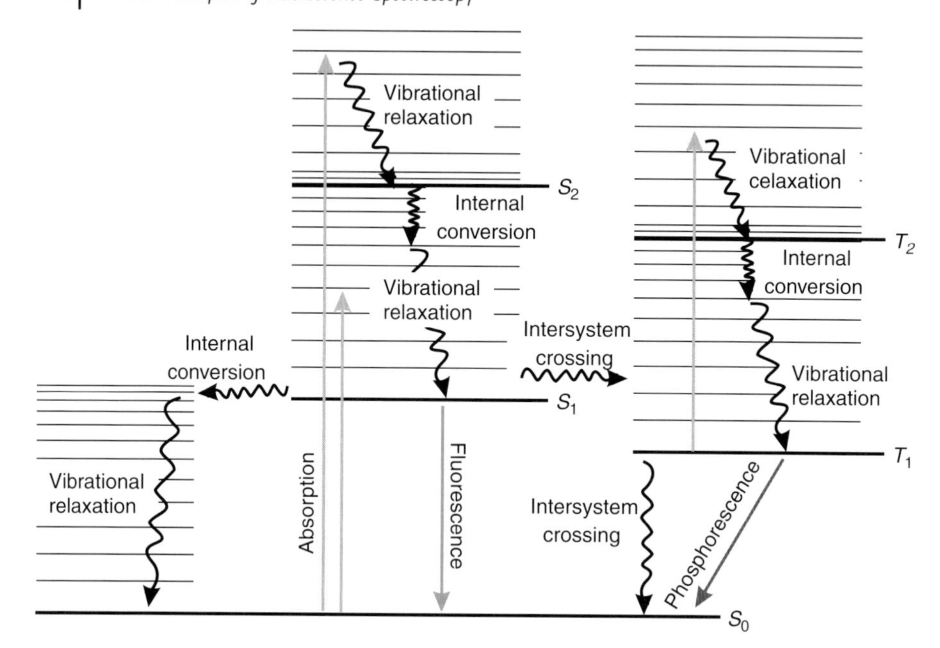

Figure 1.9 Jablonski diagram describing the electronic levels of common organic molecules and possible transitions between different singlet and triplet states.

Alternatively, the molecule might be excited into higher singlet states by absorption of a second photon. Singlet–singlet absorption ($S_1 \rightarrow S_n$) and subsequent ionization of the molecule represents a possible photobleaching pathway. The efficiency of the process depends on the absorption spectra transition strengths of the higher excited states involved, that is, whether absorption into higher excited singlet states is in resonance with the excitation light.

Depending on the molecular structure, radiative depopulation of S_1 might occur by spontaneous emission of a fluorescence photon. According to the Franck–Condon principle, the vertical transition to higher excited vibrational levels of S_0 is followed by vibrational relaxation until thermal equilibrium, according to the Boltzmann distribution, is reached. Both vibrational relaxation and internal conversion cause heating of the solvent, which thus offers an elegant method to determine the fluorescence quantum yield of fluorophores via measurement of the solvent temperature of related parameters, for example, the refractive index. Generally speaking, loose and floppy molecules exhibiting several rotational and vibrational degrees of freedom will seriously exhibit lower fluorescence intensity.

The spin of an excited electron can also be reversed by intersystem crossing, usually leaving the molecule in the first excited triplet state, T_1. In most organic dyes intersystem crossing is fairly inefficient as a spin forbidden process, even though the triplet state is of lower electronic energy than the excited singlet state. The probability of intersystem crossing increases if the vibrational levels of the two states overlap. For example, the lowest singlet vibrational level can overlap one of the higher vibrational levels of the triplet state. Overall, the intersystem crossing efficiency strongly depends on the nature of the fluorophore and the transition probabilities and is generally not predictable. However, it is known [31] that the presence of heavy atoms can substantially increase the intersystem crossing rate constant.

A molecule in a high vibrational level of the triplet state can lose energy through collisions with solvent molecules (vibrational relaxation), leaving it at the lowest vibrational level of the triplet state. It can then again undergo intersystem crossing to a high vibrational level of the electronic singlet ground state, where it returns to the lowest vibrational level through vibrational relaxation. As in the case of singlet states, triplet states can be excited into higher excited triplet states, T_n, by absorption of a second photon. Because the triplet/singlet transition is also spin forbidden, triplet-state lifetimes can last up to 100 s, in comparison with the 10^{-7}–10^{-9} s average lifetime of an excited singlet state. Because internal conversion and other radiationless transfers of energy compete so successfully with radiative deactivation, phosphorescence is usually seen only at low temperatures or in highly viscous media. Owing to the long lifetime of the triplet state and distinct overlap of the $T_1 \rightarrow T_n$ and the $S_0 \rightarrow S_1$ absorption spectra of most organic dye molecules, triplet states are most probably involved in photobleaching pathways [32, 33]. Photobleaching denotes the loss of fluorescence properties of a dye due to an irreversible reaction that dramatically changes the absorption and emission capabilities. Furthermore, as higher excited states are in general more reactive than their underlying ground states, they are the most likely to be involved in photobleaching pathways. Therefore, different strategies have been developed to increase the photostability of common fluorophores, especially in single-molecule experiments, with respect to the total number of photons that can be emitted and also with respect to the fluorescence emission rate itself [34].

An interesting example of a molecule with high triplet yield presents benzophenone, with a triplet yield of 100% [35]. Generally, rhodamine, oxazine, or carbocyanine derivatives are used in applications requiring high sensitivity. In air-saturated ensemble solutions, the triplet-state lifetimes, τ_T, of such dyes vary between less than 1 μs up to several μs, with intersystem crossing rates, k_{isc}, ranging from 10^5 to $10^8\ s^{-1}$ [36–40]. The efficiency of $T_1 \rightarrow T_n$ absorption strongly depends on (i) the intersystem crossing rate, k_{isc}, that is, the probability of finding the molecules in the triplet state, (ii) the extinction coefficient $\varepsilon_T(\nu)$ at the excitation wavelength, and (iii) the triplet-state lifetime, τ_T. To minimize triplet–triplet absorption, the triplet state has to be depopulated by addition of triplet quenchers such as cyclooctatetraene (COT) or molecular oxygen. Both COT and O_2 exhibit an energetically low lying triplet state that acts as an efficient acceptor for triplet–triplet energy transfer. Thus, the dye will be transferred into the singlet ground state upon contact formation with triplet quenchers. Unfortunately, to ensure high efficient triplet quenching, mM concentrations of triplet quenchers have to be added, which renders the applicability of the method more difficult. On the other hand, triplet quenchers such as anthracence, stilbene, or naphthalene derivatives (all exhibiting low lying triplet states) can be

Table 1.1 Overview of possible depopulation pathways of excited fluorophores.

Internal conversion	$S_n \rightarrow S_1$, $T_n \rightarrow T_1$	k_{ic}	10^{10}–10^{14} s^{-1}
Internal conversion	$S_1 \rightarrow S_0$	k_{ic}	10^{6}–10^{7} s^{-1}
Vibrational relaxation	$S_{1,\nu=n} \rightarrow S_{1,\nu=0}$	k_{vr}	10^{10}–10^{12} s^{-1}
Singlet–singlet absorption	$S_1 \rightarrow S_n$	k_{exc}	10^{15} s^{-1}
Fluorescence	$S_1 \rightarrow S_0$	k_f	10^{7}–10^{9} s^{-1}
Intersystem crossing	$S_1 \rightarrow T_1$, $S_n \rightarrow T_n$, $T_n \rightarrow S_n$	k_{isc}	10^{5}–10^{8} s^{-1}
Phosphorescence	$T_1 \rightarrow S_0$	k_p	10^{-2}–10^{3} s^{-1}
Triplet–triplet absorption	$T_1 \rightarrow T_n$	k_{exc}	10^{15} s^{-1}

coupled directly to the fluorophore via a short aliphatic chain, ensuring highly efficient intramolecular triplet–triplet energy transfer [33].

Emission of a fluorescence photon from the vibrational ground state of the first excited singlet state constitutes a spontaneous process that contains information about the environment of the fluorophore and its interactions. For example, the fluorescence emission spectrum and its maximum contain information about the polarity of the solvent, whereas the fluorescence lifetime and fluorescence quantum yield directly reflect transient quenching interactions of the fluorophore with other molecules. Table 1.1 summarizes radiative and nonradiative reaction pathways in common organic dyes and corresponding time scales, neglecting quenching by external molecules.

1.8
Stokes Shift, Solvent Relaxation, and Solvatochroism

As organic dyes consist of many atoms (typically 50–100) they thus show a manifold and complex vibrational spectrum. Accordingly, the fluorophore has a large number of energetically different transition possibilities to the vibrational ground state after excitation with light of appropriate wavelength. Owing to the solvation shell and corresponding interactions between fluorophores and solvent molecules, the resulting vibrational transitions are considerably broadened at room temperature. The complete shift of the fluorescence emission band compared with the absorption band, due to the radiationless deactivation processes, is called the *Stokes Shift*. Because the electron distribution changes upon excitation, different bonding forces and dipole moments arise. Therefore, the solvent molecules experience a new equilibrium configuration, which they adjust to within several picoseconds at room temperature. The kinetics of dielectric relaxation of solvent molecules can be followed by monitoring the time-dependent shift in the fluorescence emission spectrum with picosecond time-resolution [41–46]. If a fluorescing molecule is excited into a more polar excited state, the electronic polarization of the solvent molecules adjusts instantaneously to the new electron distribution in the molecule. In contrast, the orientational polarization of the solvent molecules does not change instantaneously with the excitation. Therefore, the orientational polarization is not in equilibrium with the excited molecule. This means that the solvent molecules have to react by dielectric relaxation until the equilibrium configuration of the corresponding excited state is reached. The same happens upon subsequent fluorescence emission, that is, the orientational polarization remains conserved during the quasi instantaneous optical transition. Thus, the fluorescence spectrum experiences a time-dependent red-shift, which is expressed as function $C(t)$.

$$C(t) = \frac{\nu(t) - \nu(\infty)}{\nu(0) - \nu(\infty)}$$

where

$\nu(0)$, $\nu(t)$, and $\nu(\infty)$ denote the frequencies of the fluorescence emission maximum immediately after excitation, at time t, and after complete relaxation has occurred, respectively.

Typically, dielectric relaxation is completed in 10 ps. Assuming monoexponential solvent relaxation times, time-resolved fluorescence measurements at different detection wavelengths deliver the following trend: at shorter detection wavelengths ($\lambda_{blue} < \lambda$) the measured fluorescence is dominated by those molecules whose solvation shell is incompletely relaxed, whereas at longer detection wavelengths ($\lambda < \lambda_{red}$) fluorescence is controlled by molecules with completely relaxed solvation shells. Thus, the fluorescence lifetime of fluorophores always exhibits the trend $\tau_{blue} < \tau < \tau_{red}$.

Furthermore, owing to the different properties of the ground and excited states, the dipole moment changes upon excitation $\Delta\mu = \mu_e - \mu_g$, which is reflected in a shift in the absorption and emission band, which is dependent on solvent polarity (*Solvatochromism*). Therefore, charge transfer transitions, for example, in coumarin dyes, show pronounced solvatochromism effects. On the other hand, distinct shifts in absorption and emission of suitable candidates can be used advantageously for the definition of new solvent polarity parameters [47]. For a complete description of solvatochromic effects, the refractive index n and the dielectric constant ε_s of the solvent, in addition to the change in dipole moment of the fluorophore upon excitation, $\Delta\mu$, have to be considered. Using the Lippert equation [48, 49]

$$\nu_{abs} - \nu_{em} = \frac{2\left(\mu_e - \mu_g\right)^2}{cha^3}\left[\frac{2(\varepsilon_s - 1)}{(2\varepsilon_s + 1)} - \frac{2(n^2 - 1)}{(2n^2 + 1)}\right]$$

the Stokes shift ($n_{abs} - n_{em}$) can be expressed as a function of solvent properties (n, ε_s) and the dipole moments of the fluorophore in the ground, μ_g, and excited states, μ_e, where c is the speed of light, h is Planck's constant, and a the *Onsager radius* of the fluorophore in the respective solvent ($a \sim 60\%$ of the longitudinal axis of the fluorophore). While for coumarin dyes the Stokes shift generally increases with solvent polarity (i.e., the emission maximum shifts further bathochromically than the absorption maximum), rhodamine derivatives show negligible solvatochromism.

1.9
Fluorescence Quantum Yield and Lifetime

Both fluorescence quantum yield and lifetime are among the most important selection criteria for fluorophores in single-molecule fluorescence spectroscopy. The fluorescence quantum yield Φ_f of a fluorophore is the ratio of fluorescence photons emitted to photons absorbed. According to the following equation

$$\Phi_f = \frac{k_r}{k_r + k_{nr}}$$

the quantum yield can be described by two rate constants, the radiative rate constant, k_r, and the nonradiative rate constant, k_{nr}, comprising all possible competing deactivation pathways, such as internal conversion, intersystem crossing, or other intra- and intermolecular quenching mechanisms. Even though complete prediction of the fluorescence quantum yield of a certain fluorophore (which would theoretically allow us to design fluorophores with ideal properties) is impossible, some requirements can be formulated that advance high fluorescence quantum yields: (i) the fluorophore should exhibit a rigid structure to minimize radiationless deactivation due to rotation or vibration of fluorophore side groups, (ii) to ensure a low intersystem crossing rate constant strong spin–orbit coupling, for example, due to heavy atoms, should be avoided, and (iii) charge transfer transitions due to conjugated electron donor and acceptor groups, as for example in coumarin dyes, often show bright fluorescence. However, there are many other nonradiative processes that can compete efficiently with the emission of light and thus reduce the fluorescence quantum yield. The reduction in fluorescence efficiency depends in a complicated fashion on the molecular structure of the dye [50].

Neglecting all radiationless deactivation processes, the rate constant for radiative deactivation of the first excited singlet state, k_f, can be approximated using the *Strickler–Berg* relationship [51]:

$$k_f = 2.88 \times 10^9 \, \nu_0^{-2} \, n^2 \frac{g_g}{g_e} \int \varepsilon \, d\nu$$

where

n denotes the refractive index of the solvent
g_g and g_e are the degeneracy of the ground and first excited states, respectively
ε is the extinction coefficient
ν_0 is the wavenumber of the absorption maximum (the integration is executed over the entire absorption spectrum).

Thermal blooming is commonly applied to determine the absolute fluorescence quantum yield of a dye solution. A thermal blooming measurement is, in essence, a calorimetric determination of the very small temperature gradients induced by the absorption of light energy. The technique can be extremely sensitive and allows one to measure exceptionally weak absorption [52–56]. The basic idea involving power conservation is very simple. The laser power that is incident on any sample must be equal to the sum of the power transmitted plus the power emitted as fluorescence plus the power degraded to heat. Heat generation in the region of the absorption increases the local temperature, modifies the refractive index, and induces what is, in fact, an optical lens, which is negative for most liquids. The thermal lens develops over a period of a few tenths of seconds. During that time, the laser beam is observed as an increasing ("blooming") spot on a plane located a few meters behind the sample. Appropriate observation of the spot size with time enables the determination of the absolute fluorescence quantum yield of the sample.

The easiest way to determine the fluorescence quantum yield is the measurement of fluorescence efficiency relative to that of a standard solution. However, relative fluorescence quantum yield measurements are aggravated by many technical problems. Firstly, a fluorescence standard absorbing and emitting in the wavelength range of the dye to be investigated must be found. Unfortunately, only a few examples, such as quinine sulfate ($\Phi_f = 70\%$ in 0.1 N H_2SO_4 [57]), fluorescein ($\Phi_f = 95\%$ in 0.1 N aqueous sodium hydroxide [56]), and rhodamine 6G in addition to rhodamine 101 ($\Phi_f = 90$–100% [50]) at concentrations below 10^{-5} M have been characterized so far with high precision. Furthermore, as the measurements require comparison of the fluorescence efficiency of an unknown with a standard, careful attention must be given to corrections for differences in solvent, temperature, the wavelength response of monochromators and detectors, polarization effects, and so on.

The average time a molecule spends in its excited singlet state S_1 before spontaneous emission occurs is denoted as fluorescence lifetime, τ_f.

$$\tau_f = \frac{1}{k_r + k_{nr}}$$

The fluorescence lifetime of a fluorophore can be described as the decrease in the number of excited fluorophores $[F(t)^*]$ with time following optical excitation with a infinitesimally short light pulse (δ-pulse).

$$\frac{d[F(t)^*]}{dt} = -(k_r + k_{nr})[F(t)^*]$$

As the number of excited fluorophores $[F(t)^*]$ is proportional to the fluorescence intensity $I(t)$, integration between $t = 0$ and t yields a single exponential function similar to a radioactive decay (Figure 1.10).

$$I(t) = I_0 \exp\left(-\frac{t}{\tau_f}\right) \tag{1.1}$$

That is, the fluorescence lifetime can be determined by measuring the time it takes the fluorescence intensity to reach $1/e$ of its original value I_0 at $t = 0$ upon optical excitation with a δ-pulse. In the case of heterogeneous samples, for example, due to different interaction possibilities between the biomolecule and the attached fluorescent label, or for mixtures of fluorophores, the fluorescence decay has to be

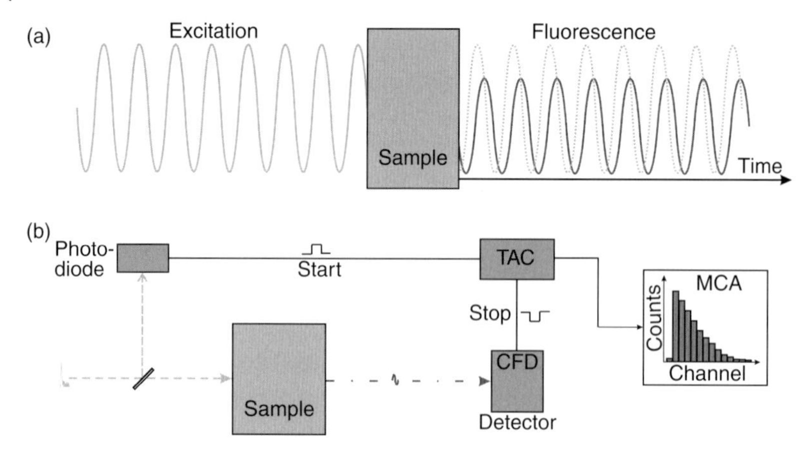

Figure 1.10 Principle of (a) phase modulation and (b) time-correlated single-photon counting (TCSPC) for the measurement of fluorescence decays.

described by applying a multiexponential model.

$$I(t) = \sum_i a_i \exp\left(-\frac{t}{\tau_{i,f}}\right)$$

$$\sum_i a_i = 1$$

Thus, time-resolved fluorescence spectroscopy enables the quantitative assignment of the relative contribution, a_i, of the i-th component of the sample with a characteristic fluorescence lifetime, $\tau_{i,f}$.

In principle, two popular methods for the determination of fluorescence lifetimes exist, the pulse or photon-counting and the phase modulation method (Figure 1.10) [58]. Instead of pulsed excitation with light pulses of duration substantially shorter than the fluorescence lifetime of the sample, the phase modulation method excites the sample with light whose intensity is modulated sinusoidally. The forced fluorescence emission of the sample directly follows the excitation modulation. Because of the finite fluorescence lifetime, emission modulation is delayed in phase by an angle relative to the excitation. In addition, fluorescence emission is demodulated due to the fact the fluorescence quantum yield of common fluorophores, Φ_f is less than 100%. Thus, fluorescence lifetime information can be extracted from both the degree of demodulation and the phase angle. However, as phase modulation methods are inoperative for fluorescence lifetime measurements at the single-molecule level, the pulse method is mainly used. The pulse method, that is, time-correlated single-photon counting (TCSPC) [59–61], features high sensitivity and the ability to deal with low photon count rates with a time resolution down to the ps region, that is, typical parameters to be handled in single-molecule experiments.

TCSPC is based on the ability to detect and count individual photons. As it is a counting process it is inherently digital in nature. Essentially, the TCSPC technique is a start–stop or "stopwatch" technique. Usually, the excitation light pulse is split such that a photodiode is triggered at the same time that the sample is excited. It is as if a stopwatch is started at this point. When the first fluorescence photon is detected by a photomultiplier tube (PMT), microchannel plate photmultiplier (MCP), or an avalanche photodiode (APD) the stopwatch is stopped and the time lag measured is collected. This experiment is repeated several times and the start/stop time lags are plotted as a histogram to chart the fluorescence decay. The TCSPC measurement relies on the concept that the probability distribution for emission of a single photon after an excitation yields the actual intensity versus time distribution of all photons emitted as a result of the excitation. Ideally, the fluorescence intensity emitted by the sample is very low, that is, one fluorescence photon should be observed every few hundred excitation laser pulses, to prevent so-called pile-up effects. Pile-up results from the fact that a TCSPC experiment can record only one photon per excitation pulse. At high photon detection count rates, that is, when more than one fluorescence photon is produced per excitation cycle, the probability of detecting "early" emitted photons (photons with shorter arrival times) is significantly higher, thus shortening the measured fluorescence decay. To prevent pile-up effects in TCSPC measurements, the power of the excitation light should be adequately reduced.

More precisely, as the exact arrival time of a fluorescence photon at the detector is crucial for the whole TCSPC measurement, it is determined by the use of a constant fraction discriminator (CFD), which sends a precisely timed signal to *start* the charging of a linear voltage ramp in the time-to-amplitude converter (TAC). The charging linear voltage ramp of the TAC is *stopped* by the regular electronic output of the photodiode, which represents the highly stable and exact repetition rate of the optical excitation. Subsequently, a pulse is output from the TAC, the amplitude of which is proportional to the charge on the ramp, and, hence, the time between *start* and *stop*. It has to be pointed out here that the TAC is run in the inverted mode, so that each photon that is detected is counted. It takes a finite time to reset the voltage ramp and if the TAC were to be started by each laser trigger, many counts would be lost while the TAC was being reset. The pulse height is digitized by an analog-to-digital converter and a count is stored in a multi-channel analyzer(MCA) in an address corresponding to that number. These components are contained on a PC card. The MCA uses a variable number of channels (usually 512–4096 channels) that determine the time/channel. The experiment is repeated as described until the histogram of number of counts against address number corresponds to the required precision, given as a fixed number of counts at the maximum channel, the decay curve of the sample. Depending on the number of channels used, at least several thousand counts should be accumulated at the peak (Figure 1.11). The resolution of the TCSPC measurements is limited by the spread of the transit times in the detector, by the timing accuracy of the discriminator that receives the detector pulses, and by the accuracy of the time measurements. With an MCP-PMT the width of the instrument response function is of the order of 25–30 ps. The width of the time channels of the histogram can be made to be less than 1 ps.

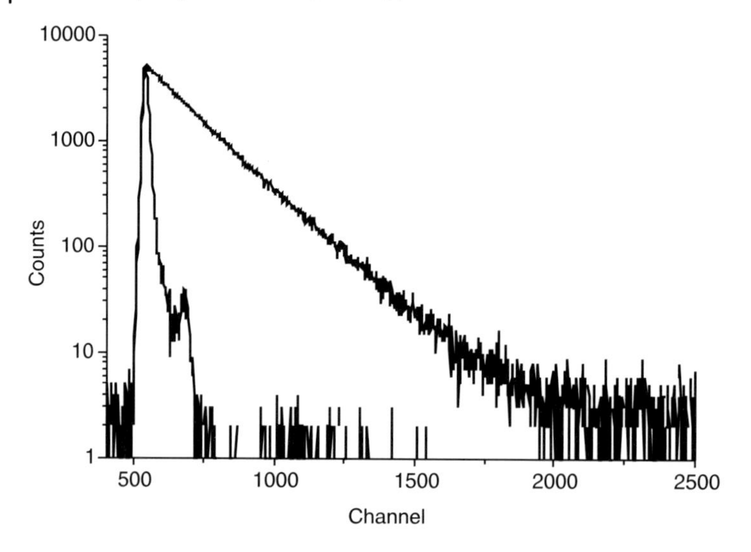

Figure 1.11 Fluorescence decay curve of a red-absorbing fluorophore and instrument response function (IRF) with a full width half maximum (FWHM) of 220 ps measured by time-correlated single-photon counting (TCSPC). For excitation, a pulsed diode laser with a repetition rate of 10 MHz and a pulse length of 200 ps (FWHM) at 635 nm is used. The IRF is measured at the excitation wavelength using a scattering solution. The logarithmic plot of the fluorescence intensity measured at 680 nm (photon counts) versus time (4096 channels, 12 ps/channel) shows a straight line indicating monoexponential fluorescence decay behavior.

One limitation to the analysis of TCSPC measurements is the fact that, in general, the length of the laser pulse can not be neglected (Figure 1.11). Thus, the total observed fluorescence decay $R(t)$ is represented as a convolution of the instrument response function (IRF) $G(t)$ with the impulse response of the sample, which would be obtained by applying an infinitesimally small δ-pulse, $F(t)$. $G(t)$ contains all of the relevant correction factors for the detector and the complete TCSPC system:

$$R(t) = \int_0^t G(t^*)F(t-t^*)dt^* \tag{1.2}$$

The inverse process of deconvolution is mathematically difficult. Therefore, most approaches fit a sum of exponentials convoluted to an instrument response function to experimental data through an iterative convolution. In single-molecule experiments, photon statistics are generally low, which renders the application of deconvolution procedures more difficult and time-consuming. Therefore, alternative strategies appear to be more suitable. To fully maximize the information content of a TCSPC single-molecule fluorescence data set, statistical estimator techniques have been shown to give the best results when dealing with the stochastic nature of single-molecule fluorescence data. In particular, for identifying molecules based on their fluorescence lifetime, pattern matching using the maximum likelihood estimator (MLE) technique has been shown to give the theoretically best possible identification for typical fluorophores with nanosecond fluorescence lifetimes [62–65]. In addition, the use of neuronal networks for the identification of single molecules according to their fluorescence lifetime has been shown successfully [66]. For ideal single-molecule data, it was found that the neural networks and the MLE perform almost equally well.

Most approaches to the determination of the fluorescence lifetime of single-molecule data were realized using a monoexponential MLE-algorithm [67, 68]:

$$1+\left(e^{\frac{T}{\tau}}-1\right)^{-1}-m\left(e^{\frac{mT}{\tau}}-1\right)^{-1}=N^{-1}\sum_{i=1}^{m} iN_i$$

where

T is the width of each channel
m the number of utilized time channels
N the number of photon counts taken into account
N_i is the number of photon counts in time channel i.

Typical parameters used in single-molecule data analysis are $m=50$ and $T=0.2$ ns [64, 65, 69]. The left-hand side of the equation is independent of the measured experimental data and only determined by the fluorescence lifetime, τ.

However, it should be considered that in most single-molecule experiments it is not necessary to determine the fluorescence lifetime of individual unknown species precisely. Rather, the different species and their fluorescence lifetimes are known a priori and have thus to be identified and discriminated relative to each other. For the identification of different fluorescent molecules it is not necessary to collect as many photons as are necessary for an exact lifetime measurement with the TCSPC technique. Instead, a much smaller number of photons is sufficient [70–73]. The discrimination is achieved by comparing the raw data with the expected fluorescence decays measured in concentrated solutions with high precision.

1.10 Fluorescence Anisotropy

When fluorophores are excited by polarized light (e.g., by a laser), molecules whose transition dipole moments are oriented parallel to the electric field vector E will be excited preferentially. This *photo selection* similarly results in polarized fluorescence emission, which can seriously influence lifetime measurements. As transition dipole moments for absorption and emission have fixed orientations within a fluorophore, excitation of fluorophores oriented in a rigid matrix (an organic solvent at low temperature or a polymer at room temperature) with polarized light leads to significant polarization effects. For excitation with linearly polarized light the fluorescence anisotropy, *r*, and polarization, *P*, are defined by

$$r = \frac{I_{vv} - I_{vh}}{I_{vv} + 2I_{vh}}$$

$$P = \frac{I_{vv} - I_{vh}}{I_{vv} + I_{vh}}$$

where

I_{vv} and I_{vh} are the fluorescence intensities measured upon vertical (v) excitation under parallel (vv) or perpendicular (horizontal) (vh) oriented polarizers.

If the rotational correlation time ϕ (the characteristic lifetime of rotational diffusion) is much faster than the fluorescence lifetime, τ, of the fluorophore ($\phi \ll \tau$), emission of the sample is completely depolarized ($I_{vv} = I_{vh}$ and $r = 0$). With typical fluorescence lifetimes of organic fluorophores in the nanosecond range and rotational correlation times of the order of 100 ps, fluorescence emission of organic fluorophores in low viscosity solvents (e.g., water) is depolarized. On the other hand, if ϕ is slower than τ, for example, for fluorophores conjugated to large biomolecules, measurements at lower temperatures or in solvents of high viscosity, the emission is strongly polarized. However, the maximum anisotropy, r_0, which corresponds to the limit of a transparently frozen solvent, can only be measured infrequently because of reabsorption and the different energy transfer processes that promote depolarization. Therefore, the technique of fluorescence anisotropy can be used advantageously to study protein–protein interactions or interactions between nucleic acids and proteins in homogeneous solution.

Complete loss of anisotropy occurs when detection is performed below an angle of 54.7°, the *magic angle*, relative to the direction of polarization of the excitation light. That is, anisotropy is eliminated when the angle between the emission dipole of the fluorophore, γ, and the incident excitation light corresponds to 54.7°.

$$r = \frac{3\cos^2\gamma - 1}{2}$$

Time-resolved fluorescence anisotropy measurements, that is, recording of the fluorescence decay behind a polarizer in parallel and perpendicular positions relative to the linear polarization of the pulsed excitation light, enables the construction of the time-resolved fluorescence anisotropy decay, $r(t)$.

$$r(t) = r_0\, e^{-\frac{t}{\phi}}$$

For spherically symmetrical molecules the anisotropy decay can be described by a monoexponential model. Multiexponential anisotropy decays imply that the molecule under investigation exhibits unsymmetrical geometry. As rotational correlation times of small organic fluorophores are of the order of 100 ps, a short excitation pulse and deconvolution methods are even more important than for time-resolved fluorescence measurements. With the aid of the rotational correlation time, measured from time-resolved anisotropy measurements, the rotational volume of the molecule V in the solvent of viscosity η at temperature T can be determined using the gas constant, R.

$$\phi = \frac{\eta V}{RT}$$

References

1 Schäfer, F.P. (1973) Principles of dye laser operation, in *Topics in Applied Physics "Dye Lasers"*, vol. 1 (ed. F.P. Schäfer) Springer-Verlag Berlin, Heidelberg, New York, pp. 1–83.

2 Kuhn, H. (1948) *J. Chem. Phys.*, **16**, 840–841.

3 Kuhn, H. (1949) *J. Chem. Phys.*, **17**, 1198–1212.

4 Scheibe, G. (1941) *Angew. Chem.*, **49**, 567.

5 Jelley, E.E. (1936) *Nature*, **138**, 1009–1010.

6 Herz, A.H. (1977) *Adv. Colloid Interface Sci.*, **8**, 237–298.

7 Kasha, M., Rawls, H.R., and El-Bayoumi, M.A. (1965) *Pure Appl. Chem.*, **11**, 371–392.

8 Czikklely, V., Försterling, H.D., and Kuhn, H. (1970) *Chem. Phys. Lett.*, **6**, 207–210.

9 Scheibe, G. (1948) *Z. Elektrochem.*, **52**, 283–292.

10 Möbius, D. (1995) *Adv. Mater.*, **7**, 437–444.

11 Rabinowitch, E. and Epstein, L. (1941) *J. Am. Chem. Soc.*, **63**, 69.

12 Förster, T. and König, E. (1957) *Z. Elektrochem.*, **61**, 344–348.

13 Rösch, U., Yao, S., Wortmann, R., and Würthner, F. (2006) *Angew. Chem. Int. Ed.*, **45**, 7026–7030.

14 Doose, S., Neuweiler, H., and Sauer, M. (2005) *ChemPhysChem.*, **6**, 1–10.

15 Heinlein, T., Knemeyer, J.P., Piestert, O., and Sauer, M. (2003) *J. Phys. Chem. B*, **107**, 7957–7964.

16 Seidel, C.A.M., Schulz, A., and Sauer, M. (1996) *J. Phys. Chem.*, **100**, 5541–5553.

17 Mataga, N. and Kubota, T. (1970) Chs 3 and 7, in *Molecular Ineractions and Electronic Spectra*, Marcel Dekker, New York.

18 Ketelaar, J.A.A., Van de Stolpe, C., Goudsmit, A., and Dzcubas, W. (1952) *Recl. Trav. Chim. Pays-Bas*, **71**, 1104.

19 Neuweiler, H., Schulz, A., Böhmer, A., Enderlein, J., and Sauer, M. (2003) *J. Am. Chem. Soc.*, **125**, 5324–5330.

20 Marmé, N., Knemeyer, J.P., Wolfrum, J., and Sauer, M. (2003) *Bioconjugate Chem.*, **14**, 1133–1139.

21 Internationla Union of Pure and Applied Chemistry, (1997) *IUPAC Compendium of Chemical Terminology*, 2nd edn, Blackwell Scientific, Oxford.

22 Franck, J. (1926) *Trans. Faraday Soc.*, **21**, 536–542.

23 Condon, E. (1926) *Phys. Rev.*, **28**, 1182–1201.

24 Condon, E. (1928) *Phys. Rev.*, **32**, 858–872.

25 Shpolski, E.V. (1962) *Usp. Fiz. Nauk.*, **77**, 321.

26 Moerner, W.E. and Carter, T.P. (1987) *Phys. Rev. Lett.*, **59**, 2705.

27 Tamarat, Ph., Maali, A., Lounis, B., and Orrit, M. (2000) *J. Phys. Chem. A*, **104**, 1–16.

28 Moerner, W.E. and Kador, L. (1989) *Phys. Rev. Lett.*, **62**, 2535.

29 Orrit, M. and Bernard, J. (1990) *Phys. Rev. Lett.*, **65**, 2716.

30 Wild, U.P., Güttler, F., Priotta, M., and Renn, A. (1992) *Chem. Phys. Lett.*, **193**, 451.

31 Kasha, M. (1952) *J. Phys. Chem.*, **20**, 71.

32 Labhart, H. and Heinzelmann, W. (1973) *Organic Molecular Photophysics*, vol. 1 (ed. J.B. Birks) John Wiley & Sons, Ltd, Chichester, pp. 297–355.

33 Liphardt, B. and Lüttke, W. (1982) *Chem. Ber.*, **115**, 2997–3010.

34 Tsien, R.Y. and Waggoner, A. (1995) Fluorophores for confocal microscopy – photophysics and photochemistry, In: *Handbook of Biological Confocal Microscopy*, 2nd edn, (ed. J.B. Pawley) Plenum Press, New York, p. 267.

35 Moore, W.M.L. and Hammond, G.S. (1961) *J. Am. Chem. Soc.*, **83**, 2789–2794.
36 Widengren, J. and Schwille, P. (2000) *J. Phys. Chem. A*, **104**, 6416.
37 Asimov, M.M., Gavrilenko, V.N., and Rubinov, A.N. (1990) *J. Lumin.*, **46**, 243.
38 Menzel, R. and Thiel, R. (1998) *Chem. Phys. Lett.*, **291**, 237.
39 Menzel, R., Bornemann, R., and Thiel, E. (1999) *Phys. Chem. Chem. Phys.*, **1**, 2435.
40 Tinnefeld, P., Herten, D.P., and Sauer, M. (2001) *J. Phys. Chem. A*, **105**, 7989–8003.
41 Maroncelli, M. and Fleming, G.R. (1987) *J. Chem. Phys.*, **86**, 6221–6239.
42 Maroncelli, M., MacInnis, J., and Fleming, G.R. (1989) *Science*, **243**, 1674–1681.
43 Jarzeba, W., Walker, G.C., Johnson, A.E., and Barbara, P.F. (1988) *J. Phys. Chem.*, **92**, 7039–7041.
44 Jarzeba, W., Walker, G.C., Johnson, A.E., and Barbara, P.E. (1991) *Chem. Phys.*, **152**, 57–68.
45 Akesson, E., Walker, G.C., and Barbara, P.F. (1991) *J. Chem. Phys.*, **95**, 4188–4194.
46 Jiang, Y., McCarthy, P.K., and Blanchard, G.J. (1994) *Chem. Phys.*, **183**, 249–267.
47 Reichardt, C. (1979) *Angew. Chem.*, **91**, 119–131.
48 Lippert, E. (1957) *Z. Elektrochem.*, **61**, 962–975.
49 Suppan, P. (1990) *J. Photochem. Photobiol. A: Chem.*, **50**, 293–330.
50 Drexhage, K.H. (1973) Structure and properties of laser dyes, In: *Topics in Applied Physics "Dye Lasers"*, vol. 1 (ed. F.P. Schäfer) Springer-Verlag, Berlin, Heidelberg, New York, pp. 144–179.
51 Strickler, S.J. and Berg, R.A. (1962) *J. Chem. Phys.*, **37**, 814–822.
52 Demas, J.N. and Crosby, G.A. (1971) *J. Phys. Chem.*, **75**, 991.
53 Flu, C. and Whinnery, J.R. (1973) *Appl. Opt.*, **12**, 72.
54 Long, M.E., Swofford, R.L., and Albrecht, A.C. (1976) *Science*, **191**, 183–185.
55 Twarowskl, A.J. and Kliger, D.S. (1977) *Chem. Phys.*, **20**, 259.
56 Brannon, J.H. and Magde, D. (1978) *J. Phys. Chem.*, **82**, 705–709.
57 Scott, T.G., Spencer, R.D., Leonard, N.J., and Weber, G.J. (1970) *J. Am. Chem. Soc.*, **92**, 687–695.
58 Lakowicz, J.R. (1999) *Principles of Fluorescence Spectroscopy*, 2nd edn, Plenum Press, New York.
59 Bollinger, T. (1961) *Rev. Sci. Instrum.*, **32**, 1044.
60 O'Connor, D.V. and Phillips, D. (1984) *Time-Correlated Single Photon Counting*, Academic Press, London.
61 Becker, W. (2005) Advanced time-correlated single photon counting techniques, In: *Springer Series in Chemical Physics*, Springer-Verlag, Berlin Heidelberg.
62 Enderlein, J. and Sauer, M. (2001) *J. Phys. Chem. A*, **105**, 48–53.
63 Maus, M., Cotlet, M., Hofkens, J., Gensch, T., De Schryver, F.C., Schaffer, J., and Seidel, C.A.M. (2001) *Anal. Chem.*, **73**, 2078.
64 Herten, D.P., Tinnefeld, P., and Sauer, M. (2000) *Appl. Phys. B*, **71**, 765.
65 Sauer, M., Angerer, B., Han, K.T., and Zander, C. (1999) *Phys. Chem. Chem. Phys.*, **1**, 2471.
66 Bowen, B.P., Scruggs, A., Enderlein, J., Sauer, M., and Woodbury, N. (2004) *J. Phys. Chem. A*, **108**, 4799–4804.
67 Tellinghuisen, J. and Wilkerson, C.W. Jr. (1993) *Anal. Chem.*, **65**, 1240–1246.
68 Enderlein, J., Goodwin, P.M., Van Orden, A., Ambrose, W.P., Erdmann, R., and Keller, R.A. (1997) *Chem. Phys. Lett.*, **270**, 464–470.
69 Sauer, M., Arden-Jacob, J., Drexhage, K.H., Göbel, F., Lieberwirth, U., Mühlegger, K., Müller, R., Wolfrum, J., and Zander, C. (1998) *Bioimaging*, **6**, 14–24.
70 Kullback, S. (1959) *Information Theory and Statistics*, John Wiley & Sons, Inc., New York.
71 Köllner, M. and Wolfrum, J. (1992) *Chem. Phys. Lett.*, **200**, 199–204.
72 Köllner, M. (1993) *Appl. Opt.*, **32**, 806–820.
73 Köllner, M., Fischer, A., Arden-Jacob, J., Drexhage, K.H., Seeger, S., and Wolfrum, J. (1996) *Chem. Phys. Lett.*, **250**, 355–360.

2
Fluorophores and Fluorescent Labels

2.1
Natural Fluorophores

Substances that display significant fluorescence generally possess delocalized electrons, formally present in conjugated double bonds. Most proteins and all nucleic acids are colorless in the visible region of the spectrum. However, they exhibit absorption and emission in the ultraviolet (UV) region. Natural fluorophores in tissue include the reduced form of nicotinamide adenine dinucleotide (NADH) and flavin adenine dinucleotide (FAD), structural proteins such as collagen, elastin, and their crosslinks, and the aromatic amino acids, each of which has a characteristic wavelength for excitation with an associated characteristic emission. Within the proteins the aromatic amino acids tryptophan, tyrosine, and phenylalanine are responsible for the fluorescence signal emitted (Figure 2.1) [1].

Among the three aromatic amino acids, tryptophan is the most highly fluorescent. Owing to differences in fluorescence quantum yield and resonance energy transfer from proximal phenylalanine to tyrosine or tyrosine to tryptophan, fluorescence of proteins is usually dominated by tryptophan fluorescence. The tryptophan residues of proteins generally account for about 90% of the total fluorescence of proteins. A natural fluorophore is highly sensitive to the polarity of its surrounding environment. In general, proteins absorb light at 280 nm, and fluorescence emission maxima range from 320 to 350 nm. The fluorescence quantum yield for tryptophan in different proteins is virtually unpredictable, and may lie anywhere in a range varying from <0.01 to around 0.35, with lifetimes well below 1 ns up to $\sim$7 ns. One serious limitation of native fluorescence detection of aromatic amino acids is their low photostability under one- and two-photon excitation conditions, which renders the application of native fluorescence for highly sensitive detection schemes, for example, single-molecule fluorescence spectroscopy, almost impossible [2, 3]. In addition, the intrinsic fluorescence of tryptophan or tyrosine residues is generally much weaker compared with conventional fluorescent dyes.

Nucleotides and nucleic acids are generally nonfluorescent at room temperature ($\Phi_f < 10^{-4}$) in aqueous solvents, but show a slightly increasing fluorescence yield with decreasing temperature [4, 5]. However, some modified derivatives, such as

Handbook of Fluorescence Spectroscopy and Imaging. M. Sauer, J. Hofkens, and J. Enderlein

ISBN: 978-3-527-31669-4

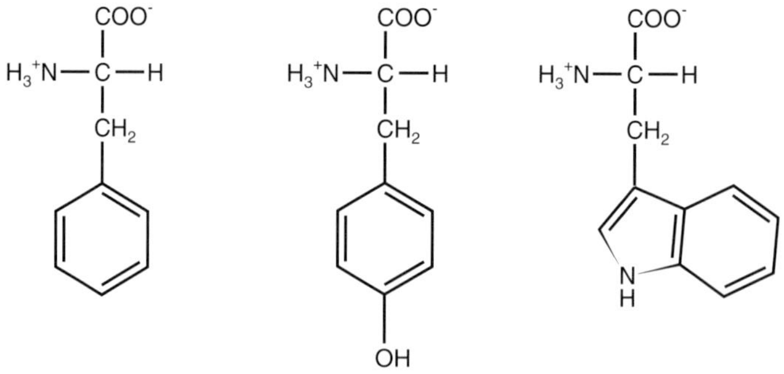

Phenylalanine Tyrosine Tryptophan

Figure 2.1 Molecular structures of the aromatic amino acids.

7-methylguanosine and 7-methylinosine are strongly fluorescent at room temperature [5]. Also deoxyadenosine derivatives bearing alkenyl side chains at C(8) are highly fluorescent upon excitation at ~280 nm with emission maxima at ~400 nm [6].

Nicotinamide adenine dinucleotide (NADH) and flavin adenine dinucleotide (FAD) play important roles in the energy metabolism of cells. The reduced cofactor NADH is highly fluorescent, with absorption and emission maxima at 340 and 450 nm, respectively. On the other hand, the oxidized form NAD^+ is nonfluorescent. The fluorescent group in NADH is the reduced nicotinamide ring, and its fluorescence is partially quenched by collisions with the adenine moiety. Upon binding to proteins, the quantum yield of NADH generally increases drastically, thus enabling the relatively sensitive detection of native proteins carrying an NADH residue. Since the isolation of the *Old Yellow Enzyme* by Warburg and Christian in 1932, the number of known flavoproteins has increased considerably [7, 8]. Flavoenzymes, that is, enzymes that contain the naturally fluorescent flavin cofactor as the redox-active prosthetic group, are involved in numerous redox processes in metabolic oxidation–reduction, photobiology, and biological electron transport [9]. Flavin cofactors are derivatives of riboflavin, a compound better known as vitamin B_2 (Figure 2.2).

Riboflavin is synthesized by bacteria and plants but has to be absorbed by higher organisms. The enzymes flavokinase and FAD can convert riboflavin into flavin mononucleotide (FMN) and flavin adenine dinucleotide (FAD), the cofactors commonly found in flavoproteins. The essential part of the flavin cofactor is the isoalloxazine ring (Figure 2.2). FMN and FAD exhibit a characteristic yellow color, that is, they absorb light in the visible range at ~450 nm, and emit around 515 nm. The fluorescence spectra as well as the fluorescence quantum yield of flavins strongly depend on the environment. In aqueous solution, riboflavin and FMN possess a rather high fluorescence quantum yield, Φ_f, of 0.26 [10]. On the other hand, FAD exhibits a much smaller fluorescence quantum yield, Φ_f, of 0.03, because of the formation of an intramolecular quenching complex between the flavin and adenine moiety [10]. As FAD is only naturally fluorescent in its oxidized form, fluctuations in

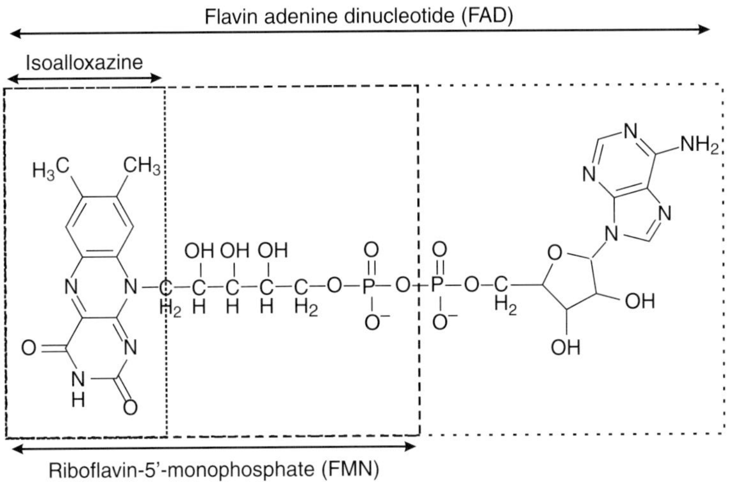

Flavin adenine dinucleotide (FAD)

Isoalloxazine

Riboflavin-5'-monophosphate (FMN)

Figure 2.2 Molecular structures of FMN and FAD in the oxidized state.

fluorescence intensity of single flavoenzymes can be used directly to follow the oxidation of substrates by oxygen. That is, fluctuations in the fluorescence of their active sites can be used to examine enzymatic turnovers of single flavoenzyme molecules [11].

Phycobiliproteins derived from cyanobacteria and eukaryotic algae constitute another class of water soluble natural fluorophores [12]. They are found in light harvesting structures (phycobilisomes) and are used as accessory or antenna pigments for photosynthetic light collection. They absorb light in the wavelength range of the visible spectrum that is poorly utilized by chlorophyll and, through fluorescence energy transfer, convey the energy to the membrane-bound photosynthetic reaction centers, where fast electron transfer occurs with high efficiency, converting solar energy into chemical energy. Phycobiliproteins are classified on the basis of their color into phycoerythrins (red) and phycocyanins (blue), with absorption maxima lying between 490 and 570 nm, and 610 and 665 nm, respectively. Energy transfer proceeds successively from phycoerythrin via phycocyanin and allophycocyanin to chlorophyll *a* with an overall efficiency of almost 100% [13, 14]. A minor competing process to the deactivation of excited pigments is the emission of chlorophyll *a* fluorescence. At room temperature, most fluorescence is emitted by chlorophyll *a* of photosystem II, with a sharp peak around 685 nm and a broad shoulder at about 740 nm [15].

The phycobilisomes allow the various pigments to be arranged geometrically in a manner that helps to optimize the capture of light and transfer of energy. Accordingly, they have a very low fluorescence yield *in vivo*, which increases enormously on extraction when transfer to chlorophyll is prevented. Phycobiliproteins are composed of a number of subunits, each having a protein backbone to which linear tetrapyrrole chromophores are covalently bound. All phycobiliproteins contain several phycocyanobilin or phycoerythrobilin chromophores [12–14, 16]. Each bilin has unique

Table 2.1 Spectroscopic characteristics (absorption maximum, λ_{abs}, emission maximum, λ_{em}, and molecular weight, *MW*) of some prominent phycobiliproteins in aqueous buffer.

Pigment	λ_{abs} (nm)	λ_{em} (nm)	*MW* (kDa)
R-Phycoerythrin	565 (495)	575	240
B-Phycoerythrin	545	575	240
C-Phycocyanin	615	647	220
R-Phycocyanin	617 (555)	637	100
Allophycocyanin	652	660	100

spectral characteristics, which may be further modified by interactions of the subunits and of the chromophore with the apoprotein (Table 2.1).

B-phycoerythrin is compromised of three polypeptide subunits forming an aggregate containing a total of 34 bilin chromophores. It exhibits an absorption cross-section equivalent to that of ~20 rhodamine 6G chromophores and has the highest fluorescence quantum yield of all phycobiliproteins, 0.98, with a fluorescence lifetime of 2.5 ns [17–19]. B-phycoerythrin was the first species to be detected at the single-molecule level using laser-induced fluorescence [19–21]. However, single allophycocyanin molecules with a smaller fluorescence quantum yield of 0.68 [22] can also be easily visualized when immobilized on a cover glass surface under aqueous buffer [23]. Therefore, phycobiliproteins are often used as fluorescent labels in bioanalytical applications requiring high sensitivity.

Phytochromes are biliprotein photosensors that regulate many physiological processes in green plants, enabling them to adapt to fluctuating light environments. A red, far-red reversible chromoprotein, phytochrome, was the first photoreceptor to be identified [24, 25]. Phytochromes are large proteins with covalently bound linear tetrapyrrole, that is, bilin, chromophores that transduce light signals by reversibly photointerconverting between red-light-absorbing and far-red-light-absorbing species, a process that typically initiates a transcriptional signaling cascade [26]. Because of efficient double-bond isomerization in the bilin chromophores, the excited state is rapidly depopulated. Therefore, phytochromes are comparatively poorly fluorescent biliproteins [27]. However, by introduction of bilin analogs that lack the photoisomerizing double bond, strongly yellow-orange fluorescent holoproteins (phytofluors) have been produced [28]. The ability to tag proteins of interest through fusion with an apophytochrome gene and to produce phytofluors within living cells has attracted a great deal of interest and considerable effort has been spent on the investigation of the fluorescence properties of red-emitting phytofluors using fluorescence correlation spectroscopy (FCS) [29]. A comparative study with standard organic fluorophores demonstrated that a specific mutant of a phytochrome (PR1: phytofluor red 1) enables even single-molecule detection upon excitation at 632.8 nm using a conventional helium:neon laser [30, 31].

While phytochromes monitor the red and far-red regions of the electromagnetic spectrum, UV-A/blue light perception is mediated by the cryptochromes, photoreceptors regulating stomatal aperture in response to blue light, and the chromoprotein encoded by the NPH1 gene. NPH1 apoprotein noncovalently binds FMN to form the holoprotein nph1. The N-terminal region of the protein contains two domains of about 110 amino acids (LOV1 and LOV2), which are regulated by environmental factors that affect their redox status: light, oxygen, or voltage (LOV) [32–34]. Moreover, both domains show bright fluorescence at ~500 nm upon excitation at ~450 nm and might therefore be used as efficient alternatives to green fluorescent proteins for genetic labeling.

2.2 Organic Fluorophores

Organic fluorophores or fluorescent dyes are characterized by a strong absorption and emission band in the visible region of the electromagnetic spectrum. The long-wavelength absorption band of a fluorophore is attributed to the transition from the electronic ground state S_0 to the first excited singlet state S_1. As the transition moment for this process is typically very large, the corresponding absorption bands exhibit oscillator strengths of the order of unity. The reverse process $S_1 \rightarrow S_0$ is responsible for spontaneous emission known as fluorescence and for stimulated emission. Much of the knowledge we have today stems from the time when new organic fluorescent dyes were developed as amplifying media for dye lasers [35].

However, since the use of organic fluorescent dyes for qualitative and quantitative determination of analyte molecules (especially for automated DNA sequencing in the 80th), their importance for bioanalytical applications has increased considerably [36–39]. Owing to the enhanced demand for fluorescent markers, the development of new fluorescent dyes has increased significantly in recent years [40–45]. Using fluorescent dyes, extremely high sensitivity down to the single-molecule level can be achieved. Furthermore, time- and position-resolved detection without contact with the analyte is possible. Typically, the fluorescent probes or markers are identified and quantified by their absorption and fluorescence emission wavelengths and intensities. The sensitivity achievable with a fluorescent label is directly proportional to the molar extinction coefficient for the absorption and the quantum yield of the fluorescence. The extinction coefficient typically has maximum values of about $10^5\,l\,mol^{-1}\,cm^{-1}$ in organic fluorophores and the quantum yield may approach values close to 100%. Absorption and emission spectra, in addition to the fluorescence quantum yield and lifetime, are dependent on environmental factors. Furthermore, for labeling of biological compounds, for example, antibodies or DNA/RNA, the dye must carry a functional group suitable for a mild covalent coupling reaction, preferentially with free amino or thiol groups of the analyte. In addition, the fluorophore should be as hydrophilic as possible to avoid aggregation and nonspecific binding in aqueous solvents.

Suitable organic fluorescent dyes are distinguished by a high fluorescence quantum yield. That is, upon excitation into the excited singlet state and subsequent

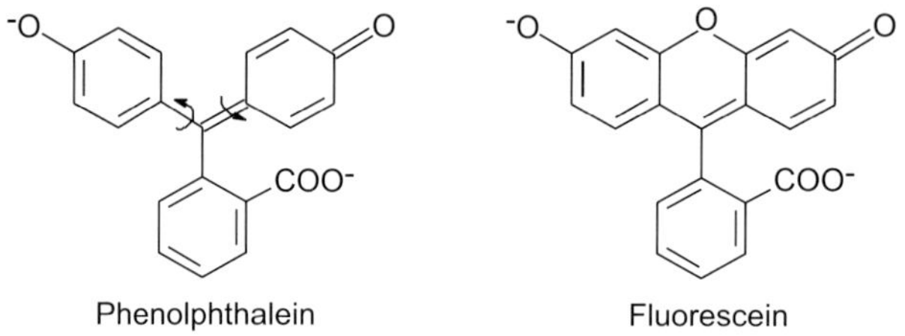

Phenolphthalein Fluorescein

Figure 2.3 Molecular structures of phenolphthalein and fluorescein in basic solution, pH 9.0.

relaxation to the lowest vibronic level of S_1, the radiative decay to the singlet ground state S_0 is the preferred deactivation pathway. However, there are many nonradiative processes that can compete efficiently with light emission, and thus reduce the fluorescence efficiency to a degree that depends, in a complicated fashion, on the molecular structure of the dye. Internal conversion, that is, the nonradiative decay of the lowest excited singlet state S_1 directly to the ground state S_0, is mainly responsible for the loss of fluorescence efficiency of organic dyes [46]. One important factor that substantially lowers the fluorescence quantum yield of organic dyes is structural flexibility. For example, phenolphthalein is practically nonfluorescent in alkaline solution due to the rotational mobility of the phenyl rings (Figure 2.3). Therefore, the introduction of an oxygen bridge causes a strong increase in fluorescence quantum yield. The resulting fluorescein dye exhibits a strong fluorescence intensity in basic solutions (Figure 2.3).

Since the introduction of fluorescein isocyanate for immunofluorescence by Coons *et al.* [47], it has been the fluorophore of choice in many applications. Fluorescein exhibits a high fluorescence quantum yield of 0.90 and a Stokes shift of 22 nm ($\lambda_{abs} = 494$ nm; $\lambda_{em} = 526$ nm) at pH 9.0. Furthermore, the dye is fairly hydrophilic. Therefore, it shows only a slight nonspecific affinity to biological material. Subsequently, the isocyanate has been replaced as the active intermediate for covalent coupling by isothiocyanate and *N*-succinimidyl esters (NHS), being more convenient and safe derivates. On the other hand, the extinction coefficient and the fluorescence quantum yield and lifetime are strongly pH dependent, and decrease with deceasing pH value. Furthermore, fluorescein derivatives exhibit a comparably low photostability in aqueous solvents, the preferred solvent for biological applications.

Another pathway of internal conversion which is, to a first approximation, independent of temperature and solvent viscosity, and occurs efficiently in certain dyes even if their chromophore is fully rigid and planar, is hydrogen vibrations [46]. The process involves the conversion of the lowest vibronic level of the excited state to a higher vibronic level of the ground state. It can be expected that only those hydrogen atoms that are directly attached to the chromophore of the dye will influence the nonradiative deactivation. Indeed, replacement of hydrogen by deuterium, that is, by dissolving the dye in monodeuterated methanol, reduces the rate of nonradiative decay and thus increases the fluorescence efficiency [48]. The mechanism is of minor importance in dyes that emit in the visible range, but can seriously reduce the fluorescence efficiency of infrared dyes. Therefore, long-wavelength (>700 nm) absorbing fluorophores generally exhibit only very poor fluorescence intensities, especially in aqueous surrounding.

Besides fluorescence quenching via photoinduced electron transfer (PET) or fluorescence resonance energy transfer (FRET), a molecule excited to S_1 may enter a triplet state and relax to the lowest level T_1. The occupation of triplet states is undesirable with respect to single-molecule fluorescence applications for various reasons. Firstly, owing to the relatively long lifetimes of triplet states, the chromophore might be excited into higher excited triplet states and undergo irreversible chemical reactions, that is, it might photobleach. Secondly, provided the dye resides in the triplet state, no fluorescence photon can be detected, thus decreasing the number of detected fluorescence photons. On the other hand, photoinduced reverse intersystem crossing might occur, which induces complicated photophysics and renders the interpretation of single-molecule data more difficult. That is, excitation into higher excited triplet states might repopulate the singlet manifold, because singlet and triplet energies are better matched in higher excited states, and/or some of the restrictions of intersystem crossing are relaxed due to the different nature of the triplet symmetry [49, 50]. Therefore, in some organic fluorescent dyes the triplet lifetime can be reduced by increasing the excitation intensity, that is, with increasing excitation intensity the dye is pumped into higher excited triplet states, thus inducing reverse intersystem crossing [51–54].

Furthermore, the intrinsic intersystem crossing rate can be enhanced if the dye is substituted with heavier elements, which increase the spin–orbit coupling [46]. This effect can be demonstrated by comparing the fluorescence quantum yield of some fluorescein derivatives (Figure 2.4).

While fluorescein has a triplet yield of ~0.03 in basic solution, the triplet yield of eosin, a fluorescein derivative bearing four bromine substituents was found to be 0.76 [55]. On the other hand, substitution of bromine by chlorine atoms has very little effect on the intersystem crossing rate of fluorescein. However, replacement of the oxygen atoms at the 3- and 6-positions by sulfur yields the dye dithiofluorescein, a fluorescein derivative with an absorption maximum in basic ethanol at 635 nm, which is absolutely nonfluorescent [46].

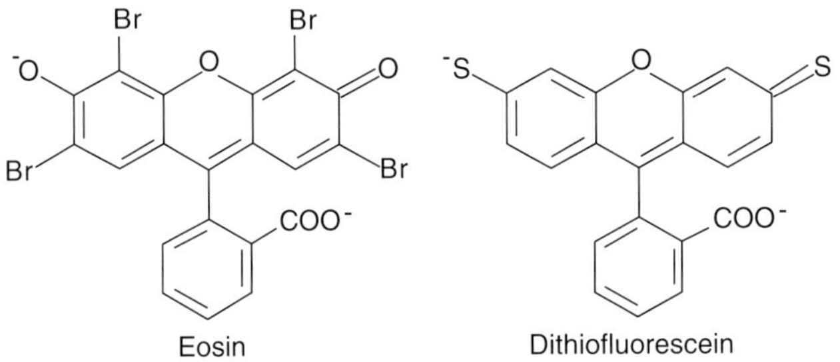

Br Br $^{-}$O O O Br Br COO^{-} Eosin

$^{-}$S O S COO^{-} Dithiofluorescein

Figure 2.4 Molecular structures of eosin and dithiofluorescein in basic solution, pH 9.0.

2.3
Different Fluorophore Classes

Coumarin derivatives were used successfully as efficient laser dyes because of their marked Stokes shift, a property that was used to achieve a wide tuning range in dye lasers. Coumarins can essentially be described by two mesomeric forms, one nonpolar form with a low dipole moment and a more polar form with a higher dipole moment where a positive charge is located on the nitrogen atom and a negative charge is on the oxygen atom (Figure 2.5).

In the electronic ground state S_0 of coumarins, the nonpolar mesomeric structure is predominant and the polar form makes only a minor contribution to the actual π-electron distribution. The more polar mesomeric form is stabilized if the dye molecule is surrounded by polar solvent molecules. Therefore, the absorption maximum of coumarin dyes is generally shifted to longer wavelengths with increasing solvent polarity. In the electronic excited state S_1, the more polar mesomeric form is predominant. That is, the electric dipole moment in coumarin dyes increases upon optical excitation. This induces the rearrangement of the surrounding solvent molecules and stabilizes the excited state, which lowers the energy of the excited state considerably. Therefore, coumarin derivatives exhibit a large Stokes shift as compared with, for example, rhodamine or oxazine dyes. Figure 2.6 shows the molecular structure of some prominent coumarin derivatives. Usually, coumarin derivatives are coupled covalently to biomolecules using activated carboxyl functions, as for example in the case of 7-diethylaminocoumarin-3-acetic acid (Figure 2.6) [56]. As electron donating alkyl groups stabilize the positive charge on the nitrogen atom in the more polar mesomeric form, a shift in the absorption band to longer wavelengths occurs from Coumarin 120 ($\lambda_{abs} = 351$ nm, $\lambda_{em} = 440$ nm) via Coumarin 1 ($\lambda_{abs} = 373$ nm, $\lambda_{em} = 460$ nm) to Coumarin 102 ($\lambda_{abs} = 390$ nm, $\lambda_{em} = 480$ nm). Finally, it was found that the absorption and emission characteristics of coumarin derivatives can be further extended towards longer wavelengths if a heterocyclic substituent is introduced into the 3-position (see Coumarin 6 with $\lambda_{abs} = 455$ nm, $\lambda_{em} = 540$ nm in methanol). As a typical example of Coumarin dyes, Coumarin 1 exhibits a molar extinction coefficient of 23 500 M^{-1} cm^{-1} at 373.25 nm [57] and a fluorescence quantum yield of 0.73 [58].

It has been known for more than two decades that the fluorescence of 7-aminocoumarins is quenched by a variety of organic electron donors and acceptors [59]. In general, donors with half-wave oxidation potentials that are less positive

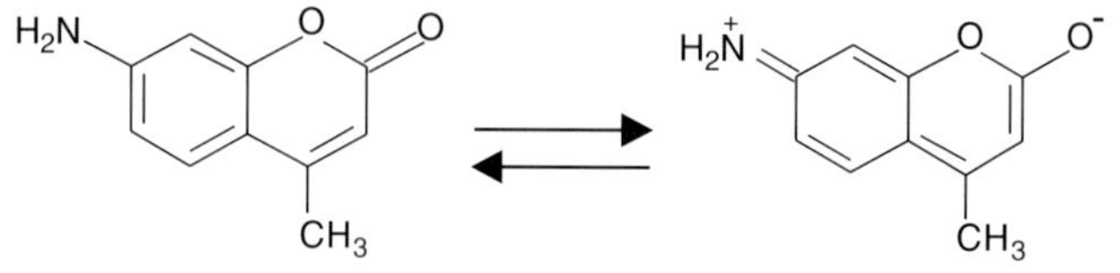

Figure 2.5 Two mesomeric forms of Coumarin 120. In methanol Coumarin 120 exhibits an absorption and an emission maximum of 351 and 440 nm, respectively.

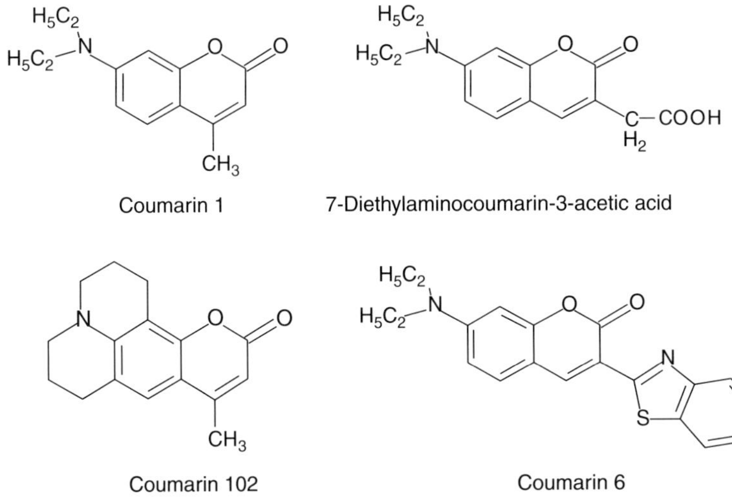

Coumarin 1 7-Diethylaminocoumarin-3-acetic acid

Coumarin 102 Coumarin 6

Figure 2.6 Molecular structures of different coumarin derivatives (Coumarin 1, Coumarin 102, Coumarin 6, and 7-diethylaminocoumarin-3-acetic acid).

than 1.0 V versus SCE (standard calomel electrode) and acceptors with reduction potentials less negative than −1.5 V versus SCE are candidates for diffusion limited quenching of coumarin singlet states. However, it was discovered only recently that coumarin derivatives are quenched via photoinduced electron transfer (PET) by the four different DNA bases in a specific manner. Therefore, coumarin dyes are sometimes denoted as "intelligent" dyes [60]. An intelligent dye is one that has a fluorescence lifetime that depends on the DNA base to which it is bound. The shift in lifetime is caused by excited-state interactions between the fluorescent dye and the DNA base. The base-specific fluorescence quenching efficiency results in different fluorescence lifetimes, which can be used for identification of the base type. A dye that is appropriate for this purpose is Coumarin 120. Phosphothioate modified nucleotides labeled with C-120 influence the fluorescence lifetime and quantum yield of Coumarin 120 in a peculiar manner. The four conjugates have fluorescence lifetimes between 5.3 and 1.9 ns for the Coumarin 120 adenosine and guanosine conjugate, respectively. Depending on the redox properties of the DNA base, the dye is reduced or oxidized in its excited state. The measured fluorescence quantum yield and lifetime strongly depend on the DNA base, and on the length and type of linker connecting the base and the chromophore [61].

Unfortunately, most coumarins have a very low photochemical stability, which renders their use in single-molecule spectroscopic applications more difficult. The quantum yield of photobleaching under moderate one-photon excitation (OPE) conditions is of the order of 10^{-3}–10^{-4}, which is two orders of magnitude larger than the photobleaching yield of rhodamine dyes [62]. Brand and coworkers studied fluorescence bursts from single Coumarin 120 molecules using OPE at 350 nm and two-photon excitation (TPE) at 700 nm in aqueous solution [63]. Their results give clear evidence that OPE at a high irradiance results in two-step photolysis via the first electronic excited singlet and triplet states, S_1 and T_1, producing dye radical ions and solvated electrons. Hence, this additional photobleaching pathway limits the applicable irradiance for OPE. Using coherent TPE for single-molecule detection, saturation of the fluorescence was observed for a high quasi-CW (continuous wave) irradiance ($10^8\,\mathrm{W\,cm^{-2}}$), which may also be caused by photobleaching. Furthermore, TPE is deteriorated by other competing nonlinear processes (e.g., continuum generation in the solvent), which only occur above a certain threshold irradiance. They concluded that the single-molecule detection sensitivity of Coumarin 120 molecules is enhanced substantially by using TPE, primarily due to the higher background with OPE at UV wavelengths.

Alexa Fluor 350 and 430, and the two dyes ATTO 390 and ATTO 425, belong to a class of commercially available coumarin derivatives for biological labeling applications. Their absorption maxima are reflected in their names, that is, the absorptions maxima are at ~350, 430, 390, and 425 nm, respectively. The emission maxima of the three coumarins are located around 445, 545, 480, and 485 nm, respectively. They exhibit extinction coefficients in the range of 20 000–50 000 $\mathrm{M^{-1}\,cm^{-1}}$ with fluorescence quantum yields of up to 0.90 and lifetimes of between 3 and 4 ns.

Today, most (bio)analytical applications requiring high sensitivity use xanthene dyes that absorb and emit in the wavelength region from 500 to 700 nm. Owing to their structural rigidity, xanthene dyes show high fluorescence quantum yields. As with fluorescein dyes, xanthenes exhibit a small Stokes shift of about 20–30 nm. They are applied as complementary probes together with other fluorophores in double-label staining, as energy donors and acceptors in FRET experiments, and as fluorescent markers in DNA sequencing and immunoassays. Xanthene dyes are in general far more stable than fluorescein and coumarin derivatives under aqueous conditions [45, 64]. As most commercially available ATTO and Alexa dyes absorbing and emitting in the wavelength range between 480 and 630 nm belong to the class of xanthene dyes, this fluorophore class will be introduced in more detail. To increase the water solubility of the fluorophores for bioanalytical applications, the xanthene chromophores are often modified by the attachment of sulfonate groups.

The π-electron distribution in the chromophore of the xanthene dyes can be described approximately by two identical mesomeric structures, in which the positive charge is located on either of the two nitrogen atoms (Figure 2.7). Unlike the coumarin dyes, the two forms have the same weight, and thus in xanthene dyes there is no static dipole moment parallel to the long axis of the molecule in either the ground or excited states.

The transition moment of the main long-wavelength absorption band is oriented parallel to the long axis of the molecule. Interestingly, some transitions at shorter

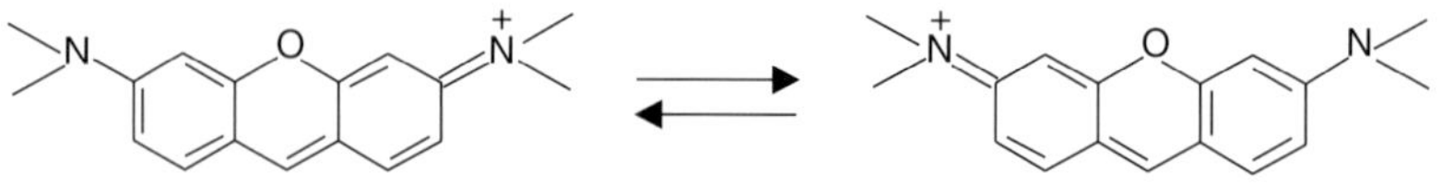

Figure 2.7 Two identical mesomeric structures of xanthene dyes.

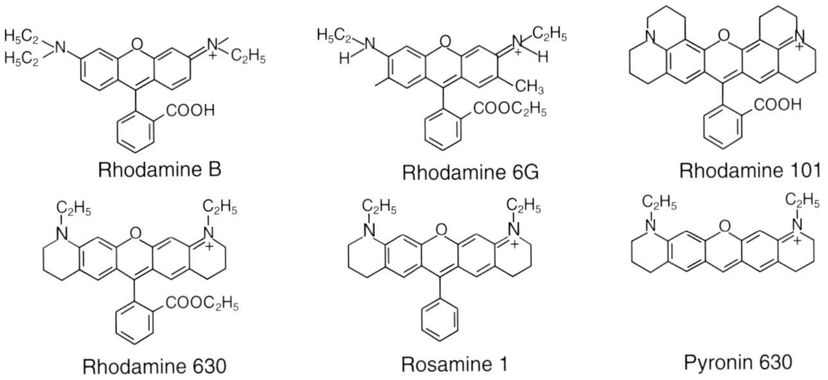

Rhodamine B Rhodamine 6G Rhodamine 101

Rhodamine 630 Rosamine 1 Pyronin 630

Figure 2.8 Molecular structures of Rhodamine B, Rhodamine 6G, Rhodamine 101, Rhodamine 630, Rosamine 1, and Pyronin 630.

wavelength are oriented perpendicular to the long axis [65]. The absorption spectrum of xanthene dyes is determined by the symmetrical π-electron system extending across the diaminoxanthene frame. Because the dipole moment does not change upon excitation, the absorption maximum shows only little dependence on the polarity of the solvent. On deprotonation of the carboxyl group, for example, in Rhodamine B or Rhodamine 101 (Figure 2.8), a small hypsochromic shift to shorter wavelengths occurs [46]. In contrast to fluorescein derivatives, the esterified xanthene derivatives, such as Rhodamine 630 and Rhodamine 6G, do not show a pH-dependent absorption or emission spectrum.

The fluorescence efficiency of xanthene dyes shows a peculiar dependence on the substitution pattern of the amino groups. If the amino groups are fully alkylated, as in the case of Rhodamine B, the fluorescence efficiency is strongly dependent on solvent viscosity and temperature. These effects can be attributed to some type of mobility of the diethylamino groups in the excited state, which is enhanced by increasing temperature and reduced by increasing viscosity. However, the decrease in fluorescence lifetime by changing from ethanol to water, a solvent with nearly similar viscosity, demonstrates that other solvent properties, such as solvent polarity, also influence the excited-state lifetime and thereby the fluorescence efficiency (Table 2.2). Although only of minor importance for xanthene dyes absorbing and emitting between 500 and 600 nm, it strongly reduces the fluorescence lifetime and quantum yield of xanthene and oxazine derivatives absorbing and emitting above 600 nm. On the other hand, if the amino groups are only partially alkylated or incorporated in six-membered rings, for example, in the case of Rhodamine 630, Rhodamine 101, and Rhodamine 6G, the fluorescence efficiency is close to unity and virtually independent of solvent polarity and temperature (Table 2.2).

Although Rhodamine B and Rhodamine 101 carry an unesterified carboxyl group, this group is not amenable to covalent coupling to analyte molecules, due to steric hindrance. Usually, xanthene dyes are coupled to analytes via an additional reactive

Table 2.2 Spectroscopic characteristics (absorption maximum, λ_{abs}, emission maximum, λ_{em}, and fluorescence lifetime, τ) of various xanthene derivatives in aqueous buffer and ethanol at 25 °C. Fluorescence quantum yields are given only for ethanolic dye solutions. To ensure protonation of the *o*-carboxyl group in Rhodamine B and Rhodamine 101, measurements were performed in aqueous buffer at pH 3.0 and upon addition of a drop of trifluoroacetic acid to 1 ml of alcoholic dye solution, respectively. Typically, xanthene dyes exhibit extinction coefficients in the range of 1×10^5 to $1.3 \times 10^5\ M^{-1}cm^{-1}$ in alcoholic solutions.

Xanthene derivative	λ_{abs} (nm)	λ_{em} (nm)	τ (ns)	Φ_f (in ethanol)
Rhodamine B	557/552	578/579	1.43/2.28	0.55
Rhodamine 6G	526/530	556/556	3.89/3.79	0.95
Rhodamine 101	579/574	608/601	4.17/4.28	0.92
Rhodamine 630	564/563	588/587	4.04/4.06	0.97
Rosamine 1	565/562	592/589	2.68/3.76	0.88
Pyronin 630	559/559	579/579	3.69/3.72	0.92

carboxyl group (typically activated as NHS) attached to one of the amino groups or to the carboxyphenyl ring. As the carboxyphenyl substituent is held in a position almost perpendicular to the xanthene moiety, it is not part of the chromophore system. Hence it has only minor influence on the absorption and emission spectrum of the dye.

Removal of the bulky *o*-carboxyl group leads to so-called rosamines, which show reduced quantum efficiency in fluid solvents such as ethanol, methanol, or water [44]. In Rosamine 1 (Figure 2.8) the steric hindrance for torsion of the phenyl group is reduced, leading to a configuration in which the planes of the phenyl substituents and the xanthene ring system can nearly approach coplanarity in the first excited state. This process can result in a state with charge transfer character and a reduced $S_1 - S_0$ energy gap. Hence, the internal conversion rate increases as demonstrated by a reduced fluorescence efficiency and lifetime (Table 2.2). In this double-potential minimum model, the height of the potential barrier and the energy difference between the initial and the final states are controlled by solvent viscosity and polarity as well as by steric and electronic properties of the chromophore and the phenyl substituents [44]. Therefore, rosamine derivatives with strong electron donating or accepting phenyl substituents, for example, with *p*-amino or *p*-nitro phenyl substituents, show only weak fluorescence efficiency.

Rhodamine dyes that carry a free *o*-carboxyl group can exist in several forms. The deprotonation is enhanced by dilution or by adding a small amount of a base. The deprotonated zwitterionic form exhibits an absorption maximum shifted to shorter wavelengths (3–10 nm). In nonpolar solvents such as acetone, the zwitterionic form is not stable and forms an intramolcular lactone in a reversible fashion. The lactone is colorless because the π-electron system of the dye is interrupted. A further interesting characteristic of xanthene derivatives consists in the fact that they are selectively quenched upon contact formation with the DNA base guanine and the amino acid tryptophan via photoinduced electron transfer [66–74]. Both guanine and tryptophan act as efficient electron donors quenching the excited singlet state of xanthene derivatives efficiently.

There are several functional xanthene derivatives commercially available for biological labeling: ATTO 488, ATTO 520 and ATTO 532 related to Rhodamine 6G, ATTO 550 related to Rhodamine B, ATTO 565 related to Rhodamine 630, ATTO 590, and ATTO 594. The molecular structures of the Alexa derivatives Alexa 488, Alexa 532, Alexa 546, Alexa 568, Alexa 594, and Alexa 633 are also based on the xanthene chromophores shown in Figure 2.8 [75].

Owing to the small number of compounds that demonstrate intrinsic fluorescence above 600 nm, the use of near-infrared (NIR) fluorescence detection in bioanalytical samples is a desirable alternative to visible fluorescence detection. This fact has prompted current efforts towards the use of NIR dyes for bioanalytical applications. However, there are few chromophores that show sufficient fluorescence quantum yields in the NIR region, especially in aqueous surroundings, and that can be coupled covalently to analyte molecules [76]. This finding is due to the fact that at longer wavelengths the fluorescence efficiency tends to decrease with decreasing energy difference between S_1 and S_0. As already mentioned, hydrogen vibrations in particular lead to a decreased quantum yield in the red-near-IR region. The quanta of hydrogen vibrations have the highest energies for organic compounds. Thus, hydrogen vibrations are most likely to contribute to internal conversion between S_1 and S_0. This mechanism, which is expected only for those hydrogens that are attached directly to the chromophore, seriously reduces the fluorescence efficiency of infrared dyes [45, 46]. Therefore, only a limited number of fluorescent dyes with sufficient quantum yields that absorb at wavelengths above 620 nm are known, and fewer still are available in reactive form for conjugation with biomolecules.

There are several other advantages to using fluorescent dyes that absorb in the red over those that absorb at shorter blue and green wavelengths. The most important of these advantages is the reduction of the background signal, which ultimately improves the sensitivity achievable. There are three major sources of background: (i) elastic scattering, that is, Rayleigh scattering, (ii) inelastic scattering, that is, Raman scattering, and (iii) fluorescence from impurities. The efficiency of both Rayleigh and Raman scattering are dramatically reduced by shifting to longer wavelength excitation (scales with $1/\lambda^4$). Likewise, the number of fluorescent impurities is significantly reduced with longer excitation and detection wavelengths.

One class of red-absorbing fluorophores constitute cyanine dyes [77–80]. Cyanine derivatives belong to the class of polymethine dyes, that is, planar fluorophores with conjugated double bonds where all atoms of the conjugated chain lie in a common plane linked by σ-bonds (see Section 1.1). Cyanine dyes can be best described by two identical mesomeric structures $R_2N[CH{=}CH]_nCH{=}N^+R2 \leftrightarrow R_2N^+{=}CH[CH{=}CH]_nNR_2$, where n is a small number that defines the longest wavelength absorption, and the nitrogen and part of the conjugated chain usually form part of a heterocyclic system, such as imidazole, pyridine, pyrrole, quinoline, or thiazole. Figure 2.9 gives the molecular structure of some commercially available symmetric (e.g., Cy5) and asymmetric (e.g., Dy-630) cyanine derivatives. As in the case of

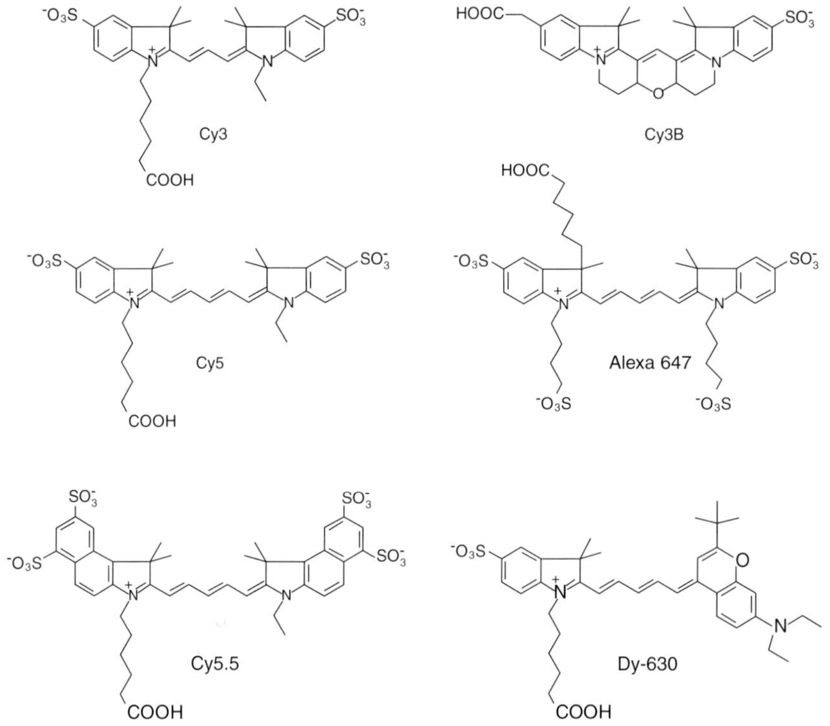

Figure 2.9 Molecular structures of the cyanine derivatives Cy3, Cy3B, Cy5, Alexa 647, Cy5.5, and Dy-630.

xanthene dyes, there is no static dipole moment parallel to the long axis of symmetric cyanine derivatives in either the ground or the excited states, because the two mesomeric forms have the same weight.

Cyanine dyes are widely used in ultrasensitive imaging and spectroscopy, especially for biological applications. The absorption can be tuned through the visible and near-infrared region by variation of the length of the polymethine chain joining the two heads of the cyanine dye. As can be seen in Table 2.3 each additional double bond shifts the absorption maximum by roughly 100 nm towards the red (see Cy3 $\rightarrow$ Cy5 $\rightarrow$ Cy7). The molar extinction coefficients of cyanine derivatives are comparably high and lie between 1.2 and $2.5 \times 10^5\,M^{-1}\,cm^{-1}$ [81–83]. On the other hand, the fluorescence quantum yield only varies between 0.04 and 0.4 with lifetimes in the range of a few hundred picoseconds (0.2–1.0 ns). In addition, cyanine dyes are less photostable than xanthene dyes, especially under single-molecule conditions [84] and some derivatives (Cy5) are destroyed by environmental ozone [85].

Fluorescence experiments have revealed several expected and also unexpected photophysical phenomena of cyanine derivatives (especially Cy5), such as *cis–trans* isomerization, off-states in addition to triplet formation, and complex photobleaching pathways including nonfluorescent intermediates that still absorb light in the

Table 2.3 Spectroscopic characteristics (absorption maximum, λ_{abs}, emission maximum, λ_{em}) of various cyanine derivatives in aqueous buffer at 25 °C. Cy5, for example, exhibits a fluorescence quantum yield of 0.27, a lifetime of 0.91 ns, and a molar extinction coefficient of $2.5 \times 10^5\ M^{-1}cm^{-1}$.

Cyanine derivative	λ_{abs} (nm)	λ_{em} (nm)
Cy3	548	562
Cy3B	558	572
Cy3.5	581	596
Cy5	647	664
Alexa 647	649	666
Cy5.5	673	692
Cy7	747	774
Dy-630	627	651
Dy-640	627	667
Dy-650	646	670

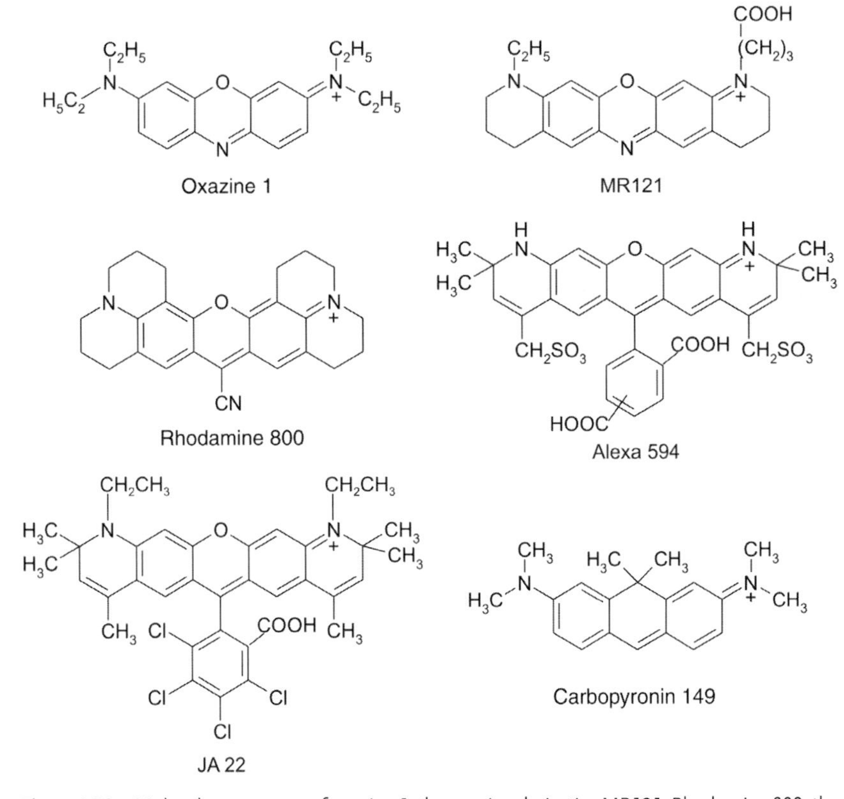

Figure 2.10 Molecular structures of oxazine 1, the oxazine derivative MR121, Rhodamine 800, the Rhodamine derivatives Alexa 594 and JA 22, and ATTO 611x .

visible range [86–89]. A fluorescence correlation spectroscopy study of Cy5 showed that for most excitation conditions that are relevant for ultrasensitive fluorescence spectroscopy, a photostationary equilibrium is established between the *cis*- and *trans*-forms, where approximately 50% of the Cy5 dye molecules can be expected to be in their nonfluorescent *cis*-states [87]. The rate constant for *cis–trans* isomerization in water was measured to be $2.5 \times 10^7\,s^{-1}$ under moderate excitation conditions (>10 $kW\,cm^{-2}$). To reduce *cis–trans* isomerization in Cy3B, the polymethine structure has been made more rigid by the incorporation of three 6-membered rings (Figure 2.9). Therefore, Cy3B exhibits a fluorescence quantum yield of 0.70 and a prolonged fluorescence lifetime of 2.8 ns.

However, a method of stabilizing normal cyanine dyes in solution has been developed and is used extensively in single-molecule experiments, by applying an oxygen scavenging system and thus retracting the main reason for the photobleaching of the cyanine dyes [90]. The oxygen scavenging system is composed of phosphate-buffered saline (PBS), pH 7.4, containing 10% (wt/vol) glucose, 12.5% (vol/vol) glycerin, 50–100 $\mu g\,ml^{-1}$ glucose oxidase, 100–200 $\mu g\,ml^{-1}$ catalase, and 0.4–0.8 mM DTT. To quench the lifetime of the triplet states in the absence of oxygen, 100 mM β-mercaptoethylamine (MEA) is added as a triplet quencher. Consequently, cyanine derivatives have emerged as a set of standard dyes in many multicolor single-molecule assays.

Recently [91, 92], it was demonstrated that the commercially available cyanine derivatives Cy5 and Alexa 647 can intriguingly act as efficient reversible single-molecule photoswitches, and that the fluorescent states of these, after apparent photobleaching by 633 nm excitation, can be restored by irradiation in the range of ~300–532 nm. Besides the importance of single-molecule photoswitches, for example, for optical data storage, this finding implies limitations for the use of cyanine dyes such as Cy5 and Alexa 647 as acceptors in single pair fluorescence resonance energy transfer (sp-FRET) experiments.

In addition to cyanine derivatives, xanthene derivatives are also available with absorption and emission wavelengths above 620 nm. For example, it has been known for a long time that if the central carbon group of a pyronin or rhodamine dye is replaced by a nitrogen atom, a compound is obtained whose absorption and emission are shifted by about 100 nm to longer wavelengths [46]. Such planar oxazine derivatives are rigid and exhibit suitable spectroscopic characteristics in the wavelength region 580–700 nm. The exchange of the carboxyphenyl substituent by an electron-accepting group at the central carbon has an effect similar to the introduction of a nitrogen atom (see Figure 2.10 and compare with the data in Table 2.4). In the case of a cyano group, the resulting rhodamine derivative exhibits absorption and emission spectra shifted by about 100 nm. The squares of the frontier orbital coefficients, calculated with semiempirical methods, show that the electron density at the central carbon of a pyronin chromophore is zero in the HOMO but high in the LUMO. By introduction of an electron acceptor at this position, as in case of Rhodamine 800, or exchange of the methane group by a more electronegative atom, as in case of oxazine 1, the energy of the LUMO decreases, resulting in a decreased excitation energy.

Table 2.4 Spectroscopic characteristics (absorption maximum, λ_{abs}, emission maximum, λ_{em}, and fluorescence quantum yield Φ_f) of various red-absorbing and emitting dyes in aqueous buffer (or ethanol[a]) at 25 °C. Fluorescence lifetimes in water vary between ~0.5 ns for ATTO 725 and ~4 ns for ATTO 590 or JA 22.

Fluorophore	λ_{abs} (nm)	λ_{em} (nm)	Φ_f
Boidpy FL	504	513	
Bodipy 630/650	632	640	
Rhodamine 800	682[a]	698[a]	0.19[a]
Oxazine 1	645[a]	667[a]	
MR121	661	673	
JA 22	621[a]	642[a]	0.90[a]
Alexa 594	590	617	
ATTO 590	594	624	0.80
ATTO 594	601	627	0.85
Alexa 610	612	628	
ATTO 610	615	634	0.70
ATTO 611x	611	681	0.35
ATTO 620	619	643	0.50
Alexa 633	632	647	
ATTO 633	629	657	0.64
ATTO 635	635	659	0.25
ATTO 637	635	659	0.29
ATTO 647N	644	669	0.65
ATTO 655	663	684	0.30
Alexa 660	663	690	
Alexa 680	679	702	
ATTO 680	680	700	0.30
Alexa 700	702	723	
ATTO 700	700	719	0.25
ATTO 725	725	752	0.10
ATTO 740	740	764	0.10
Alexa 750	749	775	

Another possibility for shifting the absorption maximum towards longer wavelengths is the addition of two double bonds in the nitrogen-containing rings of xanthene derivatives. This strategy is used to shift the absorption spectrum of organic fluorescent dyes such as Alexa 568, Alexa 594, Alexa 633, and ATTO 590. Owing to the two additional double bonds, the absorption and emission maxima shifts to longer wavelengths are of about 30 nm (Table 2.4). Alternatively, the oxygen atom in rhodamine derivatives can be exchanged by a tetrahedral carbon atom. The absorption maxima of the resulting carbopyronin or carborhodamine dyes (e.g., ATTO 611x or ATTO 635) exhibit a 40–80 nm shift towards longer wavelengths, as compared with the corresponding oxygen-bridged dyes. On the other hand, ATTO 611x exhibits a remarkable Stokes shift of 70 nm in aqueous solvents (Table 2.4).

To summarize, red-absorbing rhodamine, oxazine, and carborhodamine or carbopyronin derivatives exhibit a high fluorescence quantum yield of up to 0.90 with lifetimes of 2.0–4.0 ns in alcoholic solutions. Usually, the fluorescence lifetime of xanthene derivatives absorbing and emitting above 620 nm shows a marked decrease on changing from ethanol to an aqueous environment [45]. Exceptions are dyes derived from JA 22 (Figure 2.9). This rhodamine derivative JA 22, with the tetrachlorocarboxyphenyl substituents, also shows a fluorescence efficiency of close to unity in an aqueous buffer. Finally, it has to be noted that all long-wavelength absorbing rhodamine, oxazine, and carborhodamine or carbopyronin derivatives show a much higher photostability than their related cyanine derivatives.

Finally, bora-diaza-indacene derivatives have to be considered, which span the visible spectrum from 500 to ~650 nm. These so-called Bodipy dyes are unusual in that they are relatively nonpolar, they exhibit only a small Stokes shift, and the chromophore is electrically neutral. Therefore, they tend to bind nonspecifically to biomolecules in aqueous surroundings. On the other hand, they exhibit extinction coefficients and quantum yields comparable to xanthene derivatives.

Interestingly, some of the red-absorbing xanthene derivatives show a loss of absorption and emission under basic conditions, for example, while coupling the activated dye to amino functions of a biomolecule at pH ~9.0. Unfortunately, all triphenylmethane and related dyes show a tendency to react at the central carbon with nucleophiles, for example, hydroxide ions, if this carbon is sterically available. Therefore, some of the above mentioned derivatives become colorless on addition of a base to their aqueous solutions, due to the formation of a so-called pseudobase. In the pseudobase the π-electron system is interrupted and therefore the long-wavelength absorption is lost. Although this process is reversible, subsequent reactions, for example, with oxygen, may lead to an irreversible destruction of the dye. It has to be pointed out that the tendency for the formation of the pseudobase is strongly controlled by the structure of the dye. Furthermore, other destructive, that is, irreversible, reactions also tend to increase with increasing absorption maximum. In the case of "normal" rhodamines, the central carbon of the chromophore is protected by the carboxyphenyl substituent. Therefore, rhodamine derivatives, such as JA 22 or Alexa 594, are completely stable even in strong basic solutions. Due to the decrease in fluorescence quantum yield with increasing absorption wavelength, it is advisable to work with red-absorbing fluorescent dyes whose absorption and emission maxima lie between 620 and 700 nm.

Owing to their outstanding chemical and photochemical stabilities and their high fluorescence quantum yields, rylene derivatives have been established as alternative fluorophores, especially in single-molecule fluorescence spectroscopy. The comparably high photostability of rylene dyes immobilized in different polymeric matrices enables the observation of photophysical processes, for example, photon antibunching and electron or energy transfer, at the single-molecule level over extended periods of time [93–98]. Perylenetetracarboxdiimide (PDI) represents the key structure from

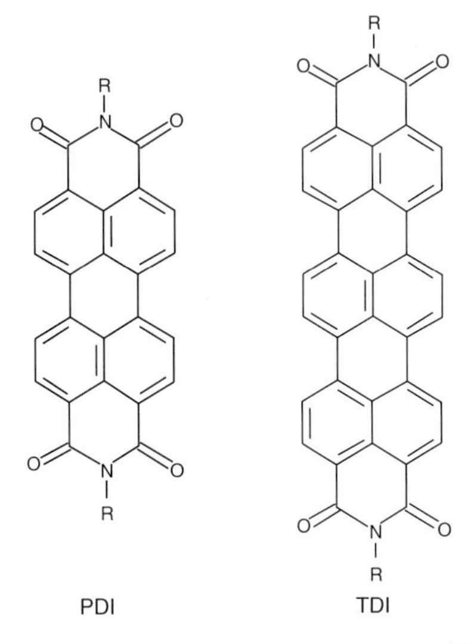

Figure 2.11 Basic molecular structures of the perylene dyes perylenetetracarboxdiimide (PDI) and terrylenediimide (TDI).

which new types of fluorophores, important intermediates, and various functional perylene dyes are derived. Depending on the substitution pattern, PDI exhibits an absorption maximum between 520 and 580 nm with a typical Stokes shift of 30–40 nm. The fluorescence quantum yield is close to unity with the fluorescence lifetime in the range of 4–5 ns. Extension of PDI by one naphthalene results in terrylenediimide derivatives (TDI) (Figure 2.11) with absorption maxima shifted by about 100 nm to the red. By extending the aromatic π-system, the molar extinction coefficient also increases from 60 000 $M^{-1}\,cm^{-1}$ (PDI) to 93 000 $M^{-1}\,cm^{-1}$ (TDI). One limitation of perylene derivatives is represented by their low water solubility. Therefore, strategies based on the incorporation of sulfonated phenoxy groups have been developed to make new red-absorbing water-soluble TDI derivatives available for biological labeling applications [93].

2.4 Multichromophoric Labels

As an alternative to conventional organic fluorescent dyes, fluorescent nano- and microspheres imbedded with organic fluorescent dyes can be used. They have been developed and then conjugated to antibodies or streptavidin for immunological studies using flow cytometry, fluorescence microscopy, and ELISAs (enzyme-linked immunosorbent assay) [99–102]. Furthermore, fluorescent spheres or beads can be synthesized with other surface functional groups, for example, carboxy-, aldehyde-, sulfate-, or amino-modified, to enable conjugation to the molecule of interest. These brightly fluorescent nano- and microspheres are available with diameters from 20 nm to several μm. With respect to the fluorescent dyes embedded, they absorb and emit in the visible range from 400 to 700 nm. Owing to the incorporation of the fluorophores into polystyrene or silica, fluorescent spheres posses many advantages, such as high photostability and signal-to-noise ratio, and environmental insensitivity. Furthermore, homogeneously fluorescing spheres can be used advantageously for calibration and alignment of fluorescence microscopes and flow cytometers. However, even the smallest fluorescent spheres with a size of 20–40 nm are much larger than organic fluorophores with a size of $\leq$1 nm, which seriously deteriorates the binding and interaction characteristics of labeled biomolecules, for example, antibodies with a size of $<$10 nm (Figure 2.12).

Instead of the organic fluorescent dyes, luminescent dyes with long lifetimes in the region of hundreds of nanoseconds to milliseconds can be incorporated into micro- and nanospheres. Besides rare earth metal chelates such as Eu(III)- or Tb (III)-complexes, Pt- and Pd-porphyrin complexes, and ruthenium, osmium or rhenium complexes with polypyridyl ligands are useful long-lived luminescent dyes [103–107]. In combination with time-resolved or time-gated detection schemes, these long-lived luminescent dyes, with lifetimes much longer than that of conventional fluorescent dyes, and autofluorescence are suited to the highly sensitive detection of target molecules in biological samples, for example, tissue. In addition, some of these dyes feature long-wavelength emission at ~600 nm, which

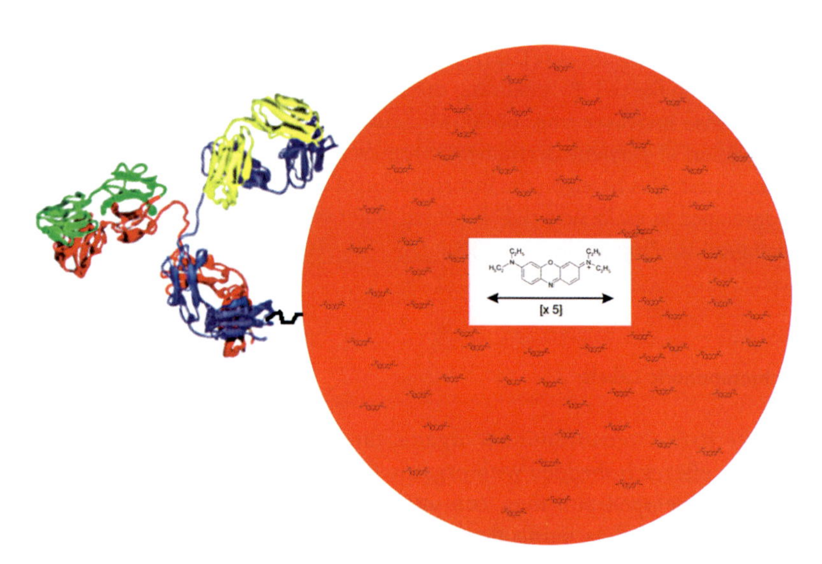

Figure 2.12 A IgG antibody labeled with a fluorescent nanosphere with a size of 20 nm are shown to scale. Comparison of the size of the fluorescent dyes embedded (~1 nm) and the sphere (20 nm) clarifies the different proportions.

is well separated from the excitation peak at <400 nm. On the other hand, these compounds are prone to quenching by oxygen. Therefore, these materials have to be encapsulated in material that is impermeable to oxygen. An additional limitation to the use of luminescent nano- or microspheres is their long lifetime, which seriously limits the maximum number of photons detectable within a given time.

The need to detect as many different analyte molecules as possible simultaneously has driven the development of multicolor or multiparameter strategies, that is, the development of fluorescent labels that can be excited at single wavelengths with comparable efficiency, but emit at different wavelengths. This can be achieved, for example, by the incorporation of two or more differently absorbing and emitting organic fluorophores into microspheres that can undergo fluorescence resonance energy transfer (FRET). Each microsphere contains one class of dyes that absorbs at the desired excitation wavelength. In addition, each microsphere contains one or more longer-wavelength dyes, which are carefully chosen to create an energy transfer cascade that guarantees efficient energy transfer from the initially excited dye to the longest-wavelength dye. Ideally, the excitation energy is transferred quantitatively from dye to dye so that only the longest-wavelength dye in the cascade emits significant fluorescence. As these microspheres exhibit a large Stokes shift, they are ideally suited to detection in samples with significant Rayleigh or Raman scattering or strong autofluorescence. Depending on the dye composition, 40 nm nanospheres (so-called TransFluoSpheres) can be prepared that can be excited efficiently at 488 nm but emit at 560, 605, 645, 685, or 720 nm, respectively. Besides applications in multiparameter experiments, such as energy transfer, nano- or microspheres can be used advantageously in high-resolution multicolor colocalization experiments [108]. Furthermore, bioimaging with luminescent nanoparticle probes has recently attracted widespread interest in biology and medicine. Because luminescent nanoparticles are better in terms of photostability and sensitivity, they are suitable for real time tracking and monitoring of biological events at the cellular level, which may not be accomplished using regular fluorescent dyes.

Alternatively, so-called tandem dyes or resonance energy transfer (RET) dyes have been synthesized to increase the Stokes shift of common fluorophores. The conjugates rely on efficient fluorescence resonance energy transfer between two fluorophores. In most cases tandem dyes based on phycobiliproteins, serving as the energy transfer donors, have been developed, particularly for immunofluorescence applications that apply flow cytometry [109, 110]. For the sake of completeness, the development of light harvesting dendrimers has to be discussed. Tree-like multichromophoric dendrimers with spectrally different fluorophores featuring various absorption and emission spectra have been synthesized and studied at the single-molecule level. The multichromophoric dendrimers show strong absorption over the whole visible spectral range, but only the longest-wavelength absorbing fluorophore shows strong fluorescence [111, 112]. Unfortunately, the hitherto synthesized energy transfer dendrimers are not suitable for mild covalent labeling of biomolecules.

2.5
Nanocrystals

One class of fluorescent nanoparticles with great expectations are semiconductor nanocrystals (NC), such as core-shell CdSe/ZnS NCs [113]. Their unique optical properties – tunable narrow emission spectrum, broad excitation spectrum, high photostability and long fluorescence lifetime (of the order of tens of nanoseconds) – make these bright probes attractive in experiments involving long observation times, multicolor and time-gated detection. Furthermore, the relatively long fluorescence lifetime of CdSe nanocrystals, in the region of several tens of nanoseconds, can be used advantageously to enhance the fluorescence biological imaging contrast and sensitivity using time-gated detection [114].

In contrast to metals, semiconductors exhibit a significant energy gap, E_G, typically in the range of a few eV, between the fully occupied valence and the empty conduction band (Figure 2.13). Classical semiconductor materials are based on the electronic structure of the elements of the fourth row (IV) of the Periodic Table. Therefore, semiconductors based on the element combinations of groups III–V, II–VI, and I–VII are frequently found.

Although most of the outer shell electrons are located in the valence band, even at room temperature, there is a certain fraction that is excited into the conduction band and is responsible for the conductive properties of semiconductors. Each electron, e^- in the conduction band leads to a reduced screening of the nuclear charge in the valence band. Therefore, the missing electron is denoted as a "hole," h^+ (Figure 2.13). Generally, electrons and holes can move freely within the bands. However, owing to their opposite charges they can also interact to form an electron–hole (*e–h*) pair or "exciton." The exciton is a quasi-particle with paired spin, that is, a Boson, and can be regarded as a state where the electron and hole orbit around their center-of-mass, similar to the description of a hydrogen atom (Figure 2.13b). The *e–h* binding energy, E_B, is given for

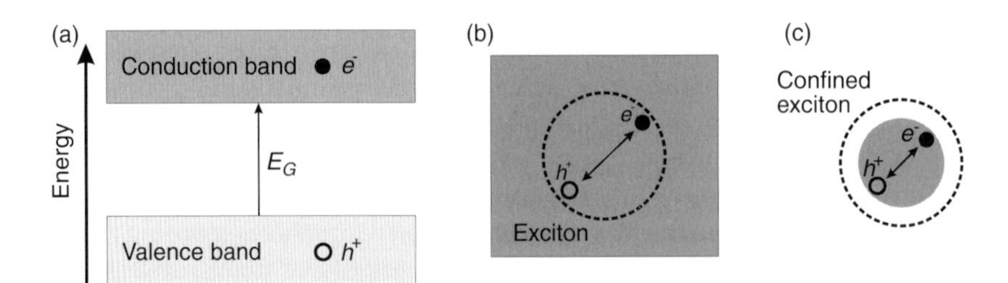

Figure 2.13 (a) Semiconductor bands with fully occupied valence and empty conduction bands separated by the band gap energy, E_G. Excitation of an electron, e^- generates a hole, h^+, in the valence band. (b) In bulk semiconductor an exciton can be formed as an *e–h* pair with a characteristic radius (dashed circle). (c) In the case of a nanocrystal the semiconductor is smaller than the typical radius, thus confining the *e–h* pair.

the nth state by

$$E_B = \frac{\mu^* e^4}{2(4\pi\varepsilon_0\varepsilon\hbar)^2}\frac{1}{n^2}$$

where

μ^* denotes the reduced mass of the electron and hole
ε is the electric permittivity of the semiconductor.

The average distance between e and h, that is, the exciton radius, a_X, in the ground state, is much larger than the Bohr radius of hydrogen of $a_H = 0.5 \times 10^{-10}$ m, and lies in the region of 1–13 nm (e.g., $a_X = 1.7$, 6.1, and 12.5 nm for ZnS, CdSe, and GaAs, respectively).

Significant deviations from bulk semiconductor properties occur when the crystal size is reduced in one or more dimensions to the diameter of the exciton radius (Figure 2.13c) [115–117]. The strong influence of the reduction in size on the semiconductor properties can be understood by considering that both the electron and the hole can theoretically only exist within the semiconductor material, which means, its interface constitutes an infinite potential. For such a particle within a boundary, that is, a particle in the box, the allowed energy states can be calculated according to (Section 1.1)

$$E_c = \frac{h^2}{8\mu^* a^2}\frac{1}{n^2}$$

where

a is the "length" of the box.

While for sizes $a \gg a_X$ this is of minor importance for the energy levels of the e–h pair, in the case of $a \leq a_X$, the confinement regime, the exciton energy is forced into a higher energy level and discrete absorption bands appear (Figure 2.14). Thus, the exciton energy can be controlled by size reduction. In principle, confinement is possible in all three dimensions. Therefore, different structures exist, such as "quantum wells" (1-dimensional confinement), "quantum wires" (2-dimensional confinement), and "quantum dots" (3-dimensional confinement). Although all three types of confinement are important, it is mainly quantum dots (QDs), that is, colloidal nanocrystals (NCs) that are used for external labeling of analyte molecules.

The crucial step for fabrication of such small particles is the controlled growing procedure, which ideally leads to crystals of similar composition, similar structure and nearly equal size (monodisperse particles). Growing of particles starts from an atomic precursor of both materials involved. As soon as clusters have been formed, there are two competing reactions in the supersaturated solution: growth of initially formed clusters and formation of new nuclei. As the latter will lead to increasing heterodispersity, the nucleation process has to be interrupted instantaneously, for example, by a sudden change in temperature. Once a sufficient number of nuclei have been produced, further growth can continue under milder conditions, where

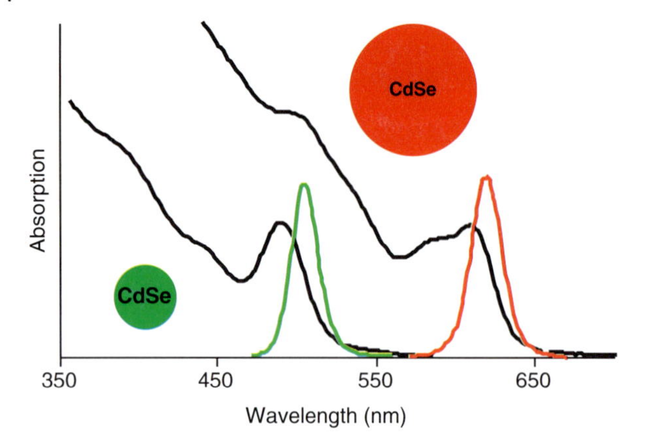

Figure 2.14 Absorption (black) and emission spectra of CdSe NCs of various sizes (green: 2.2 nm; red: 5 nm). The emission maximum is only slightly red-shifted to the longest-wavelength absorption shoulder (corresponding to the ground state of the exciton). The absorption in NCs stems from the photoinduced creation of discrete *e–h* pair combinations. Hence, the absorption would probably be expected to consist of narrow bands instead of a broad band distribution. However, owing to the slight heterodispersity of the sample, inhomogeneous line width broadening results. The overlap of the broadened bands with different absorption cross-sections results in the observed absorption spectra.

no new nucleation occurs [113,118] . For the most widely used material, CdSe, the precursor substances generally consist of dimethylcadmium (Me_2Cd) and tri-*n*-octylphosphine selenide [$(C_8H_{17})_3PSe$], both of which rapidly disintegrate into non-reactive organic compounds and Cd and Se atoms upon heating to 250–300 °C in tri-*n*-octylphosphine oxide, TOPO [118]. Cooling of the reaction mixture to temperatures <200 °C prevents further nucleation, and growth continues for several hours under heating at 250 °C. The progress of the reaction can be easily followed by absorption spectroscopy. A more sophisticated procedure is size-selective precipitation, where, under careful conditions, only the largest NCs in solution are precipitated and removed. This method can be successfully used to synthesize CdSe NCs with a diameter of 1.2–11.5 nm, corresponding to photoluminescence emission in the range of 400–700 nm. NCs synthesized by this method exhibit a wurtzite-like crystal structure and are more or less spherical. However, because the *e–h* pairs created upon excitation strongly interact with the environment, resulting in increased non-radiative deactivitation via "traps", the photoluminescence quantum yield is less than 0.10. Therefore, the "core" of a NC is nowadays coated with a "shell," that is, an insulator material, to yield a stabilized core–shell NC with high quantum yield [119, 120]. The best material for this purpose has been found to be ZnS, which can be grown onto the NCs in an analogous manner to the NC preparation, by addition of precursors (dimethylzinc, Me_2Zn, and hexamethyldisilathiane, $[Me_3Si)_2S]$ to a heated NC solution in TOPO. As ZnS has a much higher bandgap energy than CdSe (3.91 versus 2.51 eV), interactions of the carriers with the environment are greatly reduced, resulting in photoluminescence quantum yields in solution of >0.50.

For biological applications the water solubility of the rather hydrophobic NCs has to be increased. For example, surface modification with thiols with charged residues are the simplest approach to increasing the water solubility. On the other hand, thiol coatings exhibit only a limited stability, which renders their application for long-term studies in living systems difficult [121, 122]. Other materials used to increase the water solubility, and through this enabling biological imaging applications to be made, include coating with silanes [113, 123], peptides [124–126], and ambiphilic polymers [127–130].

Although reports showing NC biocompatibility actually appeared a few years ago [113, 121], their breakthrough for biological targeting was only very recent [130–136]. At present, NCs can be easily functionalized, for example with streptavidin, to facilitate mild coupling to biotinylated biomolecules (Figure 2.15). Immunofluorescent labeling of cancer markers and other cellular targets on the surface of fixed and live cancer cells in addition to staining of actin and microtubule fibers in the cytoplasm, and the detection of nuclear antigens inside the nucleus, have been successfully demonstrated [136]. Therefore, NCs might also be useful for *in vivo* single-molecule studies and the first successful results have already been presented [137, 138].

However, NCs also posses serious limitations. One disadvantage of NCs is the difficulty to engineer them with single binding sites that can be specifically conjugated to only one molecule of interest. Instead, during the labeling step, NCs tend to bind to several molecules simultaneously. Another problem is blinking, which is strongly controlled by the excitation intensity (Figure 2.15b), and often obeys a power law [139, 140]. Although blinking of NCs is efficiently reduced in the presence of 1–10 mM mercaptoethanol, the addition of reducing agents is not compatible with live cell imaging [141]. In addition, not all semiconductor particles are generally active, that is, luminescent, but they can be photoactivated [142]. To explain the strong blinking phenomenon it is assumed that trapping of an electron occurs, induced by photon absorption. Trapping means that the electron is emitted into the surrounding of the NC, thus leaving a charged NC behind. The charged NC can still be excited,

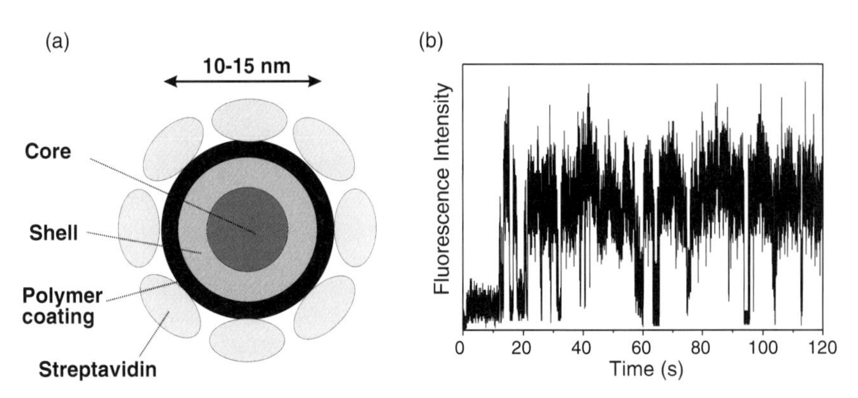

Figure 2.15 (a) Assembling of water soluble quantum dots carrying a streptavidin layer for labeling of biotinylated biomolecules. (b) Fluorescence trajectory of a single QD605 immobilized on a dry cover slide surface (1 ms bin^{-1}).

resulting in the formation of an *e–h* pair, which is efficiently quenched by Auger interactions. After a certain time, the electron can be released back into the NC, neutralizing the charge and resulting in an emissive NC. Typical photoluminescence lifetimes measured for core-shell CdSe or CdTe NCs are in the range of 10–80 ns [143, 144]. A systematic study of the radiative and non-radiative rate fluctuations in single NCs revealed a correlation of photoluminescence emission maxima, lifetime, and intensity of individual NCs with millisecond time resolution [144]. Furthermore, the technique that was applied enabled the exact determination of the photoluminescence quantum yield of single NCs. In this study, average quantum yields of 0.82 for QD 705 and 0.85 for QD 605 were obtained. Very recently [145], it was shown that single NCs show multiexciton emission under high power excitation. This means that it is possible to create more than one *e–h* pair per NC. In contrast to organic dyes *e–h* pairs can coexist in NCs because they are Bosons.

Another issue is whether such particles, which are composed of seemingly toxic material, are well suited for *in vivo* studies and whether they retain biological functionality. For example, besides their core–shell structure, commercially available NCs have a third layer – an organic surface coating – to provide chemical and photophysical stability, inertness in different environments, buffer solubility, and to introduce reactive groups for linking to biomolecules. Ultimately, this results in particle sizes of 15–25 nm in diameter (Figure 2.15a), that is, about 15–25 times larger than conventional fluorophores used to tag biomolecules (for comparison see Figure 2.12).

A new class of water-soluble nanocrystals or nanodots made from small numbers of gold or silver atoms could be the basis for a new biological labeling system with narrower excitation spectra, smaller particle size, and fluorescence comparable to systems based on semiconductor nanocrystals. Nontoxic noble metal nanoclusters composed of only a few atoms also show very strong, robust, discrete, size-dependent emission but with much smaller sizes than those of semiconductor nanocrystals [146–149]. For example, gold nanodots are made up of 5, 8, 13, 23 or 31 atoms, each size fluorescing at a different wavelength between ~350 and ~850 nm to produce ultraviolet, blue, green, red and infrared emissions, respectively [149, 150]. The fluorescence energy varies according to the radius of the crystal, with the smallest structures being the most efficient at light emission.

References

1 Teale, F.W.J. and Weber, G. (1957) *Biochem. J.*, **65**, 476–482.

2 Lippitz, M., Erker, W., Decker, H., van Holde, K.E., and Basché, T. (2002) *Proc. Natl. Acad. Sci. USA*, **99**, 2772–2777.

3 Schüttpelz, M., Müller, C., Neuweiler, H., and Sauer, M. (2006) *Anal. Chem.*, **78**, 663–669.

4 Vigny, P. and Favre, A. (1974) *Photochem. Photobiol.*, **20**, 345–349.

5 Morgan, J.P. and Daniels, M. (1980) *Photochem. Photobiol.*, **31**, 101–113; Kulikowska, E., Bzowska, A., Wierzchowski, J., and Shugar, D. (1986) *Biochim. Biophys. Acta*, **874**, 355–363.

6 Seela, F., Zuluaf, M., Sauer, M., and Deimel, M. (2000) *Helv. Chim. Acta*, **83**, 910–927.

7 Warburg, O. and Christian, W. (1932) *Biochem. Z.*, **254**, 438.

8 Wellner, D. (1967) *Annu. Rev. Biochem.*, **36**, 669–690.

9 Müller, F. (ed.) (1991) *Chemistry and Biochemistry of Flavoenzymes*, vol. I, II, III, CRC Press, Boca Raton.

10 Weber, G. (1950) *Biochem. J.*, **47**, 114–121.

11 Lu, H.P., Xun, L., and Xie, X.S. (1998) *Science*, **282**, 1877–1882.

12 MacColl, R. and Guard-Friar, D. (1987) *Phycobiliproteins*, CRC Press, Boca Raton, Florida.

13 Glazer, A.N. (1982) *Annu. Rev. Microbiol.*, **36**, 173–198.

14 Glazer, A.N. (1984) *Biochim. Biophys. Acta*, **768**, 29–51.

15 Krause, G.H. and Weis, E. (1991) *Annu. Rev. Plant Physiol. Plant Mol. Biol.*, **42**, 313–349.

16 Kronick, M.N. (1986) *J. Immunol. Methods*, **92**, 1–13.

17 Mathies, R.A. and Peck, K. (1990) *Anal. Chem.*, **62**, 1786.

18 Wehrmeyer, W., Wendler, J., and Holzwarth, A.R. (1985) *Eur. J. Cell. Biol.*, **36**, 17.

19 Wu, M., Goodwin, P.M., Ambrose, W.P., and Keller, R.A. (1996) *J. Phys. Chem.*, **100**, 17406–17409.

20 Nguyen, D.C., Keller, R.A., Jett, J.H., and Martin, J.C. (1987) *Anal. Chem.*, **59**, 2158.

21 Peck, K., Stryer, L., Glazer, A.N., and Mathies, R.A. (1989) *Proc. Natl. Acad. Sci. USA*, **86**, 4087.

22 Oi, V., Glazer, A.N., and Stryer, L. (1982) *J. Cell Biol.*, **93**, 981.

23 Ying, L. and Xie, X.S. (1998) *J. Phys. Chem. B*, **102**, 10399–10409.

24 Butler, W.L., Norris, K.H., Siegelman, H.A., and Hendricks, S.B. (1959) *Proc. Natl. Acad. Sci. USA*, **45**, 1703–1708.

25 Briggs, W.R. and Olney, M.A. (2001) *Plant Physiol.*, **125**, 85–88.

26 Rockwell, N.C., Su, Y.S., and Lagarias, J.C. (2006) *Annu. Rev. Plant Biol.*, **57**, 837–858.

27 Braslavsky, S.E. (2003) *Photochroism: Molecules and Systems* (eds. H. Dürr and H. Bouas-Laurent) Elsevier Science, Amsterdam, pp. 738–755.

28 Murphy, J.T. and Lagarias, J.C. (1997) *Curr. Biol.*, **7**, 870–876.

29 Bose, G., Schwille, P., and Lamparter, T. (2004) *Biophys. J.*, **87**, 2013–2021.

30 Fischer, A.J. and Lagarias, J.C. (2004) *Proc. Natl. Acad. Sci. USA*, **101**, 17334–17339.

31 Miller, A.E., Fischer, A.J., Laurence, T., Hollars, C.W., Saykally, R.J., Lagarias, J.C., and Huser, T. (2006) *Proc. Natl. Acad. Sci. USA*, **103**, 11136–11141.

32 Christie, J.M., Salomon, M., Nozue, K., Wada, M., and Briggs, W.R. (1999) *Proc. Natl. Acad. Sci. USA*, **96**, 8779–8783.

33 Kottke, T., Heberle, J., Hehn, D., Dicks, B., and Hegemann, P. (2004) *Biophys. J.*, **84**, 1192–1201.

34 Kottke, T., Dick, B., Hegemann, P., and Heberle, J. (2006) *Biopolymers*, **82**, 373–378.

35 Schäfer, F.P. (ed.) (1973) *Topics in Applied Physics "Dye Lasers"*, vol. 1, Springer-Verlag, Berlin, Heidelberg, New York.

36 Smith, L.M., Fung, S., Hunkapillar, M.W., and Hood, L.E. (1985) *Nucleic Acids Res.*, **13**, 2399.

37 Prober, J.M., Trainer, G.L., Dam, R.J., Hobbs, F.W., Robertson, C.W., Zagursky, R.J., Cocuzza, A.J., Jensen, M.A., and Baumeister, K. (1987) *Science*, **238**, 336.

38 Smith, L.M. (1991) *Nature*, **349**, 812.

39 Ansorge, W., Sproat, B., Stegemann, J., Schwager, C., and Zenke, M. (1987) *Nucleic Acids Res.*, **15**, 4593.

40 Hemmilä, I.A. (1989) *Appl. Fluoresc. Technol.*, **1**, 1–16.

41 Whitaker, J.A., Haugland, R.P., Ryan, D., Hewitt, P.C., and Prendergast, F.G. (1992) *Anal. Biochem.*, **207**, 267–279.

42 Zen, J.M. and Patonay, G. (1991) *Anal. Chem.*, **63**, 2934–2938.

43 Ernst, L.A., Gupta, R.K., Mujumdar, R.B., and Waggoner, A.S. (1989) *Cytometry*, **10**, 3–10.

44 Sauer, M., Han, K.T., Müller, R., Schulz, A., Tadday, R., Seeger, S., Wolfrum, J., Arden-Jacob, J., Deltau, G., Marx, N.J., and Drexhage, K.H. (1993) *J. Fluoresc.*, **3**, 131–139.

45 Sauer, M., Han, K.T., Müller, R., Nord, S., Schulz, A., Seeger, S., Wolfrum, J., Arden-Jacob, J., Deltau, G., Marx, N.J., Zander, C., and Drexhage, K.H. (1995) *J. Fluoresc.*, **5**, 247–261.

46 Drexhage, K.H. (1973) Structure and properties of laser dyes, in *Topics in Applied Physics "Dye Lasers"*, vol. 1

(ed. F.P. Schäfer) Springer-Verlag, Berlin, Heidelberg, New York, pp. 144–179.

47 Coons, H., Creech, H.J., Jones, R.N., and Berliner, E. (1942) *J. Immunol. Methods*, **45**, 159.

48 Drexhage, K.H. (1973) *Laser Focus*, **9**, 35.

49 El-Sayed, M.A. (1986) *Acc. Chem. Res.*, **8**, 1.

50 Tinnefeld, P., Hofkens, J., Herten, D.P., Masuo, S., Vosch, T., Cotlet, M., Habuch, S., Müllen, K., De Schryver, F.C., and Sauer, M. (2004) *ChemPhysChem*, **5**, 1786–1790.

51 English, D.S., Harbron, E.J., and Barbara, P.F. (2000) *J. Phys. Chem. A*, **104**, 9057.

52 Widengren, J. and Seidel, C.A.M. (2000) *ChemPhysPhysChem.*, **2**, 3435.

53 Fleury, L., Segura, J.M., Zumofen, G., Hecht, B., and Wild, U.P. (2000) *Phys. Rev. Lett.*, **84**, 1148.

54 Tinnefeld, P., Buschmann, V., Weston, K.D., and Sauer, M. (2003) *J. Phys. Chem. A*, **107**, 323–327.

55 Soep, B., Kellmann, A., Martin, M., and Lindquist, L. (1972) *Chem. Phys. Lett.*, **13**, 241.

56 Corrie, J.E.T. (1994) *J. Chem. Soc. Perkin Trans. 1*, 2975–2982.

57 Reynolds, G.A. and Drexhage, K.H. (1975) *Optics Commun.*, **13**, 222–225.

58 Jones, G., Jackson, W.R., Choi, C., and Bergmark, W.R. (1985) *J. Phys. Chem.*, **89**, 294–300.

59 Jones, G., Griffin, S.F., Choi, C.Y., and Bergmark, W.R. (1984) *J. Org. Chem.*, **49**, 2705–2708.

60 Seidel, C.A.M., Schulz, A., and Sauer, M. (1996) *J. Phys. Chem.*, **105**, 5541–5553.

61 Han, K.T., Sauer, M., Schulz, A., Seeger, S., and Wolfrum, J. (1993) *Ber. Bunsenges. Phys. Chem.*, **97**, 1728–1730.

62 Eggeling, C., Brand, L., and Seidel, C.A.M. (1997) *Bioimaging*, **5**, 105–115.

63 Brand, L., Eggeling, C., Zander, C., Drexhage, K.H., and Seidel, C.A.M. (1997) *J. Phys. Chem.*, **101**, 4313–4321.

64 Davidson, R.S. and Hilchenbach, M.M. (1990) *Photochem. Photobiol.*, **52**, 431.

65 Jakobi, H. and Kuhn, H. (1962) *Z. Elektrochem. Ber. Bundenges. Phys. Chem.*, **46**, 46.

66 Knemeyer, J.-P., Marmé, N., and Sauer, M. (2000) *Anal. Chem.*, **72**, 3717–3724.

67 Piestert, O., Barsch, H., Buschmann, V., Heinlein, T., Knemeyer, J.P., Weston, K.D., and Sauer, M. (2003) *Nano Lett.*, **3**, 979–982.

68 Heinlein, T., Knemeyer, J.P., Piestert, O., and Sauer, M. (2003) *J. Phys. Chem. B*, **107**, 7957–7964.

69 Marmé, N., Knemeyer, J.P., Wolfrum, J., and Sauer, M. (2003) *Bioconjugate Chem.*, **14**, 1133–1139.

70 Sauer, M. (2003) *Angew. Chem. Int. Ed.*, **115**, 1790–1793.

71 Doose, S., Neuweiler, H., and Sauer, M. (2005) *ChemPhysChem*, **6**, 2277–2285.

72 Neuweiler, H., Schulz, A., Böhmer, A., Enderlein, J., and Sauer, M. (2003) *J. Am. Chem. Soc.*, **125**, 5324–5330.

73 Neuweiler, H. and Sauer, M. (2004) *Curr. Pharm. Biotechnol.*, **5**, 285–298.

74 Neuweiler, H., Doose, S., and Sauer, M. (2005) *Proc. Natl. Acad. Sci. USA*, **102**, 16650–16655.

75 Panchuk-Voloshina, N., Haugland, R.P., Bishop-Stewart, J., Bhalgat, M.K., Millard, P.J., Mao, F., Leung, W.L., and Haugland, R.P. (1999) *J. Histochem. Cytochem.*, **47**, 1179–1188.

76 Soper, S.A. and Mattingly, J. (1994) *J. Am. Chem. Soc.*, **116**, 3744.

77 Boyer, A.E., Lipowska, M., Zen, J.M., Patonay, G., and Tsung, V.C.W. (1992) *Anal. Lett.*, **25**, 415.

78 Southwick, P.L., Ernst, L.A., Tauriello, E.V., Parker, S.R., Mujumdar, R.B., Mujumdar, S.R., Clever, H.A., and Waggoner, A.S. (1990) *Cytometry*, **11**, 418.

79 Mujumdar, R.B., Ernst, L.A., Mujumdar, S.R., and Waggoner, A.S. (1989) *Cytometry*, **10**, 11.

80 Terpetschnig, E., Szmacinski, H., Ozinskas, A., and Lakowicz, J.R. (1994) *Anal. Biochem.*, **217**, 197.

81 Mujumdar, R.B., Ernst, L.A., Mujumdar, S.R., Lewis, C.J., and Waggoner, A.S. (1993) *Bioconjugate Chem.*, **4**, 105–111.

82 Mujumdar, S.R., Mujumdar, R.B., Grant, C.M., and Waggoner, A.S. (1996) *Bioconjugate Chem.*, **7**, 356.

83 Buschmann, V., Weston, K.D., and Sauer, M. (2003) *Bioconjugate Chem.*, **14**, 195–204.

84 Tinnefeld, P., Herten, D.P., and Sauer, M. (2001) *J. Phys. Chem A*, **105**, 7989–8003.

85 Fare, T.L., Coffey, E.M., Dai, H., He, Y.D., Kessler, D.A., Kilian, K.A., Koch, J.E., LeProust, E., Marton, M.J., Meyer, M.R., Stouhton, R.B., Tokiwa, G.Y., and Wang, Y. (2003) *Anal. Chem.*, **75**, 4672–4675.

86 Ha, T., Ting, A.Y., Liang, J., Deniz, A.A., Chemla, D.S., Schultz, P.G., and Weiss, S. (1999) *Chem. Phys.*, **247**, 107.

87 Widengren, J. and Schwille, P. (2000) *J. Phys. Chem. A*, **104**, 6416–6428.

88 Tinnefeld, P., Buschmann, V., Weston, K.D., and Sauer, M. (2003) *J. Phys. Chem. A*, **107**, 323.

89 Ha, T. and Xu, J. (2003) *Phys. Rev. Lett.*, **90**, 223002.

90 Ha, T., Rasnik, I., Chemg, W., Babcock, H.P., Gauss, G.H., Lohman, T.M., and Chu, S. (2002) *Nature*, **419**, 638.

91 Heilemann, M., Margeat, E., Kasper, R., Sauer, M., and Tinnefeld, P. (2005) *J. Am. Chem. Soc.*, **127**, 3801–3806.

92 Bates, M., Blosser, T.R., and Zhuang, X. (2005) *Phys. Rev. Lett.*, **94**, 108101.

93 Herrmann, A. and Müllen, K. (2006) *Chem. Lett.*, **35**, 978–985.

94 Hofkens, J., Maus, M., Gensch, T., Vosch, T., Cotlet, M., Kohn, F., Herrmann, A., Müllen, K., and De Schryver, F. (2000) *J. Am. Chem. Soc.*, **122**, 9278–9288.

95 Vosch, T., Hofkens, J., Cotlet, M., Kohn, F., Fujiwara, H., Gronheid, R., Van Der Biest, K., Weil, T., Herrmann, A., Müllen, K., and De Schryver, F. (2001) *Angew. Chem. Int Ed.*, **40**, 4643–4646.

96 Tinnefeld, P., Weston, K.D., Vosch, T., Cotlet, M., Weil, M., Hofkens, J., Müllen, K., De Schryver, F., and Sauer, M. (2002) *J. Am. Chem. Soc.*, **124**, 14310–14311.

97 Hofkens, J., Cotlet, M., Vosch, T., Tinnefeld, P., Weston, K.D., Ego, C., Grimsdaler, A., Müllen, K., Beljonne, D., Bredas, J.L., Jordens, S., Schweitzer, G., Sauer, M., and De Schryver, F. (2003) *Proc. Natl. Acad. Sci. USA*, **100**, 13146–13151.

98 Vosch, T., Cotlet, M., Hofkens, J., Van Der Biest, K., Lor, M., Weston, K.D., Tinnefeld, P., Sauer, M., Latterini, L., Müllen, K., and De Schryver, F. (2003) *J. Phys. Chem. A*, **107**, 6920–6931.

99 Vogt, R.F., Cross, G.D., Phillips, D.L., Henderson, L.O., and Hannon, W.H. (1991) *Cytometry*, **12**, 525–536.

100 Bhalgat, M.K., Haugland, R.P., Pollack, J.S., Swan, S., and Haugland, R.P. (1998) *J. Immunol. Methods*, **219**, 57–68.

101 Zhao, X., Hillard, L.R., Mechery, S.J., Wang, Y., Bagwe, R.P., Jin, S., and Tan, W. (2004) *Proc. Natl. Acad. Sci. USA*, **101**, 15027–15032.

102 Schaertl, S., Meyer-Almes, F., Lopez-Calle, E., Siemers, A., and Krämer, J. (2000) *J. Biolmol. Screen.*, **5**, 389–397.

103 Kürner, J.M., Klimant, I., Krause, C., Preu, H., Kunz, W., and Wolfbeis, O.S. (2001) *Bioconjugate Chem.*, **12**, 883–889.

104 Diamanidis, E.P. and Christopoulos, T.K. (1990) *Anal. Chem.*, **62**, 1149A–1157A.

105 Kessler, M.A. (1999) *Anal. Chem.*, **71**, 1540–1543.

106 Juris, A., Balzani, V., Barigelletti, F., Campagna, S., Belser, P., and Zelewsky, A.V. (1988) *Coord. Chem. Rev.*, **84**, 85–277.

107 Liebsch, G., Klimant, I., and Wolfbeis, O.S. (1999) *Adv. Mater.*, **11**, 1296–1299.

108 Lacoste, T.D., Michalet, X., Pinaud, F., Chemla, D.S., Alivisatos, A.P., and Weiss, S. (2000) *Proc. Natl. Acad. Sci. USA*, **97**, 9461–9466.

109 Roederer, M., Kantor, A.B., Parks, D.R., and Herzenberg, L.A. (1996) *Cytometry*, **24**, 191–197.

110 Gerstner, A.O., Lenz, D., Laffers, W., Hoffman, R.A., Steinbrecher, M., Bootz, F., and Tarnok, A. (2002) *Cytometry*, **48**, 115–123.

111 Cotlet, M., Vosch, T., Habuchi, S., Weil, T., Müllen, K., Hofkens, J., and De Schryver, F. (2005) *J. Am. Chem. Soc.*, **127**, 9760–9768.

112 Nantalaksakul, A., Reddy, D.R., Bardeen, C.J., and Thayumanavan, S. (2006) *Photosynth. Res.*, **87**, 133–150.

113 Bruchez, M., Moronne, M., Gin, P., Weiss, S., and Alivisatos, A.P. (1998) *Science*, **281**, 2013–2016.

114 Dahan, M., Laurence, T., Pinaud, F., Chemla, D.S., Alivisatos, A.P., Sauer, M., and Weiss, S. (2001) *Opt. Lett.*, **26**, 825–827.

115 Henglein, A. (1982) *Ber. Bunsenges.*, **86**, 301–305.

116 Ekimov, A.I. and Onushchenko, A.A. (1982) *Fiz. I Tekhnika Poluprovodnikov*, **16**, 1215–1219.

117 Efros, A.L. (1982) *Fiz. I Tekhnika Poluprovodnikov*, **16**, 1209–1214.

118 Murray, C.B., Norris, D.J., and Bawendi, M.G. (1993) *J. Am. Chem. Soc.*, **115**, 8706–8715.
119 Hines, M.A. and Guyot-Sionnest, P. (1996) *J. Phys. Chem.*, **100**, 468–471.
120 Dabbousi, B.O., Rodriguez-Viejo, J., Mikulec, F.V., Heine, J.R., Mattoussi, H., Ober, R., Jensen, K.F., and Bawendi, M.G. (1997) *J. Phys. Chem. B*, **101**, 9463–9475.
121 Chan, W.C.W. and Nile, S. (1998) *Science*, **281**, 2016–2018.
122 Aldana, J., Wang, Y.A., and Peng, X. (2001) *J. Am. Chem. Soc.*, **123**, 8844–8850.
123 Gerion, D., Pinaud, F., Williams, S.C., Parak, W.J., Zanchet, D., Weiss, S., and Alivisatos, A.P. (2001) *J. Phys. Chem. B*, **105**, 8861–8871.
124 Akerman, M.E., Chan, W.C.W., Laakkonen, P., Bhatia, S.N., and Ruoslahti, E. (2002) *Proc. Natl. Acad. Sci. USA*, **99**, 12617–12621.
125 Tsay, J.M., Doose, S., Pinaud, R., and Weiss, S. (2005) *J. Phys. Chem. B*, **109**, 1669–1674.
126 Doose, S., Tsay, J.M., Pinaud, F., and Weiss, S. (2005) *Anal. Chem.*, **77**, 2235–2242.
127 Jovin, T.M. (2003) *Nat. Biotechnol.*, **21**, 32–33.
128 Michalet, X., Pinaud, F., Bentolila, L.A., Tsay, J.M., Doose, S., Li, J.J., Sundaresan, G., Wu, A.M., Gambhir, S.S., and Weiss, S. (2005) *Science*, **307**, 538–544.
129 Han, M., Gao, X., Su, J.Z., and Nie, S. (2001) *Nat. Biotechnol.*, **19**, 631–635.
130 Larson, D.R., Zipfel, W.R., Williams, R.M., Clark, S.W., Bruchez, M.P., Wise, F.W., and Webb, W.W. (2003) *Science*, **300**, 1434–1437.
131 Dubertret, B., Skourides, P., Norris, D.J., Noireaux, V., Brivanlou, A.H., and Libchaber, A. (2002) *Science*, **298**, 1759–1762.
132 Jaiswal, J.K., Mattoussi, H., Mauro, J.M., and Simon, S.M. (2003) *Nat. Biotechnol.*, **21**, 47–51.
133 Lingerfelt, B.M., Mattoussi, H., Goldman, E.R., Mauro, J.M., and Anderson, G.P. (2003) *Anal. Chem.*, **75**, 4043–4049.
134 Goldman, E.R., Balighian, E.D., Mattoussi, H., Kuno, M.K., Mauro, J.M., Tran, P.T., and Anderson, G.P. (2002) *J. Am. Chem. Soc.*, **124**, 6378–6382.
135 Mattoussi, H., Mauro, J.M., Goldman, E.R., Anderson, G.P., Sundar, V.C., Mikulec, F.V., and Bawendi, M.G. (2000) *J. Am. Chem. Soc.*, **122**, 12142–12150.
136 Wu, X., Liu, H., Liu, J., Haley, K.N., Treadway, J.A., Larson, J.P., Ge, N., Peale, F., and Bruchez, M.P. (2003) *Nat. Biotechnol.*, **21**, 41–46.
137 Pinaud, F., King, D., Michalet, X., Doose, S., and Weiss, S. (2003) *Eur. Biophys. J.*, **32**, 261.
138 Dahan, M., Levi, S., Luccardini, C., Rostaing, P., Riveau, B., and Triller, A. (2003) *Science*, **302**, 442–445.
139 Kuno, M., Fromm, D.P., Hamann, H.F., Gallagher, A., and Nesbitt, D.J. (2000) *J. Chem. Phys.*, **112**, 3117–3120.
140 Kuno, M., Fromm, D.P., Hamann, H.F., Gallagher, A., and Nesbitt, D.J. (2001) *J. Chem. Phys.*, **115**, 1028–1040.
141 Hohng, S. and Ha, T. (2004) *J. Am. Chem. Soc.*, **126**, 1324–1325.
142 Cordero, S.R., Carson, P.J., Estabrook, R.A., Strouse, G.F., and Buratto, S.K. (2000) *J. Phys. Chem. B*, **104**, 12137–12142.
143 Brokmann, X., Coolen, L., Dahan, M., and Hermier, J.P. (2004) *Phys. Rev. Lett.*, **93**, 107403.
144 Biebricher, A., Sauer, M., and Tinnefeld, P. (2006) *J. Phys. Chem. B*, **110**, 5174–5178.
145 Fisher, B., Caruge, J.M., Zehnder, D., and Bawendi, M. (2005) *Phys. Rev. Lett.*, **94**, 087403.
146 Zheng, J. and Dickson, R.M. (2002) *J. Am. Chem. Soc.*, **124**, 13982–13983.
147 Wilcoxon, J.P., Martin, J.E., Parsapour, F., Wiedenman, B., and Kelley, D.F. (1998) *J. Chem. Phys.*, **108**, 9137–9143.
148 Rabin, I., Schulze, W., and Ertl, G. (1999) *Chem. Phys. Lett.*, **312**, 394–398.
149 Zheng, J., Petty, J.T., and Dickson, R.M. (2003) *J. Am. Chem. Soc.*, **125**, 7780–7781.
150 Zheng, J., Zhang, C., and Dickson, R.M. (2004) *Phys. Rev. Lett.*, **93**, 077402.

3
Fluorophore Labeling for Single-Molecule Fluorescence Spectroscopy (SMFS)

3.1
In Vitro Fluorescence Labeling

Depending on the fluorophore hydrophobicity and the conformational flexibility of the linker used, fluorophores tend to interact nonspecifically with the biomolecule in an unpredictable dynamic fashion. In other words, the fluorophore samples its conformational space, including unforeseeable and uncontrollable quenching interactions, through its local nanoenvironment, for example, with aromatic amino acids when attached to a protein. Therefore, highly water soluble hydrophilic fluorophores should preferably be used in combination with small and rigid linkers. Currently, most single-molecule fluorescence spectroscopy experiments are performed *in vitro*, using fluorophores introduced extrinsically after biosynthesis and purification. At present, fluorophores can be used to covalently label proteins, synthetic oligonucleotides, lipids, oligosaccharides or other biological molecules [1].

Fluorescence labeling is used to investigate localization, interactions, and movement of interesting biological molecules. Reactive groups able to couple with amine-containing molecules are by far the most common functional groups used. An amine coupling process can be used to conjugate with nearly all protein or peptide molecules and with synthetically modified oligonucleotides and other macromolecules. Most of these reactions are rapid and occur in high yield to give stable amide or secondary amine bonds. In general, amine-reactive activated fluorophores are acylating agents that form carboxamides, sulfonamides or thioureas upon reaction with amines. For labeling experiments to the amine groups of (bio)molecules it has to be considered that buffers containing free amines such as tris(hydroxymethyl)aminomethan (Tris), ammonium sulfate, and glycine must be avoided or removed before the reaction. The most significant factors affecting the reactivity of amines are class and basicity. Specific labeling of nucleic acids is easy, and several fluorophores or reactive groups can be introduced at various sites using automated solid-phase synthesis. On the other hand, site-specific labeling of proteins is very demanding with respect to site-specificity and preservation of biological functionality.

Handbook of Fluorescence Spectroscopy and Imaging. M. Sauer, J. Hofkens, and J. Enderlein

ISBN: 978-3-527-31669-4

Nearly all proteins exhibit amine groups in terms of lysine residues and at the *N*-terminus. Aliphatic amines such as the ε-amino group of lysine are moderately basic and reactive towards most acylating reagents. However, the concentration of the free base form of aliphatic amines below pH 8.0 is very low. Therefore, pH values of 8.5–9.5 are commonly applied in the modification of lysine residues [2]. In contrast, the α-amino group at the *N*-terminus of a protein can sometimes be selectively modified by reaction at a near neutral pH due to its lower pK_a value of $\sim$7. Furthermore, it has to be considered that acylation reagents tend to degrade in the presence of water with increasing pH value. Therefore, a compromise between the reactivity of the amine group and the degradation of the acylation reagent in aqueous buffers has to be found for each coupling reaction. In other words, reaction time and pH value have to be carefully optimized. Aromatic amines are very weak bases and thus they are unprotonated at pH 4.0–7.0. In aqueous solution, acylating reagents are virtually unreactive with the amino group of peptide bonds and with the side-chain amides of glutamine and asparagine residues, the guanidinium group of arginine, the imidazole group of histidine and the amines found in natural nucleotides.

Today *N*-hydroxysuccinimide (NHS) esters are most commonly used for coupling to amino groups. An NHS ester may be formed by the reaction of carboxylate with NHS in the presence of carbodiimide. To prepare stable NHS ester derivatives, the activation has to be performed in nonaqueous solvents. As exemplified in Figure 3.1, by the reaction of fluorescein-NHS with tryptophan, then the NHS or sulfo-NHS esters react with primary and secondary amines, creating stable amide and imide links, respectively. Thus, in protein molecules, NHS esters can be used to couple principally with the α-amines at the *N*-terminals and the ε-amines of lysine side chains, depending on the pH value, that is, on the degree of deprotonation of the amines. The reaction of NHS esters with thiol or hydroxyl groups does not yield stable conjugates. NHS esters can also be prepared *in situ* to react immediately with amines of the target molecules in aqueous solvents. Using the water-soluble carbodiimide EDC [1-ethyl-3-(3-dimethylaminopropyl)-carbodiimide] (carbodiimides are zero-length cross-linking agents used to mediate the formation of an amide or phosphoramidate linkage between a carboxylate and an amine or a phosphate and an amine, respectively), a carboxylate-containing fluorophore can be transformed into an active ester by reaction in the presence of NHS or sulfo-NHS (*N*-hydroxysulfosuccinimide). Sulfo-NHS esters are more water soluble than classical NHS esters, and couple rapidly with amines on target molecules with the same specificity and reactivity as NHS esters [3]. Furthermore, sulfo-NHS esters hydrolyze more slowly in water. Usually, NHS esters have a half-life of the order of hours under physiological pH conditions, but both hydrolysis and amine reactivity increase with increasing pH.

Figure 3.1 Standard amine coupling reactions used for covalent labeling of target molecules with organic fluorophores. Fluorophores can be activated as NHS esters or derivatives, isothiocyanates or sulfonyl chlorides to form carboxamides, thioureas or sulfonamides upon reaction with aliphatic amines. As an example, fluorescein-NHS and its reaction with the amino group of the aromatic amino acid tryptophan is shown in the first line.

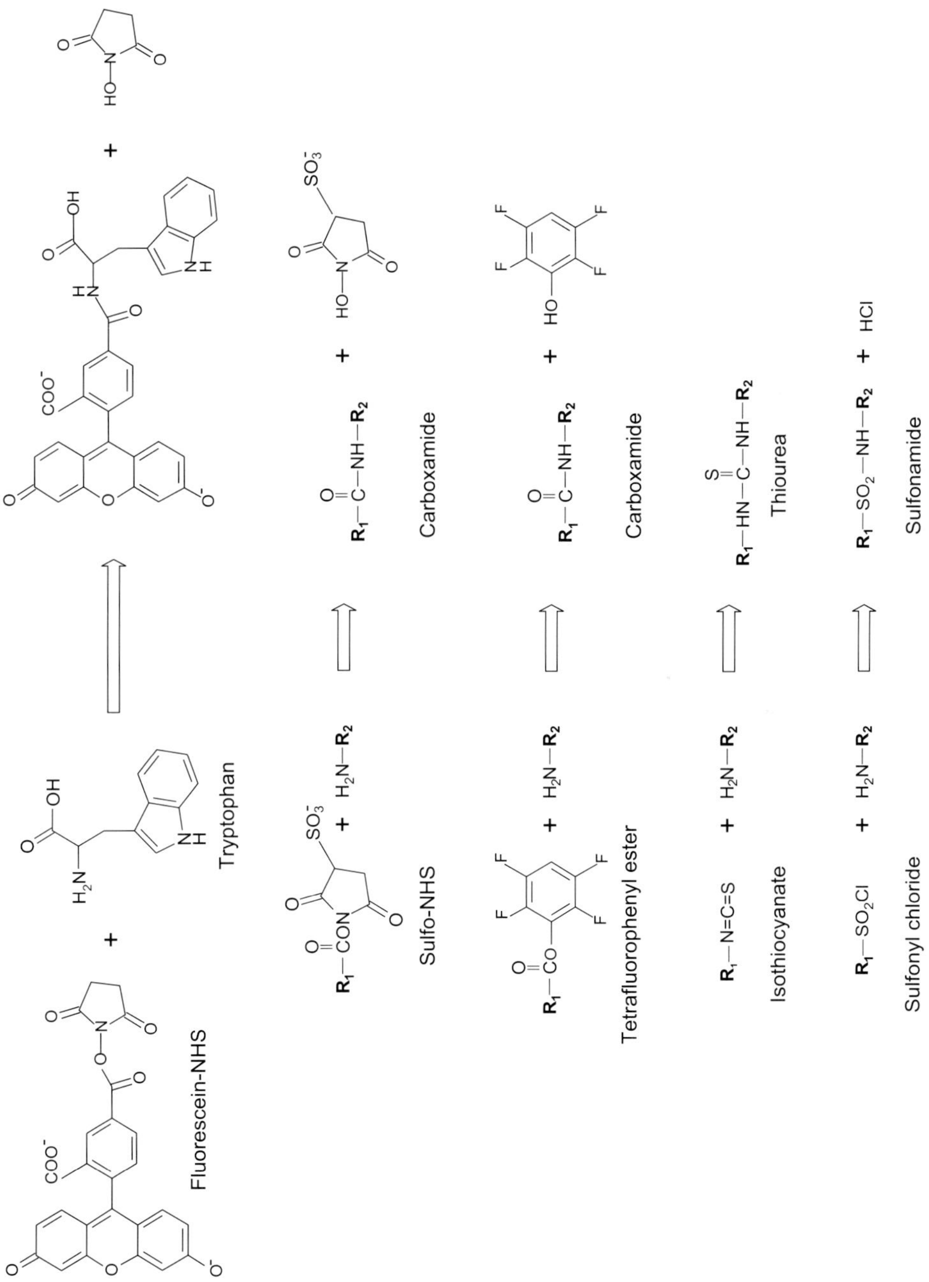

Fluorescein-NHS + Tryptophan ⇒ (fluorescein–CO–NH–tryptophan) + HO–N (N-hydroxysuccinimide)

R_1–CON (Sulfo-NHS, SO_3^-) + H_2N–R_2 ⇒ R_1–C(=O)–NH–R_2 (Carboxamide) + HO–N (SO_3^-)

R_1–CO–(tetrafluorophenyl) (Tetrafluorophenyl ester) + H_2N–R_2 ⇒ R_1–C(=O)–NH–R_2 (Carboxamide) + HO–(tetrafluorophenyl)

R_1–N=C=S (Isothiocyanate) + H_2N–R_2 ⇒ R_1–HN–C(=S)–NH–R_2 (Thiourea)

R_1–SO_2Cl (Sulfonyl chloride) + H_2N–R_2 ⇒ R_1–SO_2–NH–R_2 (Sulfonamide) + HCl

Alternatively, tetrafluorophenyl (TFP) esters can be used for covalent coupling of fluorophores to amines. NHS and TFP esters form the same strong amide bond, but TFP esters are less susceptible to hydrolysis in aqueous solvents. TFP esters are stable for several hours even under basic pH (8.0–9.0). Therefore, lower fluorophore concentrations can be used in conjugation experiments.

Besides NHS esters, fluorophores can be converted into active isothiocyanates by the reaction of an aromatic amine with thiophosgene. Isothiocyanates react with nucleophiles such as amines, thiols, and the phenolate ion of tyrosine side chains [1, 4]. The only stable product of these reactions, however, is with primary amine groups. Isothiocyanate modified fluorophores react best at alkaline pH (9.0), where the target amine groups are mainly unprotonated. On the other hand, the isothiocyanate group is relatively unstable in aqueous solution. Alternatively, isocyanates (exchanging the sulfur in an isothiocyanate by an oxygen atom) can be used to react with amines. However, the reactivity of isocyanates is even greater than that of the isothiocyanates, which renders their application more complicated due to stability and storage problems.

Reaction of a sulfonly chloride modified fluorophore with a primary amine-containing molecule proceeds with the loss of the chlorine atom and formation of a sulfonamide linker. Reaction of a sulfonyl chloride with an amine is best performed at pH 9.0–10.0. In addition, sulfonyl chlorides can be used to couple to target molecules in organic solvents. On the other hand, sulfonyl chlorides should be stored under nitrogen or in a desiccator to prevent degradation by moisture. Finally, it also has to be mentioned that fluorophores functionalized as acyl azides, aldehydes, and as epoxides can be used to label nucleophilic side groups in target molecules. However, the stability of the formed products and the specificity to react with primary amines is generally much lower.

There are fluorophore modifications that are able to couple to thiol groups, the second most common of the functional groups. Furthermore, thiol-reactive groups are frequently present on one of the two ends in heterobifunctional cross-linkers. The other end of such cross-linkers is often an amine-reactive functional group that is coupled to a target molecule before the thiol-reactive end, due to the comparable labile nature of the amine alkylation chemistries. Amine-reactive and thiol-reactive fluorophores are often used together to prepare doubly labeled fluorescent peptides, proteins, and oligonucleotides for probing biological structure, function, and interactions, using, for example, fluorescence resonance energy transfer (FRET) between a donor and an acceptor fluorophore. While polypeptides and oligonucleotides can be synthesized chemically, and thus offer the advantage of introducing side-chain protecting groups to facilitate site-specific labeling with different fluorophores, selective labeling of proteins containing more than 100 amino acids in length is difficult to achieve [5]. As each ε-amino group of lysine residues exhibits a slightly different pK_a value, site-specific labeling might be accomplished by varying the pH of the reaction. However, such approaches are cumbersome and, relatively, not very promising.

Although site-specific labeling is not always required, for example in fluorescence tracking applications, site-specificity is essential for precise distance or orientation measurements. Usually, non-specific labeling is inadequate for retrieving reliable biological information. Furthermore, one should be extremely cautious concerning the choice of labeling chemistry, optimization of labeling positions, and ensure rigorous characterization of the labeled biomolecules for labeling efficiency, site-specificity, and retention of functionality.

Because the thiol functional group present in cytosine residues is not very common in most proteins and can be labeled with high selectivity, thiol-reactive fluorophores often provide an elegant alternative for the modification of a protein at a defined site. Furthermore, many proteins are either devoid of cysteine or intrinsic cysteine residues can be removed by site-directed mutagenesis. For site-specific labeling, cysteine residues can then be introduced into the protein at carefully selected surface accessible positions for conjugation to thiol-specific fluorophores. Thiols can also be generated by selectively reducing cysteine disulfides with reagents such as dithiothreitol (DTT) or β-mercaptoethanol, each of which must then be removed by dialysis or gel filtration before reaction with the thiol-reactive fluorophore. The common thiol-reactive functional groups are primarily alkylating reagents, including maleimides, iodoacetamides, and aziridines. Reaction of these functional groups with thiols proceeds rapidly at or below room temperature in the pH range 6.5–8.0 to yield chemically stable thioethers (Figure 3.2). The high reactivity of most thiols even at pH values below 7.0 (most amino groups require higher pH values for coupling reactions with NHS esters) thus enables pH controllable sequential coupling reactions using NHS esters and maleimides for site-specific labeling of thiol- and amino-modified target molecules with different fluorophores.

Maleic acid imides (maleimides) are derivatives of the reaction of maleic anhydride with amines. The double bond of the maleimide undergoes an alkylation reaction with the thiol groups to form stable thioether bonds. Maleimide reactions are specific to thiol groups in the pH range 6.5–7.0 [6, 7]. At pH 7.0 the reaction of maleimides with thiol groups proceeds at a rate 1000 times faster than the reaction with amines [1]. At higher pH values some cross-reactivity with amino groups takes place. Maleimides do not react with methionine, histidine or tyrosine. Fluorophore-maleimides are usually synthesized in a two-step reaction. Firstly, one amino group of a diamine, for example, ethylenediamine, is converted into a maleimide by reaction with maleic anhydride. In the second reaction the maleimide is coupled to the fluorophore-NHS ester via the second amino group. Therefore, most fluorophores that are commercially available as maleimides carry a relatively long and flexible linker, for example, alkyl chains as in the case of Alexa Fluor 594, shown in Figure 3.2. Thus, the fluorophores can interact nonspecifically with the protein in an unpredictable dynamic fashion, which is often associated with quenching interactions with aromatic amino acids (see chapter 7).

Iodoacetamides readily react with thiols, including those found in peptides and proteins, to form stable thioethers (Figure 3.2). Although the primary objective of iodoacetamides is to modify the thiol groups in proteins and other molecules, the reaction is not totally specific. Iodoacetamide and the less active bromoacetamide derivatives can react with a number of functional groups within proteins: the thiol group of cysteine, both imidazolyl side chain nitrogens of histidine, the thioether of

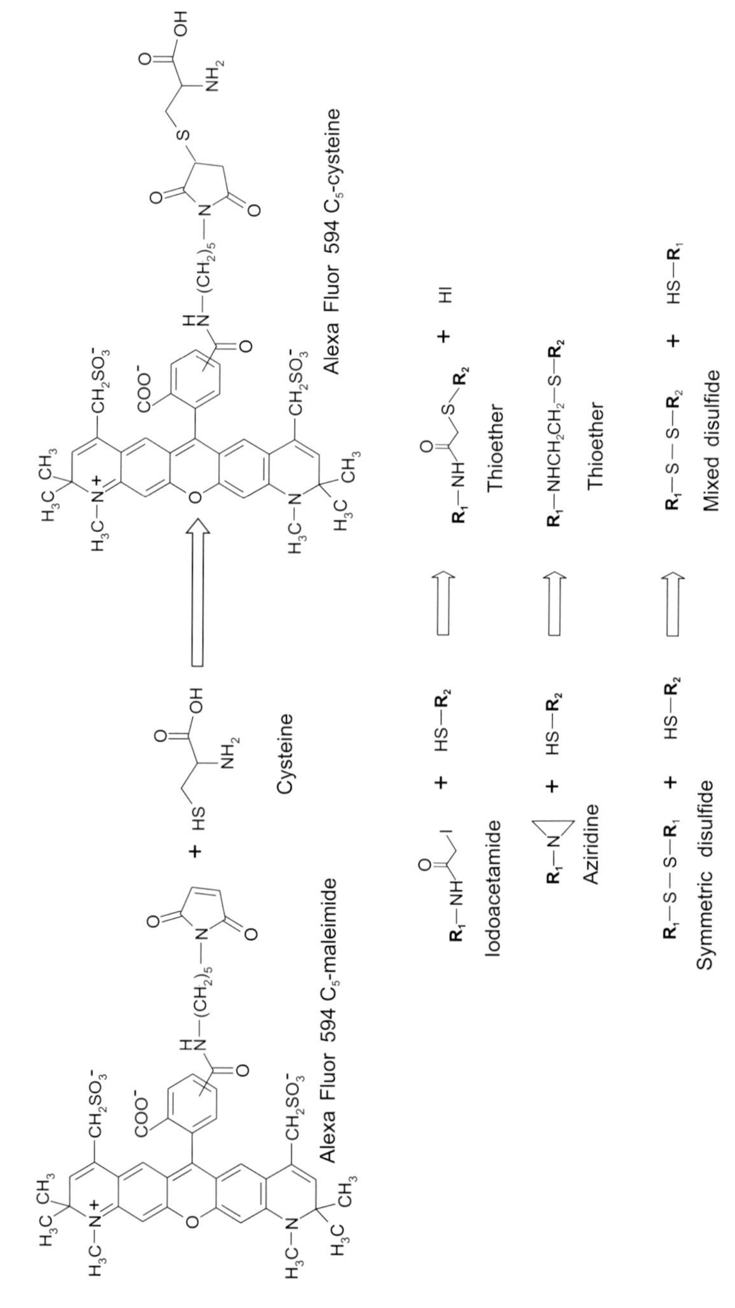

Figure 3.2 Standard thiol coupling reactions used for covalent labeling of target molecules with organic fluorophores. Fluorophores can be activated as maleimides, iodoacetamides, or aziridines to form thioethers upon reaction with thiol groups. As an example, Alexa Fluor 594 C_5-maleimide and its reaction with cysteine is shown in the first line.

methionine, and the primary ε-amino group of lysine residues along with the *N*-terminal α-amines [8]. The relative rate of reaction with each of these residues is generally dependent on the degree of ionization and thus the pH at which the modification is performed. Besides iodoacetamides, aziridines can be used to modify thiol groups in proteins. Thiols react with aziridines in a ring-opening process, forming thioether bonds (Figure 3.2). The reaction of an aziridine with a thiol group is highly specific at slightly alkaline pH values. However, in aqueous solution considerable hydrolysis occurs as an undesired side reaction. Finally, compounds containing a disulfide group can participate in disulfide exchange reactions with another thiol. The disulfide exchange process involves attack of the thiol at the disulfide, breaking the S–S bond, and subsequent formation of a new mixed disulfide (Figure 3.2). The reduction of disulfide groups to thiols in proteins using, for example, DTT, proceeds through the intermediate formation of a mixed disulfide.

Using thiol-specific fluorophores, imaging of single myosin molecules and of individual ATP turnovers, that is, cycles of adenosine triphosphate binding and hydrolysis, and imaging of single kinesin molecules as they move along microtubules, has been demonstrated [9, 10]. On the other hand, fluorophores incorporated on surface cysteine or lysine residues can also undergo noncovalent interactions with the local environment. Therefore, the orientation of the fluorophore dipole can be fixed, at least preliminarily, and thus fluorescence anisotropy experiments can give insights into the orientation and orientational dynamics of the local protein structure [11, 12]. For the permanent fixing of the transition dipole, bis-functional cysteine reactive fluorophores (Figure 3.3) can be coupled to two appropriately spaced (~16 Å) cysteine residues [13]. This intramolecular cross-linking strategy, which has only been used for proteins with an existing high-resolution structure, can be used advantageously to monitor the orientation and dynamics of protein domains or other protein structural elements [14, 15].

Cross-linking of double-cysteine proteins with homobifunctional rhodamine fluorophores (Figure 3.3) requires the application of low fluorophore-to-protein molar ratios (ideally 1: 1) to ensure stoichiometric labeling of the site. Nevertheless, the determination of the number of actual double labeled products, that is the exact fraction of cross-linked proteins, is challenging and seriously complicates the data interpretation.

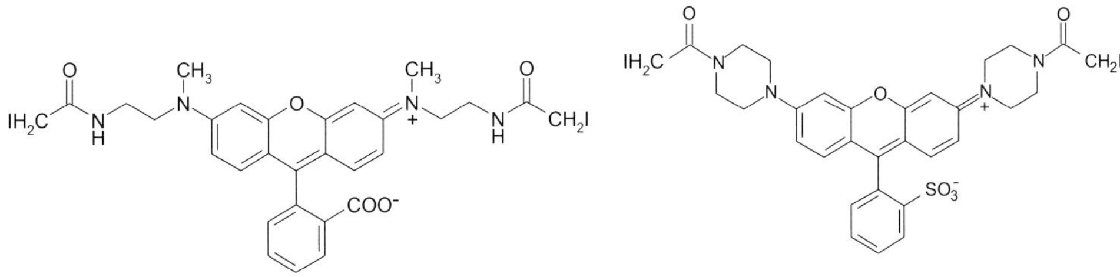

Figure 3.3 Molecular structures of two homobifunctional rhodamine derivatives (iodoacetamidotetramethylrhodamine derivatives) suitable for intramolecular cross-linking of two appropriately spaced cysteine residues on protein surfaces.

Besides these standard fluorophore coupling reactions to amino or thiol groups in biomolecules, other reactive groups can be used for bioconjugation. For example, derivatives of hydrazine can specifically react with aldehyde or ketone functional groups present in target molecules. Alternatively, reductive amination may be used to conjugate an aldehyde- or ketone-containing molecule to an amino modified fluorophore. Further detailed description of bioconjugation chemistry would go beyond the scope of this book. The interested reader is referred to the literature where excellent books about bioconjugation chemistry can be found, for example, [1].

Proteins can be site-specifically labeled with two different fluorophores, for example, a donor and an acceptor fluorophore for FRET experiments, by removing intrinsic cysteine residues by site-directed mutagenesis and reintroduction of two cysteines at carefully selected surface accessible positions. Stoichiometric labeling of the two cysteine residues with different fluorophores is then performed following a two-step reaction [15–19]. In the first reaction, the fluorophore–maleimide is added at a (sub)stoichiometric ratio to minimize double labeling of both cysteine residues. Singly modified protein molecules are subsequently separated chromatographically from unreacted or doubly labeled molecules. In the second labeling reaction, the complementary fluorophore–maleimide is coupled to the thiol group of the remaining cysteine residue. As the fluorophore can react with either of the thiol groups, a mixture of double labeled constructs cannot be circumvented. This unwanted sample heterogeneity can complicate the interpretation of FRET data, because the donor and acceptor fluorophore might exhibit slightly different spectroscopic characteristics depending on the coupling position, respectively, due to differences in local charge, pH, and hydrophobicity and the presence of neighboring quenching amino acids [20–22]. In more sophisticated multicolor FRET experiments in particular, accurate site specificity of labeling is absolutely mandatory.

To accomplish site-specific labeling of proteins carrying two cysteine residues, fluorophores modified as thioesters can be used, which react selectively with *N*-terminal cysteine residues to form amide bonds [23–25]. Other strategies involve the oxidation of an *N*-terminal serine or threonine to the corresponding aldehyde and subsequent coupling with a fluorophore modified as hydrazine [26], or the specific reaction of an *N*-terminal cysteine with aldehydes to yield thiazolidines [27, 28]. Recently [5], a method has been demonstrated that uses protein–protein interactions to site-specifically label recombinantly expressed double-cysteine proteins, without the need for extensive and time-consuming chromatography.

An alternative method to incorporate one or more distinct fluorophores within a single protein is peptide ligation [29]. This is where the full length protein is assembled from differently labeled synthetic or biosynthetic peptide fragments. The most established version of peptide ligation represents a two-step reaction between a peptide carrying a *C*-terminal thioester and another peptide containing an *N*-terminal cysteine residue, generating a peptide bond between the two peptides. There is a 50-residue limit to reliable solid-phase peptide synthesis, but peptide ligation can provide small proteins or protein domains containing up to $\sim$100 residues. Larger proteins with more than 100 amino acid residues can be labeled using peptide ligation to couple short fluorescently labeled synthetic peptides to larger recombinant proteins prepared by biosynthesis in bacteria. Proteins can also be labeled covalently with a fluorophore using cell-free RNA translation systems. The method is based on the experimental finding that a fluorescent antibiotic, puromycin, analog at lower concentrations couples efficiently to the *C*-terminus of mature proteins, using mRNA without a stop codon [30, 31]. Using synthetic amino-acylated tRNA and complementary sequences in the protein-coding DNA, unnatural amino acids carrying functional groups for fluorophore labeling can be site-specifically introduced into proteins by *in vitro* and *in vivo* transcription/translation [32, 33]. Large multiprotein complexes, so-called molecular machines composed of several interacting proteins, can be labeled by *in vitro* reconstitution (assembly) of purified and selectively fluorescently labeled subunits or components. Here it is of utmost importance to ensure that fluorescence labeling does not deteriorate the assembly of the complex [34].

3.2 Fluorescence Labeling in Living Cells

For the investigation of biologically relevant samples, target molecules have to be labeled *in vivo* with a fluorescent tag. If fluorescent labels are to be useful for the labeling of biomolecules in living cells, they have to fulfill special requirements, such as high biocompatibility, high photostability, and retention of biological function. In addition, the observation of the fluorescence signal of a single fluorophore is more complicated than *in vitro*, primarily due to strong autofluorescence, especially in the blue/green wavelength region [35, 36]. Furthermore, concentration control is difficult to perform.

The first problem, however, is site specific labeling inside a living cell and some procedures have been described in the literature [e.g., 37]. In the simplest case, the molecules to be investigated are prepared *in vitro*, utilizing standard techniques for the labeling of biomolecules. Subsequently, cell-loading is carried out via known chemical, electrical, mechanical or vehicle-based procedures (e.g., endocytosis, permeabilization, or microinjection) [38–41]. In this way the dimerization of epidermal growth factor (EGF) by monitoring FRET between donor and acceptor labeled EGF molecules has been observed on the membrane of living cells [38], and the mobility of multiple labeled single β-galactosidase molecules has been monitored in the cell nucleus [39, 40]. To measure, for example, the localization of RNA in living cells, the target sequence has to be labeled with a complementary fluorescent probe sequence. A problem that needs to be resolved in the fluorescence *in vivo* hybridization (FIVH) approach is the efficient delivery of probes to sites in a cell where they can hybridize with their target sequence. The most direct method for introducing probe molecules into a cell is by microinjection using micropipettes. Alternatively, liposomes can be used to introduce probe molecules into living cells [42].

The microinjection technique is based on the use of a micropipette with a very small diameter at the end and the application of a higher pressure for a predetermined time. As commercially available standard micropipettes for microinjection

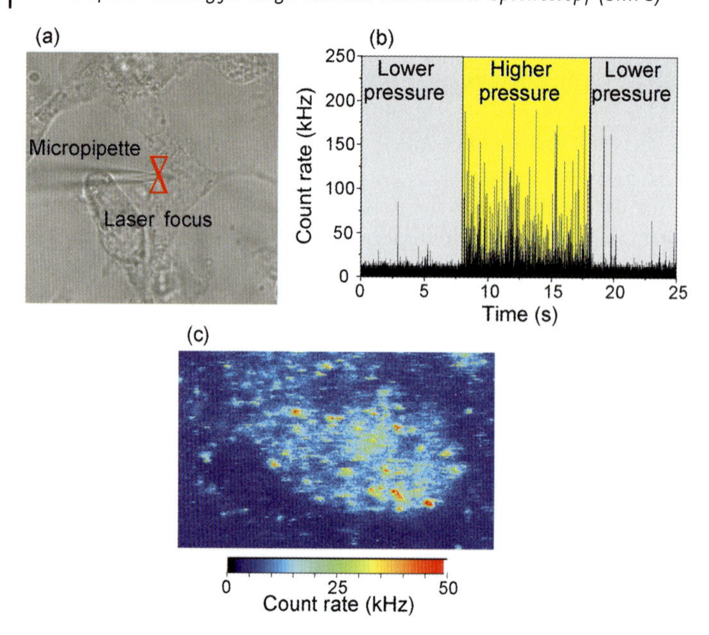

Figure 3.4 (a) Phase-contrast image of a micropipette (Femtotip I; Eppendorf-Nethler-Hinz, Hamburg, Germany) with an inner diameter of 500 ± 200 nm at the very end of the tip sticking into an adherent 3T3 mouse fibroblast. Simultaneously, a 635 nm laser diode is focused into the thin end of the micropipette. (b) Fluorescence signals observed in the pipette upon application of a higher pressure for 10 s. The micropipette was filled with a 10^{-9} M aqueous solution of oligo(dT) 43-mers labeled with a red-absorbing oxazine derivative. It is expected that oligo(dT) hybridizes to poly(A) RNA. (c) Confocal fluorescence image (25 × 12 μm, 6 ms integration time per pixel, 50 nm per pixel, 635 nm excitation wavelength with an intensity of 5 kW cm^{-2}) of a fibroblast cell in cell culture medium containing 10% (v/v) fetal calf serum, and 1 mM glutamine at room temperature (25 °C) after microinjection of ~100 fluorescently labeled oligonucleotide molecules. Some oligo(dT) molecules exhibit strongly hindered diffusion in the nucleus, indicated by blurred point-spread functions and stripes in the fluorescence intensity image; whereas, about 20% of the oligonucleotides microinjected show explicitly that they are immobile on the time scale of the measurement. They are most likely tethered to immobile elements of the transcriptional, splicing, or polyadenylation machinery [36].

typically exhibit inner diameters of 500 ± 200 nm at the very end of the tip (e.g., "Femtotips" from Eppendorf) – comparable to the diameter of a confocal detection volume – all molecules passing the micropipette and entering a living cell can thus be detected (Figure 3.4) [43–45]. Using micropipettes such as these, a well defined number of fluorescently labeled oligonucleotides were accordingly microinjected into the cytoplasm and nucleus of living 3T3 mouse fibroblast cells [36]. As a consequence, quantitative molecular information at the single cell level could be obtained (Figure 3.4). The reversible membrane permeabilization method using streptolysin O (SLO) is fast (~2 h), but as in the case of microinjection, it can only be used in *ex vivo* cellular assays, that is, when individual cells are analyzed under *in vivo* conditions.

Another very promising method is known as protein transduction. Several naturally occurring proteins have been found to enter cells easily, including the TAT protein from HIV [46, 47]. Specific short sequences within the larger molecule account for the transduction abilities of these proteins. These arginine-rich peptides allow efficient translocation through the plasma membrane and subsequent accumulation in the cell nucleus. Therefore, they could be useful vectors for the intracellular delivery of various non-permanent drugs, including antisense oligonucleotides and peptides of pharmacological interest [48]. Cellular uptake of these cationic cell-penetrating peptides have been ascribed in the literature to a mechanism that does not involve endocytosis. Living cell penetration studies without fixation using fluorescently labeled, peptidase-resistant, β-oligoarginines and HeLa cells, as well as human foreskin kerantinocytes, could demonstrate that longer-chain β-oligoarginines (8 and 10 residues) enter the cells and end up in the nuclei, particularly in the nucleoli, irrespective of temperature (37 or 4 °C) or of pretreatment with NaN_3 (Figure 3.5a,b) [49, 50]. β-Peptides have been shown to fold into stable secondary structures similar to those observed in natural peptides and proteins [51]. The finding that β-peptides are completely stable to proteolytic degradation renders them as candidates for use as peptidomimetics [52].

To circumvent washing steps in gene detection experiments in living cells, fluorescent probes have to be used that are able to recognize the target sequence with high specificity, and to exhibit a dramatic increase in fluorescence intensity only upon specific binding to their target sequence. Among the technologies currently under development for living cell gene detection and quantification, the most promising rely on the use of quenched DNA hairpin probes, for example, molecular beacons [53, 54] or smart probes [55, 56]. Both molecular beacons and smart probes form a stem–loop structure, where a fluorophore attached to one end of the stem is efficiently quenched by an external additionally attached quencher (molecular beacon) or an internal guanosine residue in the absence of a complementary target sequence. Hybridization with the mRNA target sequence initiates a conformational reorganization of the hairpin structure, that is, it opens the hairpin structure and separates the fluorophore from the quencher, which is associated with a strong increase in fluorescence intensity. However, to detect mRNA *in vivo*, one needs to deliver highly negatively charged oligonucleotide probes such as these into living cells with high efficiency. To overcome this difficulty, again short positively charged peptides that confer the ability to traverse biological membranes efficiently can be used.

Upon conjugation of one such peptide (TAT-1; N-*Tyr Gly Arg Lys Lys Arg Arg Gln Arg Arg Arg*-C) to molecular beacons yields probes that can enter into living cells with virtually 100% efficiency, fast (~30 min) delivery kinetics, and the ability to localize in the cell cytoplasm (Figure 3.5c–e) [57].

The most widespread technique for detecting specific structures or molecules in cells is immunolabeling with a primary antibody, followed by amplification with a secondary antibody conjugated to standard organic fluorophores. Alternatively, primary antibodies can be directly labeled with fluorophores and injected into living cells to bind to target molecules. When antibodies with high binding affinity are not

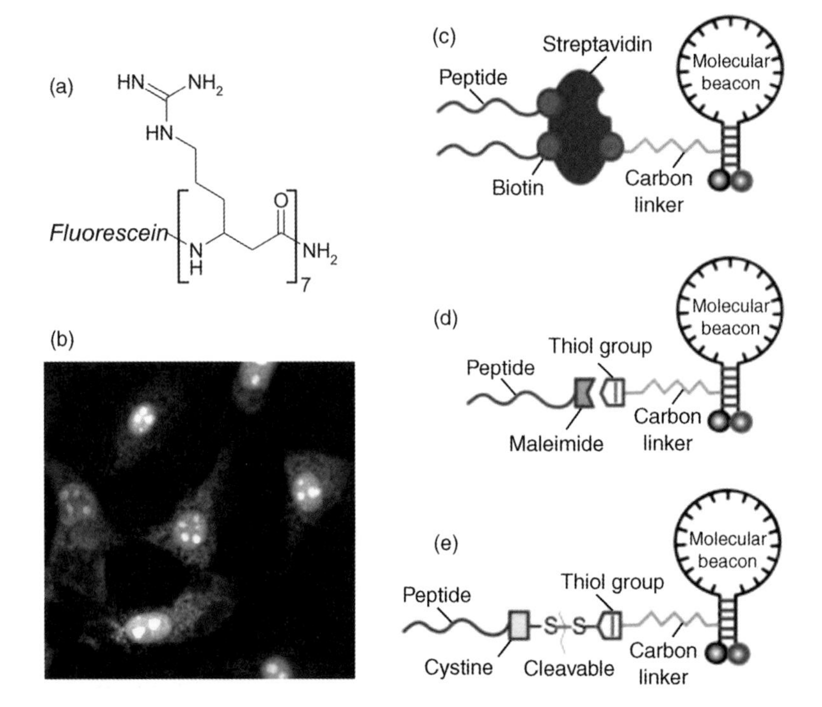

Figure 3.5 (a) Molecular structure of fluorescein labeled polycationic β-heptaarginines used to label 3T3 mouse fibroblasts. (b) Fluorescence image of 3T3 mouse fibroblast cells treated with the fluorescein labeled β-oligoarginine (1 μM) for 40 min. After incubation the cells were rinsed twice with PBS, pH 7.3 [49]. (c–d) Schematic illustration of three different conjugation schemes for linking the delivery peptide to DNA-hairpins [57]. Firstly, peptides can be linked to DNA-hairpins through a streptavidin–biotin bridge by introducing a modified oligonucleotide, biotin-dT, to the stem. As each streptavidin molecule has four binding sites, hairpins and delivery peptides can be attached on the same streptavidin molecule. Secondly, the hairpin oligonucleotide can be modified by a thiol group in the stem through a carbon linker. The thiol group then reacts with a maleimide group added to the C-terminus of the peptide to from a stable thioether. As a third approach, the TAT-1 peptide can be functionalized by adding a cysteine residue at the C-terminus, which forms a disulfide bridge with the thiol-modified hairpin. Upon entering the reducing environment of the cytoplasm the disulfide bond might be cleaved releasing the hairpin probe (Reproduced from Nitin *et al.* (2004) *Nucleic Acids Res.*, **32**, e58 [57].)

available the target can be recombinantly expressed with an epitope tag. The limitations of immunofluorescence comprise the restriction to permeabilized cells or extracellular or endocytosed proteins, in addition to problems associated with the oligomerization tendency, as a result of the multivalency of the antibodies [58]. Furthermore, as for all *in vivo* probes, the site-specificity should be exceptional to avoid nonspecific binding or incorporation into macromolecular complexes, and the probes should be nontoxic and exhibit high cell permeability. One serious disadvantage represents the fact that even after successful delivery of a fluorescently

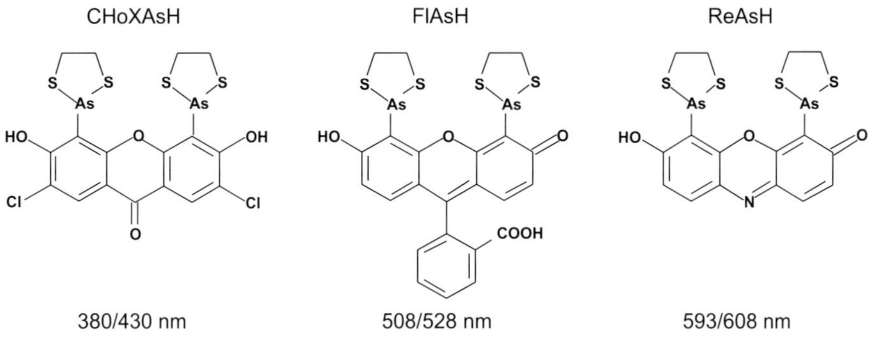

CHoXAsH FlAsH ReAsH

380/430 nm 508/528 nm 593/608 nm

Figure 3.6 Molecular structures of the three membrane permeable biarsenical fluorophores. In addition, the absorption and fluorescence emission maxima are given for aqueous solvents [103].

labeled antibody and specific high-affinity binding to the target protein, the question arises as to how free and bound fluorescently labeled antibodies can be differentiated. However, fluorescently labeled antibodies that increase fluorescence intensity only upon specific binding to their target molecules are unfortunately not yet available.

To circumvent these problems, so-called hybrid systems, composed of a small molecule that can covalently bind to genetically specified proteins inside or on the surface of living cells, have been developed [59–62]. The most promising system for covalent labeling of proteins in living cells is the tetracysteine–biarsenical system [59], which requires incorporation of a 4-cysteine α-helical motif – a 12-residue peptide sequence that includes four cysteine residues – into the target protein. The tetracysteine motif binds membrane-permeable biarsenical molecules, notably the green and red fluorophores "FlAsH" and "ReAsH" with picomolar affinity (Figure 3.6) [63].

Besides the desired characteristics, such as relatively high binding affinity and cell permeability, biarsenical fluorophores exhibit a dramatic increase in fluorescence intensity upon specific binding to the tetracysteine motif. For example, FlAsH (4′,5′-bis(1,3,2-dithioarsolan-2-yl)fluorescein) exhibits a fluorescence quantum yield of 0.49 when coupled to the tetracysteine motif, whereas the unbound form is only $\sim 5 \times 10^{-4}$ times as fluorescent [59]. Thus, similar to quenched DNA hairpin probes, in the ideal case, specific binding of the fluorophores to the target protein is reflected in the release of fluorescence intensity. In addition, some biarsenical fluorophores can be used for both fluorescence and electron microscopy (EM). Therefore, they are available as useful fluorophores for *in vitro* and *in vivo* cell staining experiments [64–66]. On the other hand, the strategy has not yet been demonstrated in intact transgenic animals, it requires the cysteine residues to be reduced for efficient labeling, and it does not permit two different proteins in the same compartment to be simultaneously labeled with different colors. Furthermore, the use of biarsenical fluorophores results in a relatively high background, due to nonspecific interactions with other molecules and reactions with other cysteine residues, and thus (as nearly all endogeneous proteins bear cysteine side chains) a subsequent increase in fluorescence intensity [58].

On the other hand, the protein of interest can be expressed when fused to a protein tag that is capable of binding a small fluorescent ligand [62]. In this way antibody tags can be fused to localization signal sequences to target hapten–fluorophore conjugates to specific subcellular compartments in living cells [67]. A 38 amino acid peptide ("fluorette"), which binds the rhodamine derivative Texas red with high affinity, can be used likewise to specifically label proteins in cells [68]. Unfortunately, a new peptide sequence has to be evolved for every new probe of interest. A generic method to selectively label proteins *in vivo* with organic fluorophores consisits in the use of fluorophores modified with a metal ion chelating nitrilotriacetate (Ni-NTA-functionalized fluorophores) moiety, which binds reversibly to engineered oligohistidine tags (hexa- or decahistidine) genetically attached to the protein of interest [69, 70]. The relatively low binding affinity can be used advantageously, for example, to exchange photobleached fluorophores in single-molecule fluorescence tracking experiments on cell membranes [71]. Analogously, bungarotoxin can be covalently labeled with an organic fluorophore. Bungarotoxin binds specifically to a 13 amino acid sequence, which can be genetically inserted into proteins [72].

Another very promising technique for the *in vivo* labeling of proteins with small organic fluorophores uses the enzymatic activity of human O^6-alkylguanine-DNA-alkyltransferase (hAGT). The enzyme hAGT irreversibly transfers the substrate alkyl group, an O^6-benzylguanine (BG) derivative, to one of its cysteine residues (Figure 3.7) [61, 73]. Kits for genetic labeling of proteins with O^6-alkylguanine-DNA-alkyltransferase are commercially available as *SNAP*-tags [74]. Thus, almost any organic fluorophore can be coupled covalently to an appropriate BG derivative, which serves as substrate for the hAGT-modified protein (Figure 3.7). Although the method seems to produce reliable results, the large size of hAGT, with a length of 207 amino acids, might induce perturbations in protein expression and functionality. Furthermore, experiments on mammalian cells would need to be performed using AGT-deficient cell lines to avoid labeling of endogeneous AGT [75]. Originally developed for the purification of proteins *Strep*-tag strategies can also be used for *in vivo* fluorescence labeling [76]. For example, *Strep*-tag II is a short peptide (eight amino acids) with highly selective but reversible binding properties for a streptavidin variant, which has been named "*Strep*-Tactin" [77]. Both interacting components have been engineered by combinatorial methods. *Strep*-tag strategies can thus also be used for specific *in vivo* fluorescence labeling of proteins using fluorophore labeled streptavidin conjugates. Each of the chemical labeling approaches has one or more substantial limitations, such as the endogenous receptor has to be knocked out in specific cell lines, the tag generates high background labeling or exhibits low cell-permeability and a half-life for labeling of several hours or more. Recently [78, 79], it has been demonstrated that trimethoprim (TMP) derivatives can be used advantageously to selectively tag *Escherichia coli* dihydrofolate reductase (eDHFR) fusion proteins in wild-type mammalian cells with minimal background and fast kinetics (Figure 3.7). Because TMP binds much more tightly to eDHFR than to mammalian forms of DHFR, the use of TMP-eDHFR does not require a knock-out or otherwise modified cell line. Furthermore, eDHFR is small (18 kDa) and TMP can be easily derivatized without substantially disrupting its binding efficiency. However, one

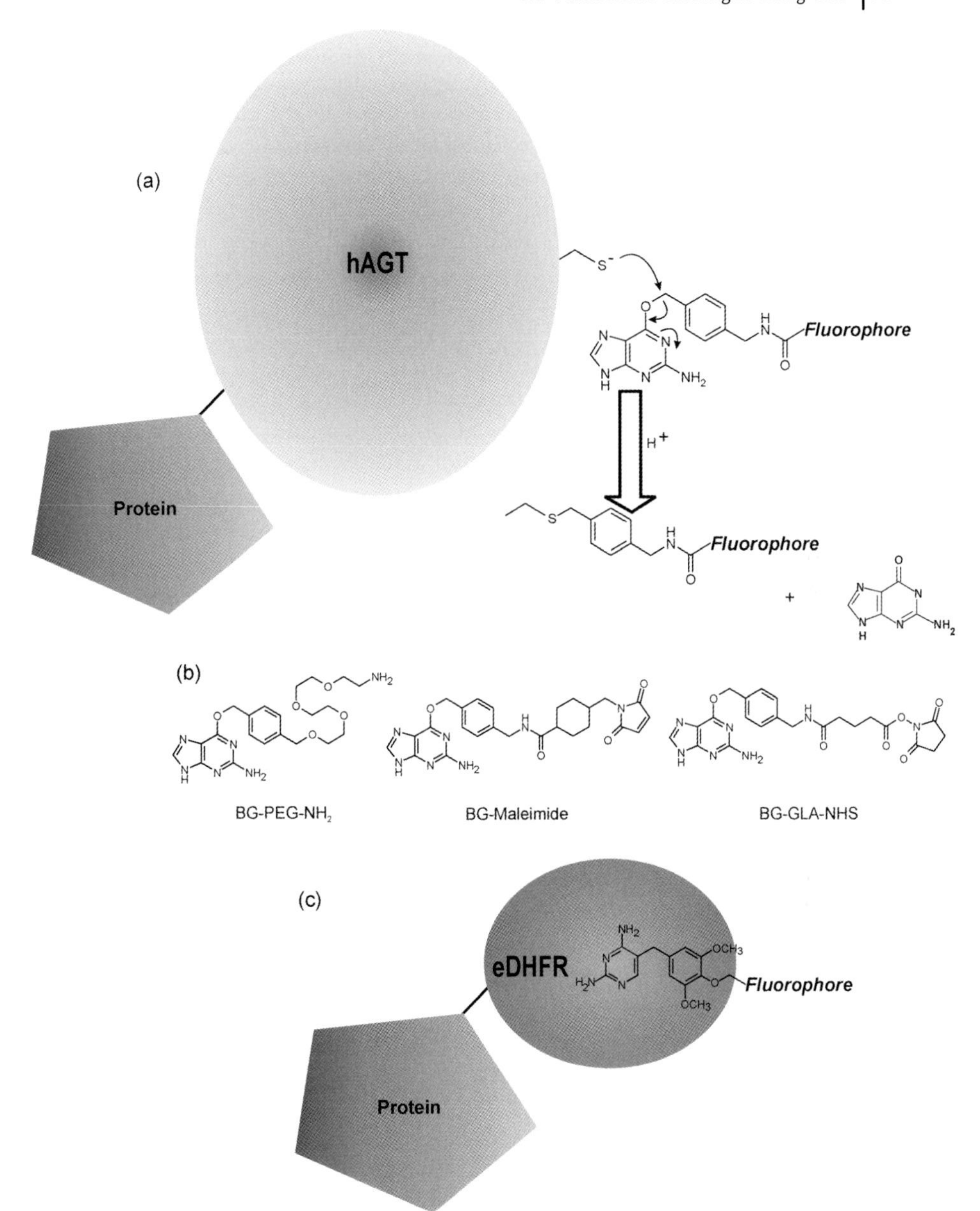

Figure 3.7 General strategy of protein labeling in living cells with fluorescent tags. (a) Covalent labeling of an hAGT fusion protein using a O^6-benzylguanine (BG) derivative. In this case a fluorophore-NHS ester is covalently coupled to $BG\text{-}NH_2$ prior to its application. (b) Alternatively, other BG derivatives can be used for the *in vivo* labeling of proteins with various small synthetic molecules [74]. (c) Strategy of protein labeling in mammalian cells with fluorescent TMP. Living cells are transfected with DNA encoding a protein of interest fused to a receptor domain, eDHFR or hAGT. Upon expression of the receptor fusion, a cell permeable small-molecule probe consisting of ligand (TMP or BG) coupled to a fluorescent tag is added to the cell growth medium.

should be aware of the fact that the labeling specificity and efficiency of the various tags in living cells is deteriorated (high degree of background staining, aggregation tendency, and low membrane permeability) due to the attachment of organic fluorophores. Thus, all tag technologies require sensitive controls to ensure specific labeling of the desired target protein with minimal perturbation.

Unfortunately, photobleaching of natural or artificial fluorophores is a serious limit observed for all living-cell applications. However, light-emitting semiconductor nanocrystals (NCs), such as core-shell CdSe–ZnS NCs (see Chapter 2), have unique optical properties – tunable narrow emission spectrum, broad excitation spectrum, high photostability, and fluorescence lifetime of the order of tens of nanoseconds – that make these bright probes attractive for use in experiments involving long observation times and multicolor and time-gated detection. Nowadays the surfaces of NCs can be modified to carry biomolecules that bind specifically to target structures in biological or biomedical applications. Thus, all the strategies used for *in vivo* fluorescence labeling with organic fluorophores can in principle be adopted for NC labeling. On the other hand, NCs also exhibit some limitations, for example, the difficulty of engineering them with single binding sites that can be specifically conjugated to just one molecule of interest. Another issue pertains to whether they retain biological functionality. Although the surfaces of NCs can be modified to achieve biological tolerance, any modification will result in an increase in particle size. Biologically compatible NCs easily reach a diameter of 20 nm, which is substantially larger than conventional fluorophores that typically have a size of $\sim$1 nm. It would appear that currently the toxicity is only of minor importance [80–82]. For example, NCs encapsulated in phospholipid micelles were injected into *Xenopus laevis* embryos, and the results obtained demonstrated that NCs are stable and non-toxic inside cytosolic compartments [83]. On binding arginine-rich peptides onto the surface of NCs (protein transduction), cellular labeling readily occurred in suspension, albeit nonspecifically [84].

However, the most elegant way to specifically label proteins *in vivo* is direct genetic labeling with fluorescence. The fluorophore used is a genetically encoded protein such as the green fluorescent protein (GFP) and its related proteins. GFP, from the bioluminescent jellyfish *Aequorea Victoria*, has revolutionized many areas of cell biology and biotechnology because it provides direct genetic encoding with strong visible fluorescence [85, 86]. GFP can function as a protein tag, as it tolerates *N*- and *C*-terminal fusion to a broad variety of proteins, many of which have been shown to retain native functions [87, 88]. According to this method, the DNA sequence coding for GFP is placed immediately adjacent to the sequence coding for the protein of interest. During biosynthesis, the protein will be prepared as a GFP-fusion protein. GFP is comprised of 238 amino acids and exhibits a barrel-like cylindrical structure where the fluorophore is highly protected, located on the central helix of the geometric center of the cylinder. These cylinders have a diameter of about 3 nm and a length of about 4 nm, that is, significantly larger than common fluorophores with a size of $\leq$1 nm (Figure 3.8). The fluorophore is a *p*-hydroxybenzylideneimidazolinone formed from residues 65–67, which are Ser-Tyr-Gly in the native protein. The fluorophore appears to be self-catalytic, requiring proper

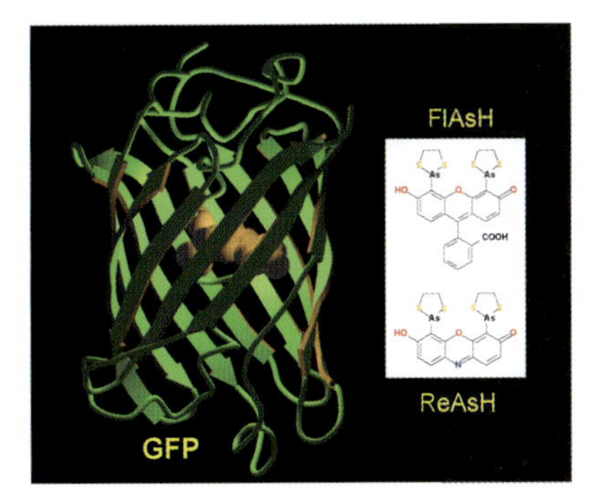

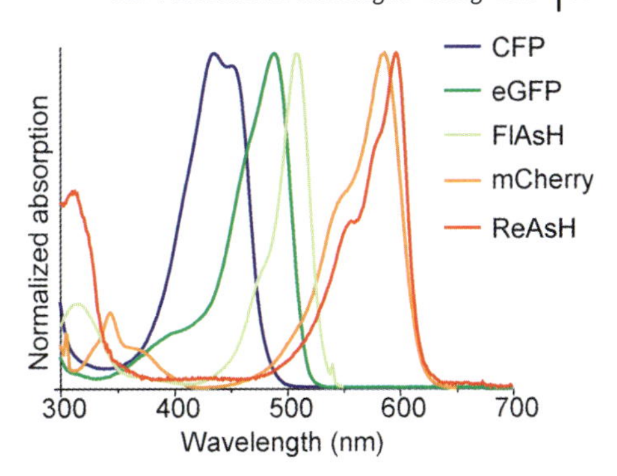

Figure 3.8 Comparison of the to-scale images of the green fluorescent protein (GFP) and the biarsenical fluorophores FlAsH and ReAsH. On the right-hand side, normalized absorption spectra of CFP, eGFP, mCherry and two biarsenical fluorophores in aqueous solvents are given.

folding of the entire structure. The protein is relatively stable, with a melting point above 65 °C.

Wild-type GFP has an extinction coefficient of 9500 l mol^{-1} cm^{-1} at 475 nm and a fluorescence quantum yield Φ_f of 0.79 compared with 53 000 l mol^{-1} cm^{-1} at 489 nm in enhanced GFP (EGFP) with $\Phi_f = 0.60$ [89]. Several spectral GFP variants with blue, cyan, and yellowish-green emission have now been successfully generated (Table 3.1).

As can be seen in Table 3.1. from the product of the extinction coefficients and the quantum yields, GFP and YFP are the brightest and most photostable candidates [90–94]. Although single-molecule studies of GFP- or yellow fluorescent protein (YFP)-fusion proteins were demonstrated in living cells, their complicated photophysics and low photostability render their application in single-molecule measurements still very difficult [95–97]. Proteins that fluoresce at red or far-red wavelengths are of specific interest because cells and tissues display reduced autofluorescence at longer wavelengths. Furthermore, red fluorescent proteins (RFPs) can be used in combination with other fluorescent proteins that fluoresce at shorter wavelengths for both multicolor labeling and fluorescence resonance energy transfer measurements. In this context, the discovery of new fluorescent proteins from nonbioluminescent

Table 3.1 Spectroscopic characteristics (absorption maximum, λ_{abs}, emission maximum, λ_{em}, extinction coefficient, ε, and fluorescence quantum yield, Φ_f, of common GFP variants [87].

GFP variant	λ_{abs} (nm)	λ_{em} (nm)	λ (l mol^{-1} cm^{-1})	Φ_f
WtGFP	475	504	9500	0.79
EGFP	489	509	53 000	0.60
GFP-S65T	489	509	55 000	0.64
EBFP	380	440	31 000	0.17
ECFP	435	478	29 000	0.40
EYFP	514	527	84 000	0.61

Table 3.2 Spectroscopic characteristics (absorption maximum, λ_{abs}, emission maximum, λ_{em}, extinction coefficient, ε, and fluorescence quantum yield, Φ_f, of fluorescent protein variants with red-shifted absorption [104].

Fluorescent protein	λ_{abs} (nm)	λ_{em} (nm)	ε (l mol^{-1} cm^{-1})	Φ_f
DsRed	558	583	75 000	0.79
mRFP1	584	607	50 000	0.25
mCherry	587	610	72 000	0.22
mOrange	548	562	71 000	0.69
mStrawberry	574	596	90 000	0.29
dTomato	554	581	69 000	0.69

Anthozoa species, in particular, the red-shifted fluorescent protein DsRed, is of general interest [98]. DsRed (drFP583) has absorption and emission maxima at 558 and 583 nm, respectively. However, several major drawbacks, such as slow maturation and residual green fluorescence, need to be overcome for the efficient use of DsRed as an *in vivo* reporter, especially in SMFS (single-molecule fluorescence spectroscopy) applications.

To improve maturation properties, and to reduce aggregation, a number of other red fluoresent proteins and variants of DsRed have been developed, for example, DsRed2, DsRed-Express, eqFP611, or HcRed1 [99–101]. The oligomerization problem of DsRed has been solved by mutagenetic means (mRFP1) [102]. Although mRFP1 overcame DsRed's tetramerization and sluggish maturation and exceeded DsRed's excitation and emission wavelengths by about 25 nm, the extinction coefficient, fluorescence quantum yield and photostability decreased (Table 3.2). Many new fluorescent proteins with different colors have been discovered in diverse anthozoan species, but most suffer from obligate tetramerization and would require efforts similar to those for the evolution of mRFP1 in order to produce a wide range of useful fusion partners [103].

An elegant alternative to common FRET studies is presented by bimolecular fluorescence complementation to study protein interactions in living cells. This approach is based on complementation between two nonfluorescent fragments of the yellow fluorescent protein (YFP) when they are brought together by interactions between proteins fused to each fragment [105].

GFP and homologous fluorescent proteins are commonly used to visualize protein localization and interactions by means of FRET in living cells. To obtain information about the movement of intracellular proteins, photobleaching techniques such as FRAP (fluorescence recovery after photobleaching) can be used [106–108]. FRAP is based on the principle of observing the rate of recovery of fluorescence due to the movement of a fluorescent marker into an area that contains this same marker, but which has been rendered non-fluorescent via an intense photobleaching pulse of laser light. The two-dimensional diffusion coefficient of the fluorophore is related to both its rate and extent of recovery. FRAP can be used to measure the lateral diffusion of various membrane or cytoplasmic constituents. However, photobleaching does

Table 3.3 Spectroscopic characteristics (absorption maximum, λ_{abs}, emission maximum, λ_{em}, extinction coefficient, ε, fluorescence quantum yield, Φ_f, and conversion wavelength, c, of monomeric photoswitchable proteins [112]. Data are given for the native/photoconverted states of the proteins.

Photoswitchable proteins	λ_{abs} (nm)	λ_{em} (nm)	ε (l mol^{-1} cm^{-1})	Φ_f	c (nm)
PA-GFP	400/504	515/517	20 700/17 400	0.13/0.79	413
PA-mRFP1-1	588/578	602/605	—/10 000	—/0.08	375–385
PS-CFP	402/490	468/511	34 000/27 000	0.16/0.19	405
PS-CFP2	400/490	468/511	43 000/47 000	0.20/0.23	405
mEosFP	505/569	516/581	67 200/37 000	0.64/0.62	400
DRONPA	503	518	95 000	0.85	490/405

not allow direct visualization of protein movement routes within a living cell. Therefore, different photoactivatable and photoswitchable fluorescent proteins have been developed that enable precise photolabeling and tracking of the protein of interest and thus give more complete information on its movement velocity and pathways (Table 3.3) [109–113]. It has to be pointed out that monomeric proteins are preferred since an oligomeric fusion tag often results in improper protein functioning and aggregation.

The monomeric photoactivatable PA-GFPs, for example, increase fluorescence intensity substantially upon irradiation at 413 nm [109], whereas the dual color monomeric photoswitchable PS-CFPs are capable of efficient photoconversion from cyan to green, changing both excitation and emission spectra in response to irradiation at 405 nm [110]. Similar to PS-CFP, EosFP is photoconverted by violet light (400 nm) but the green and orange forms are both excited by longer wavelengths (505/569 nm). The GFP-like fluorescent protein DRONPA, which was cloned from the coral Pectiniidae exhibits a reversible and highly reliable photoswitching performance [112]. A single-molecule fluorescence spectroscopic study on DRONPA molecules embedded in poly(vinyl alcohol) (PVA) demonstrated that even individual molecules can be switched between a dim and a bright state by using 488 and 405 nm laser light with a response time in the millisecond range and a repeatability of $>$100 times. This intriguing switching performance led to the term DRONPA, after "dron" a ninja term for vanishing and "pa", which stands for photoactivitation [113, 114]. Furthermore, reversible photoswitches such as DRONPA might be potentially useful for ultrahigh density optical data storage and far-field fluorescence imaging with improved optical resolution (see Chapter 8) [115, 116].

However, the use of fluorescent proteins also has limitations: they can only be introduced at the protein termini and can cause perturbations due to the large size (238 amino acids in length). Furthermore, their limited brightness, photostability, and blinking behavior, in addition to the spectral overlap of most mutants with cellular autofluorescence, complicates single-molecule fluorescence spectroscopy and imaging in living systems. Therefore, red-absorbing organic fluorophores in particular are still appealing if they could be easily attached to the target protein of interest even in living cells. On the other hand, in all *in vivo* labeling strategies introduced so far the fluorophore shows strong fluorescence upon illumination with visible light, even in the unbound state, thus deteriorating the interpretation of signals measured from living cells. Ideally, the fluorescence of the fluorophore or labeled probe would be "activated" only upon specific binding to the biomolecule, for example, the protein of interest. In some ways, biarsenical fluorophores such as FlAsH, which increase fluorescence upon binding to their tetracysteine target motif, exhibit some of the desired characteristics. However, as already demonstrated, biarsenical fluorophores have the potential to interact non-specifically with other proteins, accompanied by an increase in fluorescence intensity, that is, they lack specificity.

To realize a fluorophore labeled probe that changes its fluorescence characteristics upon specific binding to target molecules, already known quenching mechanisms might be exploited. A variety of fluorophore–quencher pairs have been identified for which photoinduced electron transfer (PET) is an exergonic process, and thus results in complete suppression of fluorescence upon molecular contact. For example, rhodamine and oxazine fluorophores are selectively quenched via PET upon contact formation with tryptophan or guanine, the most potent electron donors among the naturally occurring amino acids and DNA bases [20–22, 44, 55, 56]. Thus, the fluorophore (a rhodamine or oxazine derivative) might be modified in a way that ensures efficient fluorescence quenching via PET, for example, upon labeling to benzyl-guanine or streptavidin. In the case of streptavidin, most fluorophores are strongly quenched due to interactions with the biotin binding pockets, each of which contains three tryptophan residues. On the other hand, the quenching interactions are diminished or prevented upon biotin binding or guanine cleavage, respectively [20, 55, 117].

References

1 Hermanson, G.T. (1996) *Bioconjugate Techniques*, Academic Press, San Diego.

2 "*Introduction to amine modification*", Invitrogen detection technologies. In: Handbook of Fluorescent Probes and Research Products, 9th Edition, Richard P. Haugland (eds.), 2002, Molecular Probes.

3 Staros, J.V. (1982) *Biochemistry*, **21**, 3950–3955.

4 Podhradsky, D., Drobnica, L., and Kristian, P-. (1979) *Experientia*, **35**, 154.

5 Jäger, M., Michalet, X., and Weiss, S. (2005) *Protein Sci.*, **14**, 1–10.

6 Heitz, J.R., Anderson, C.D., and Anderson, B.M. (1968) *Arch. Biochem. Biophys.*, **127**, 627.

7 Partis, M.D., Griffiths, D.G., Roberts, G.C., and Beechey, R.B. (1983) *J. Protein Chem.*, **2**, 263–277.

8 Gurd, F.R.N. (1967) *Carboxymethylation. Methods in Enzymology*, vol. 11 (ed. C.H.W. Hirs). Academic Press, New York, p. 532.

9 Funatsu, T., Harada, Y., Tokunaga, M., Saito, K., and Yanagida, T. (1995) *Nature*, **374**, 555.

10 Vale, R.D., Funatsu, T., Pierce, D.W., Romberg, L., Harada, Y., and Yanagida, T. (1996) *Nature*, **380**, 451.

11 Warshaw, D.M., Hayes, E., Gaffeny, D., Lauzon, A.M., Wu, J., Kennedy, G., Trybus, K., Lowey, S., and Berger, C. (1998) *Proc. Natl. Acad. Sci. USA*, **95**, 8034.

12 Sase, I., Miyata, M., Ishiwata, S., and Kinosta, K. (1997) *Proc. Natl. Acad. Sci. USA*, **94**, 5646.

13 Corrie, J.E.T., Craik, J.S., and Munasinghe, V.R.N. (1998) *Bioconjugate Chem.*, **9**, 160–167.

14 Corie, J.E.T., Brandmeier, B.D., Ferguson, R.E., Trentham, D.R., Kendrick-Jones, J., Hopkins, S.C., van der Heide, U.A., Goldman, Y.E., Sabido-David, C., Dale, R.E., Criddles, S., and Irving, M. (1999) *Nature*, **400**, 425–430.

15 Peterman, E.J.G., Sosa, H., Goldstein, L.S.B., and Moerner, W.E. (2001) *Biophys. J.*, **81**, 2851–2863.

16 Sinev, M., Landsmann, P., Sineva, E., Ittah, V., and Haas, E. (2000) *Bioconjugate Chem.*, **11**, 352–362.

17 Ratner, V., Kahana, E., Eichler, M., and Haas, E. (2002) *Bioconjugate Chem.*, **13**, 1163–1170.

18 Schuler, B., Lipman, E.A., and Eaton, W.A. (2002) *Nature*, **419**, 743–747.

19 Rhoades, E., Gussakovsky, E., and Haran, G. (2003) *Proc. Natl. Acad. Sci. USA*, **100**, 3197–3202.

20 Marmé, N., Knemeyer, J.P., Wolfrum, J., and Sauer, M. (2003) *Bioconjugate Chem.*, **14**, 1133–1139.

21 Doose, S., Neuweiler, H., and Sauer, M. (2005) *ChemPhysChem.*, **6**, 2277–2285.

22 Neuweiler, H., Doose, S., and Sauer, M. (2005) *Proc. Natl. Acad. Sci. USA*, **102**, 16650–16655.

23 Dawson, P.E., Muir, T.W., Clark-Lewis, I., and Kent, S.B. (1994) *Science*, **266**, 776–779.

24 Tolbert, T.J. and Wong, C.H. (2002) *Angew. Chem. Int. Ed.*, **41**, 2171–2174.

25 Schuler, B. and Pannell, L.K. (2002) *Bioconjugate Chem.*, **13**, 1039–1043.

26 Geoghegan, K.F. and Stroh, J.G. (1992) *Bioconjugate Chem.*, **3**, 138–146.

27 Shao, J. and Tam, J.P. (1995) *J. Am. Chem. Soc.*, **117**, 3893–3898.

28 Chelius, D. and Shaler, T.A. (2003) *Bioconjugate Chem.*, **14**, 205–211.

29 Dawson, P.E. and Kent, S.B. (2000) *Annu. Rev. Biochem.*, **69**, 923–960.

30 Nemoto, N., Miyamoto-Sato, E., Husimi, Y., and Yanagawa, H. (1997) *FEBS Lett.*, **414**, 405–408.

31 Nemoto, N., Miyamoto-Sato, E., and Yanagawa, H. (1999) *FEBS Lett.*, **462**, 43–46.

32 Mendel, D., Cornish, V.W., and Schultz, P.G. (1995) *Annu. Rev. Biophys. Biomol. Struct.*, **24**, 435–462.

33 Wang, L., Brock, A., Herberich, B., and Schultz, P.G. (2001) *Science*, **292**, 498–500.

34 Mekler, V., Kortkhonjia, E., Mukhopadhyay, J., Knight, J., Revyakin, A., Kapanidis, A., Niu, W., Ebright, Y.W., Levy, R., and Ebright, R.H. (2002) *Cell*, **108**, 599–614.

35 Sauer, M., Zander, C., Müller, R., Ullrich, B., Drexhage, K.H., Kaul, S., and Wolfrum, J. (1997) *Appl. Phys. B*, **65**, 427–431.

36 Knemeyer, J.P., Herten, D.P., and Sauer, M. (2003) *Anal. Chem.*, **75**, 2147–2153.

37 Zhang, J., Campbell, R.E., Ting, A.Y., and Tsien, R.Y. (2002) *Nat. Rev. Mol. Cell Bio.*, **3**, 906–918.

38 Sako, Y., Minoguchi, S., and Yanagida, T. (2000) *Nat. Cell Biol.*, **2**, 168–172.

39 Kues, T., Peters, R., and Kubitscheck, U. (2001) *Biophys. J.*, **80**, 2954–2967.

40 Kues, T., Dickmann, A., Luhrmann, R., Peters, R., and Kubitscheck, U. (2001) *Proc. Natl. Acad. Sci. USA*, **98**, 12021–12026.

41 Byasse, T.A., Fang, M.M., and Nie, S. (2000) *Anal. Chem.*, **72**, 5606–5611.

42 Thierry, A.R. and Dritschilo, A. (1992) *Nucleic Acids Res.*, **20**, 5691–5698.

43 Sauer, M., Angerer, B., Ankenbauer, W., Foldes-Papp, Z., Gobel, F., Han, K.T., Rigler, R., Schulz, A., Wolfrum, J., and Zander, C. (2001) *J. Biotechnol.*, **86**, 181–201.

44 Zander, C., Drexhage, K.H., Han, K.T., Wolfrum, J., and Sauer, M. (1998) *Chem. Phys. Lett.*, **286**, 457–465.

45 Becker, W., Hickl, H., Zander, C., Drexhage, K.H., Sauer, M., Siebert, S., and Wolfrum, J. (1999) *Rev. Sci. Instrum.*, **70**, 1835–1841.

46 Torchilin, V.P., Rammohan, R., Weissig, V., and Levchenko, T.S. (2001) *Proc. Natl. Acad. Sci. USA*, **98**, 8786–8791.

47 Schwarze, S.R., Ho, A., Vocero-Akbani, A., and Dowdy, S.F. (1999) *Science*, **285**, 1569–1572.

48 Vives, E., Brodin, P., and Lebleu, B. (1997) *J. Biol. Chem.*, **272**, 16010–16017.
49 Rueping, M., Mahajan, Y., Sauer, M., and Seebach, D. (2002) *ChemBioChem*, **3**, 257–259.
50 Seebach, D., Namoto, K., Mahajan, Y.R., Bindschaedler, P., Sustmann, R., Kirsch, M., Ryder, N.S., Weiss, M., Sauer, M., Roth, C., Werner, S., Beer, H.-D., Munding, C., Walde, P., and Voser, M. (2004) *Chem. Biodivers.*, **1**, 65–97.
51 Seebach, D. and Matthews, J.L. (1997) *Chem. Commun.*, 2015–2022.
52 Franckenpohl, J., Arvidsson, P.I., Schreiber, J.V., and Seebach, D. (2001) *ChemBioChem*, **2**, 445–455.
53 Tyagi, S. and Kramer, F.R. (1996) *Nat. Biotechnol.*, **14**, 303–308.
54 Tyagi, S., Bratu, S.P., and Kramer, F.R. (1998) *Nat. Biotechnol.*, **16**, 49–53.
55 Knemeyer, J.P., Marmé, N., and Sauer, M. (2000) *Anal. Chem.*, **72**, 3717–3724.
56 Stöhr, K., Häfner, B., Nolte, O., Wolfrum, J., Sauer, M., and Herten, D.P. (2005) *Anal. Chem.*, **77**, 7195–7203.
57 Nitin, N., Santangelo, P.J., Kim, G., Nie, S., and Bao, G. (2004) *Nucleic Acids Res.*, **32**, e58.
58 Giepmans, B.N.G., Adams, S.R., Ellisman, M.H., and Tsien, R.Y. (2006) *Science*, **312**, 217–224.
59 Griffin, B.A., Adams, S.R., and Tsien, R.Y. (1998) *Science*, **281**, 269–272.
60 Gronemeyer, T., Godin, G., and Johnsson, K. (2005) *Curr. Opin. Biotechnol.*, **16**, 453–458.
61 Müller, L.W. and Cornish, V.W. (2005) *Curr. Opin. Chem. Biol.*, **9**, 56–61.
62 Chen, I. and Ting, A.Y. (2005) *Curr. Opin. Biotechnol.*, **16**, 35–40.
63 Adams, S.A., Campbell, R.E., Gross, L.A., Martin, B.R., Walkup, G.K., Yao, Y., Llopis, J., and Tsien, R.Y. (2002) *J. Am. Chem. Soc.*, **124**, 6063–6076.
64 Zhang, J., Campbell, R.E., Ting, A.Y., and Tsien, R.Y. (2002) *Nat. Rev. Mol. Cell. Biol.*, **3**, 906–918.
65 Andresen, M., Schmitz-Salue, R., and Jakobs, S. (2004) *Mol. Biol. Cell*, **15**, 5616–5622.
66 Dyachok, O., Isakov, Y., Sagetorp, J., and Tengholm, A. (2006) *Nature*, **439**, 349–352.
67 Farinas, J. and Verkman, A.S. (1999) *J. Biol. Chem.*, **274**, 7603–7606.
68 Marks, K.M., Rosinov, M., and Nolan, G.P. (2004) *Chem. Biol.*, **11**, 347–356.
69 Guignet, E.G., Hovius, R., and Vogel, H. (2004) *Nat. Biotechnol.*, **22**, 440–444.
70 Lata, S., Reichel, A., Brock, R., Tampe, R., and Piehler, J. (2005) *J. Am. Chem. Soc.*, **127**, 10205–10215.
71 Schreiter, C., Gjoni, M., Hovius, R., Martinez, K.L., Segura, J.-M., and Vogel, H. (2005) *ChemBioChem*, **6**, 2187–2194.
72 Sekine-Aizawa, Y. and Huganir, R.L. (2004) *Proc. Natl. Acad. Sci. USA*, **101**, 17114–17119.
73 Keppler, A., Gendreizig, S., Gronemeyer, T., Pick, H., Vogel, H., and Johnsson, K. (2003) *Nat. Biotechnol.*, **21**, 86–89.
74 Keppler, A., Pick, H., Arrivoli, C., Vogel, H., Johnsson, K. (2004) *Proc. Natl. Acad. Sci. USA*, **101**, 9955–9959.
75 Miyawaki, A., Sawano, A., and Kogure, T. (2003) *Nat. Cell Biol.*, **5**, S1–S7.
76 Schmidt, T.G.M. and Skerra, A. (1994) *J. Chromatogr. A*, **676**, 337–345.
77 Voss, S. and Skerra, A. (1997) *Protein Eng.*, **10**, 975–982.
78 Miller, L.W., Cai, Y., Sheetz, M.P., and Cornish, V.W. (2005) *Nat. Methods*, **2**, 255–257.
79 Calloway, N.T., Choob, M., Sanz, A., Sheetz, M.P., Miller, L.W., and Cornish, V.W. (2007) *ChemBioChem*, **8**, 767–774.
80 Michalet, X., Pinaud, F., Bentolila, L.A., Tsay, J.M., Doose, S., Li, J.J., Sundaresan, G., Wu, A.M., Gambhir, S.S., and Weiss, S. (2005) *Science*, **307**, 538–544.
81 Derfus, A.M., Chan, W.C.W., and Bhatia, S.N. (2004) *Nano Lett.*, **4**, 11–18.
82 Kirchner, C., Liedl, T., Kudera, S., Pellegrino, T., Javier, A.M., Gaub, H.E., Stoelzle, S., Fertig, N., and Parak, W.J. (2005) *Nano Lett.*, **5**, 331–338.
83 Dubertret, B. (2002) *Science*, **298**, 1759–1762.
84 Lagerholm, B.C., Wang, M., Ernst, L.A., Ly, D.H., Liu, H., Bruchez, M.P., and Waggoner, A.S. (2004) *Nano Lett.*, **4**, 2019–2022.
85 Tsien, R.Y. (1998) *Annu. Rev. Biochem.*, **67**, 509–544.

86 Baird, G.S., Zacharias, D.A., and Tsien, R.Y. (2000) *Proc. Natl. Acad. Sci. USA*, **97**, 11984–11989.

87 Moores, S.L., Sabry, J.H., and Spudich, J.A. (1996) *Proc. Natl. Acad. Sci. USA*, **93**, 443–446.

88 Olson, K.R., McIntosh, J.R., and Olmsted, J.B. (1995) *J. Cell. Biol.*, **130**, 639–650.

89 Patterson, G.H., Knobel, S.M., Sharif, W.D., Kain, S.R., and Piston, D.W. (1997) *Biophys. J.*, **73**, 2782–2790.

90 Harms, G.S., Cognet, L., Lommerse, P.H.M., Blab, G.A., and Schmidt, T. (2001) *Biophys. J.*, **80**, 2396–2408.

91 Harms, G.S., Cognet, L., Lommerse, P.H.M., Blab, G.A., Kahr, H., Gamsjager, R., Spaink, H.P., Soldatov, N.M., Romanin, C., and Schmidt, T. (2001) *Biophys. J.*, **81**, 2639–2646.

92 Iino, R. and Kusumi, A. (2001) *J. Fluoresc.*, **11**, 187–195.

93 Iino, R., Koyama, I., and Kusumi, A. (2001) *Biophys. J.*, **80**, 2667–2677.

94 Kubitscheck, U., Kückmann, O., Kues, T., and Peters, R. (2000) *Biophys. J.*, **78**, 2170–2179.

95 Moerner, W.E. (2002) *J. Chem. Phys.*, **117**, 10925–10937.

96 Widengren, J., Terry, B., and Rigler, R. (1999) *Chem. Phys.*, **249**, 259–271.

97 Zumbusch, A. and Jung, G. (2000) *Single Mol.*, **1**, 261–270.

98 Matz, M.V., Fradkov, A.F., Labas, Y.A., Savitsky, A.P., Zaraisky, A.G., Markelov, M.L., and Lukyanov, S.A. (1999) *Nat. Biotechnol.*, **17**, 969–973.

99 Bevis, B.J. and Glick, B.S. (2002) *Nat. Biotechnol.*, **20**, 1159.

100 Wiedenmann, J., Schenk, A., Rocker, C., Girod, A., Spindler, K.-D., and Nienhaus, G.U. (2002) *Proc. Natl. Acad. Sci. USA*, **99**, 11646–11651.

101 Gurskaya, N.G., Fradkov, A.F., Terskikh, A., Matz, M.V., Labas, Y.A., Martynov, V.I., Yanushevich, Y.G., Lukyanov, K.A., and Lukyanov, S.A. (2001) *FEBS Lett.*, **507**, 16–20.

102 Campbell, R.E., Tour, O., Palmer, A.E., Steinbach, P.A., Baird, G.S., Zacharias, D.A., and Tsien, R.Y. (2002) *Proc. Natl. Acad. Sci. USA*, **99**, 7877–7882.

103 Verkhusha, V.V. and Lukyanov, K.A. (2004) *Nat. Biotechnol.*, **22**, 289–296.

104 Shaner, N.C., Campbell, R.E., Steinbach, P.A., Giepmans, B.N.G., Palmer, A.E., and Tsien, R.Y. (2004) *Nat. Biotechnol.*, **22**, 1567–1572.

105 Hu, C.-D., Chinenov, Y., and Kerppola, T.K. (2002) *Mol. Cell*, **9**, 789–798.

106 Peters, R., Peters, J., Twes, K.H., and Bahr, W. (1997) *Biochem. Biophys. Acta*, **367**, 282–294.

107 Axelrod, D., Koppel, D.E., Schlessinger, J., Elson, E., and Webb, W.W. (1976) *Biophys. J.*, **16**, 1055–1069.

108 Edidin, M., Zagyansky, Y., and Lardner, T.J. (1976) *Science*, **191**, 466–468.

109 Patterson, G.H. and Lippincott-Schwartz, J.A. (2002) *Science*, **13**, 1873–1877.

110 Chudakov, D.M., Verkhusha, V.V., Staroverov, D.B., Souslova, E.A., Lukyanov, S., and Lukyanov, K.A. (2004) *Nat. Biotechnol.*, **22**, 1435–1439.

111 Chudakov, D.M., Belousov, V.V., Zaraisky, A.G., Novoselov, V.V., Staroverov, D.B., Zorov, D.B., Lukyanov, S., and Lukyanov, K.A. (2003) *Nat. Biotechnol.*, **21**, 191–194.

112 Chapman, S., Oparka, K.J., and Roberts, A.G. (2005) *Curr. Opin. Plant Biol.*, **8**, 565–573.

113 Ando, R., Mizuno, H., and Miyawaki, A. (2004) *Science*, **306**, 1370–1373.

114 Habuchi, S., Ando, R., Dedecker, P., Verheijen, W., Mizuno, H., Miyawaki, A., and Hofkens, J. (2005) *Proc. Natl. Acad. Sci. USA*, **102**, 9511–9516.

115 Hell, S.W. (2007) *Science*, **316**, 1153–1158.

116 Heilemann, M., Dedecker, P., Hofkens, J., and Sauer, M. (2009) *Laser & Photon. Rev.*, **3**, 180–203.

117 Buschmann, V., Weston, K.D., and Sauer, M. (2003) *Bioconjugate Chem.*, **14**, 195–204.

4
Fluorophore Selection for Single-Molecule Fluorescence Spectroscopy (SMFS) and Photobleaching Pathways

To emphasize the advantage of single-molecule over ensemble experiments, imagine that you are at a large station observing hundreds of passengers arriving on an unknown number of trains. From such an observation, you cannot answer questions about the individual routes that the trains took, how many passengers boarded the train at which station, and at what times, or how many stops, each train made. You observe simply an average and can only conclude that trains typically transport hundreds of passengers at a time. If single-molecule spectroscopy is used to monitor reactions, individual properties can be measured, whereas in standard experiments, only the overall average response is observed. The technique provides the basis for a direct comparison of models, which are usually derived by envisaging individual molecules, with solid experimental results. Furthermore, we can determine whether each molecule exhibits a different but temporally constant reaction rate (static inhomogeneity) or changes its rate with time (dynamic inhomogeneity, which can be caused by perturbations that are analogous to "elevations" on a train's route). Thus, single-molecule fluorescence spectroscopy (SMFS) is used in a range of scientific areas in physics, chemistry, and biology, and has been explained in detail in various excellent reviews [1–16].

The ongoing success of SMFS has been facilitated not only through the development of optical single-molecule techniques, but in addition through new refined organic synthesis methods and the large repertoire of molecular biology techniques. The ability to specifically label many different sites on macromolecules, for example, provides a great toolbox for the application of several spectroscopic techniques, such as fluorescence resonance energy transfer (FRET) and photoinduced electron transfer (PET). The dream of being able to detect individual molecules optically arose in conjunction with the confirmation of their existence. Nevertheless, it took more than half a century before the group working with Hirschfeld made the first successful attempts at detecting the fluorescence signal of single antibody molecules statistically labeled with 80–100 fluorophores [17]. Subsequently, at the beginning of the 1990s, Keller's group was finally able to detect a single fluorophore in a biologically relevant environment, that is, in an aqueous solvent [18]. Simultaneously, but independently, Moerner, Orrit and coworkers successfully detected single molecules at cryogenic temperatures using optical means [19, 20].

Handbook of Fluorescence Spectroscopy and Imaging. M. Sauer, J. Hofkens, and J. Enderlein

ISBN: 978-3-527-31669-4

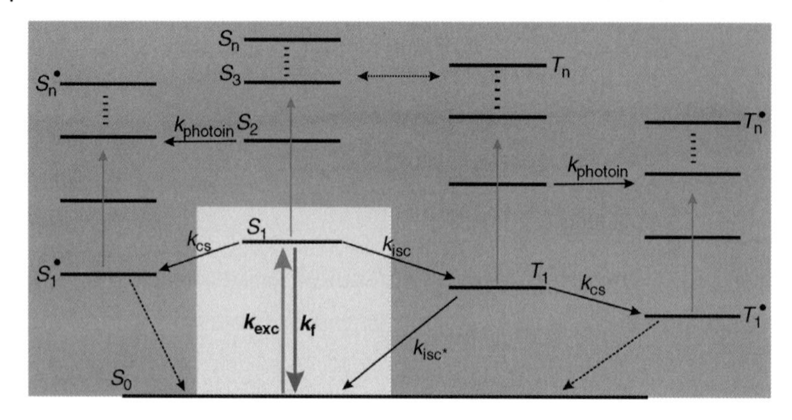

Figure 4.1 Simple photophysical model, including the main states involved in photobleaching of common organic fluorophores: excitation, k_{exc}, fluorescence emission, k_f, intersystem crossing, k_{isc}, charge separation, k_{cs}, and photoionization, $k_{photoion}$. As long as the fluorophore is cycled between the singlet ground state S_0 and the first excited singlet state S_1, it emits fluorescence photons (marked as the light-gray area).

To adequately explain the technological breakthroughs achieved in single molecule fluorescence spectroscopy (SMFS) we have to understand what the central feature is, and this is usually a fluorophore under laser irradiation (Figure 4.1). Using a laser source of appropriate wavelength, the fluorophore is excited from the singlet ground state, S_0, to the first excited singlet state, S_1. Subsequently, the fluorophore spontaneously emits a fluorescence photon to relax to the electronic ground state, S_0. Owing to the loss of vibronic excitation energy during the excitation/emission cycle, the fluorescence photon is of lower energy, that is, it is spectrally red-shifted (the so-called Stokes shift), and can be detected efficiently using appropriate optical filtering. The Stokes shift represents the basis for the high sensitivity of fluorescence spectroscopy compared with other spectroscopic methods.

A question arises with respect to the number of fluorescence photons we can expect to detect from a single fluorophore. Assuming a fluorescence quantum yield of 100%, and a typical fluorescence lifetime of 4 ns, a fluorophore would emit on average 2.5×10^6 fluorescence photons per second, if it was always excited immediately after the emission of a photon, that is, under optical saturation conditions. In other words, the fluorophore would be cycled between the singlet ground S_0 and first excited state S_1, which is highlighted light-gray in Figure 4.1, and it would never pause in any one of the states marked gray. However, owing to the presence of competing depopulation pathways from which intersystem crossing into long-lived triplet states is the most prominent, the fluorescence quantum yield never approaches 100%. Furthermore, it has to be taken into account that fluorophores undergo irreversible photobleaching upon irradiation with intense laser light because of the increased reaction capability of molecules in electronically excited states. This means that the time available to detect the presence of a single fluorophore is limited to a specific time, usually in the region of milliseconds to seconds depending on the excitation intensity.

Nevertheless, if only a small percentage of the photons emitted finally reach the detector (which is limited mainly by the collection efficiency of the applied optics and the transmission of the filters), the presence of a single fluorophore would be manifested as the detection of a few hundred thousand photons per second. Of course the quantum efficiency of typical high sensitivity CCD-cameras or avalanche-photodiodes is less than 100%, but even the human eye exhibits sufficient sensitivity, especially in the green wavelength range, to detect the fluorescence photons emitted by a single fluorophore. This demonstrates that the detection efficiency does not constitute the critical parameter in SMFS.

It is the background signal that sets the detection limit. The background stems mainly from elastic (Rayleigh) and inelastic (Raman) scattering from solvent molecules and from autofluorescent impurities. While Rayleigh scattering can be efficiently suppressed by the use of suitable optical filters, the complete suppression of Raman scattering, which is directly proportional to the number of irradiated solvent molecules, is challenging, because it occurs, at least partly, in the same spectral range as the fluorescence signal. On the other hand, autofluorescence from impurities strongly depends on the excitation and on the detection wavelength. In biological samples in particular, luminescent impurities can decrease the sensitivity or even prevent the distinct detection of individual fluorophores. As the contribution of the background signal is directly proportional to the number of molecules in the excitation volume, the reduction of the excitation/detection volume is critical for SMFS. Efficient reduction of the excitation volume can be achieved by using laser excitation in different configurations, that is, confocal, evanescent, or near-field arrangements. Although all configurations offer distinct advantages, it was confocal fluorescence microscopy, as introduced by Rigler and coworkers, which established single-molecule sensitive optical techniques as complementary standard tools for various disciplines ranging from material science to cell biology [21, 22].

Besides these general aspects, the photophysical properties of the fluorophore used in SMFS experiments are of fundamental importance. Naturally, the fluorophore should exhibit a high extinction coefficient at the excitation wavelength, ε, a high fluorescence quantum yield, Φ_f, a large Stokes shift, and a high photostability [23–25]. Regarding the brightness of a fluorophore, which is represented by the product $\varepsilon \times \Phi_f$, SMFS compatible fluorophores should exhibit values in the region of $\varepsilon > 10\,000\ \mathrm{cm^{-1}\ M^{-1}}$ and $\Phi_f > 0.10$, that is, a brightness of at least ~1 000 to facilitate their unequivocal detection by fluorescence spectroscopy at the level of individual fluorophores. For comparison, a standard rhodamine or carbopyronin dye shows a brightness of 50 000–100 000. The Stokes shift generally varies between 15 and 30 nm for the most prominent SMFS capable fluorophores, that is, xanthene, rylene, and cyanine derivatives. In addition, the fluorescence lifetime, τ, plays an important role in SMFS. On the one hand, long fluorescence lifetimes, that is, longer than a few nanoseconds, enable the efficient discrimination of inevitable autofluorescence of biological samples and inherent Raman scattering from solvent molecules using time-gated detection techniques to be carried out. On the other hand, a fluorophore has to be relaxed to the singlet ground state, S_0, before re-excitation can occur. Thus, the maximum number of photons emitted per time period is limited by the fluorescence rate, $k_f = 1/\tau$. Therefore, fluorophores with short fluorescence lifetimes, $\tau < 5$ ns, are preferable for SMFS.

These considerations are influenced by requirements concerning the intersystem crossing probability of the fluorophore and other reactions resulting in the formation of nonfluorescent intermediates, for example, triplet or radical ion states. As the lifetime of these states generally lasts for several micro- to milliseconds (or even longer depending on the environmental conditions), they severely decrease the fluorescence signal that can be detected. In addition, these fluorescence intermittencies complicate the fluorescence imaging and tracking of individual fluorophores. Therefore, fluorophores exhibiting a high triplet quantum yield or other dark-states with high yield are less suited to SMFS. However, the most important fluorophore property for SMFS is represented by photostability. Every organic fluorophore is prone to irreversible photodestruction due to the higher reactivity of the excited states and subsequent various photochemical reactions often involving photooxidation, which limits the total number of photons emitted. Unfortunately, the underlying mechanisms of photobleaching are not fully understood, and many of the photophysical parameters of fluorophores reported vary considerably between different reports and with the dye concentration levels [26, 27]. Furthermore, not only does the environment of the fluorophores but also the excitation conditions have a substantial effect on the photobleaching properties [28–31].

In general, xanthene derivatives (rhodamine, rosamine, pyronin, carbopyronin, carborhodamine, and oxazine dyes) exhibit a much higher photostability than cyanine, coumarin, and fluorescein dyes and than phycobiliproteins and other natural fluorophores (see Chapter 2). The mechanisms by which fluorophores photobleach are fairly complex, but there is consent that (i) photo-oxidation, that is, the formation of singlet oxygen through sensitization of ground-state triplet oxygen molecules by triplet states of the fluorophore, and (ii) photoionization, that is, the formation of highly reactive radical ions upon excitation to higher excited states, are most probably the predominant photobleaching pathways. Thus, to minimize photobleaching or "photodestruction" the time the fluorophore spends in any of the gray marked states in Figure 4.1 (long-lived or highly reactive states) has to be minimized. Because in most organic fluorophores the triplet $T_1 \rightarrow T_n$, and the triplet and singlet radical $T_1^\bullet \rightarrow T_n^\bullet$, and $S_1^\bullet \rightarrow S_n^\bullet$ absorption spectra, respectively, overlap considerably with their $S_0 \rightarrow S_1$ absorption spectrum, excitation in higher excited states can easily occur and initiate photobleaching. Furthermore, charge separation, that is, electron transfer between excited fluorophores and the surrounding environment can take place. The nature of the radical produced, that is, a radical cation versus a radical anion, depends on the energy levels of the ground state of the interacting molecules and on those of the excited states of the fluorophore under investigation.

In addition, it has to be considered that in FRET experiments the photobleaching efficiency of the acceptor (i) scales with the FRET efficiency, (ii) is enhanced under picosecond pulsed versus continuous wave excitation of the donor, and (iii) increases with the intensity of the donor excitation. These findings indicate that the main pathway for photobleaching of the acceptor is through absorption of a short wavelength photon from the acceptor's first excited singlet state, that is, via $S_1 \rightarrow S_n$ absorption, and subsequent irreversible reactions [29, 30]. Photobleaching can be reduced, for example using pulsed excitation with a very low repetition rate, which ensures that the fluorophore residing in a long-lived triplet state (with a lifetime of typically several microseconds) cannot be further excited via $T_1 \rightarrow T_n$, absorption [32]. However, such low repetition rates, in the region of ~100 kHz, substantially decrease the number of fluorescence photons detectable per given time span from an individual fluorophore, and renders the application for fluorescence imaging and tracking microscopy as impossible.

Molecular oxygen has a dual role, as it is primarily responsible for photobleaching via photo-oxidation, especially of cyanine and fluorescein dyes, while being an efficient triplet-state quencher. Therefore, to increase the observation time for cyanine and fluorescein derivatives, a method to remove oxygen from solution is required, together with an alternative triplet-state quencher. Most approaches use an enzymatic oxygen scavenging system, which uses a mixture of glucose oxidase and catalase that converts glucose and oxygen into gluconic acid and water in a two-step process, which results in a net loss of O_2 from the solution [33]. Although oxygen removal increases the photobleaching lifetime of Cy3 and Cy5 (typical cyanine derivatives), it introduces intermittent fluorescence signals on the millisecond time scale due to the increased triplet-state lifetime. Therefore, efficient extrinsic triplet quenchers, that is, reducing compounds such as β-mercaptoethanol (BME), β-mercaptoethylamine (MEA), or glutathione at the higher millimolar level (10–150 mM) have to be added [34–36]. Alternatively, lower concentrations of Trolox (1–2 mM) in combination with the enzymatic oxygen-scavenging system can be applied to eliminate Cy5 blinking and to reduce photobleaching [37].

To conclude, photon detectors in use at present exhibit detection efficiencies of up to 100%, the chromatic aberrations of the optical components are almost perfectly corrected, and the numerical aperture of oil-immersion objectives reaches values >1.40. In addition, white light lasers covering the whole visible range from blue, green, yellow, red, and into the infrared, have been developed. Until recently, such lasers were not available, and it was only ultrashort laser pulse technology that paved the way towards their realization. In combination with acousto-optic tunable filters (AOTFs) any desired wavelength can be selected for efficient excitation of different fluorophores. Thus, the photophysical performance of fluorophores actually represents the bottleneck for the further improvement of single-molecule fluorescence spectroscopic techniques. Vital to the detection of single fluorophores is the fact that they can be cycled efficiently between the singlet ground state S_0 and first-excited singlet state S_1 marked as the light-gray area in Figure 4.1. Thus, the occupation of long-lived and higher reactive states has to be prevented or quenched by the addition of suitable quenchers that guarantee the efficient repopulation of S_0. Furthermore, fluorophores are considerably prone to photobleaching when they are excited with light of shorter wavelength, for example, as is possible in FRET experiments, while already residing in the S_1 or T_1 state. The absorption of a second photon, that is, $S_1 \rightarrow S_n$ or $T_1 \rightarrow T_n$ absorption, generates higher excited and more reactive species, which are probably destroyed by subsequent irreversible reactions.

Recently [38], a new and universal method has been introduced that dramatically improves photostability and reduces blinking of organic fluorophore, by recovering the reactive intermediates. The method is based on the removal of oxygen and quenching of the triplet and charge separated states by electron transfer reactions. Therefore, a formula that contains reducing and oxidizing agents, that is, a reducing and oxidizing system (ROXS) has been developed. The success of the approach was demonstrated by single-molecule fluorescence spectroscopy of oligonucleotides labeled with different fluorophores, that is, cyanines, (carbo)-rhodamines, and oxazines, in aqueous solvents enabling the observation of individual fluorophores for minutes, under moderate excitation, with increased fluorescence brightness (Figure 4.2).

By using reducing and oxidizing agents simultaneously (for example methylviologen and ascorbic acid at ~1 mM concentration) fluorophores that enter the triplet state are efficiently reduced by the reducing compound and a radical anion is formed. The radical anion is then quickly reoxidized by the oxidizing compound to repopulate the singlet ground state. Alternatively, the fluorophore is oxidized by the oxidant to form a radical cation and is subsequently returned to the ground state by the reducer.

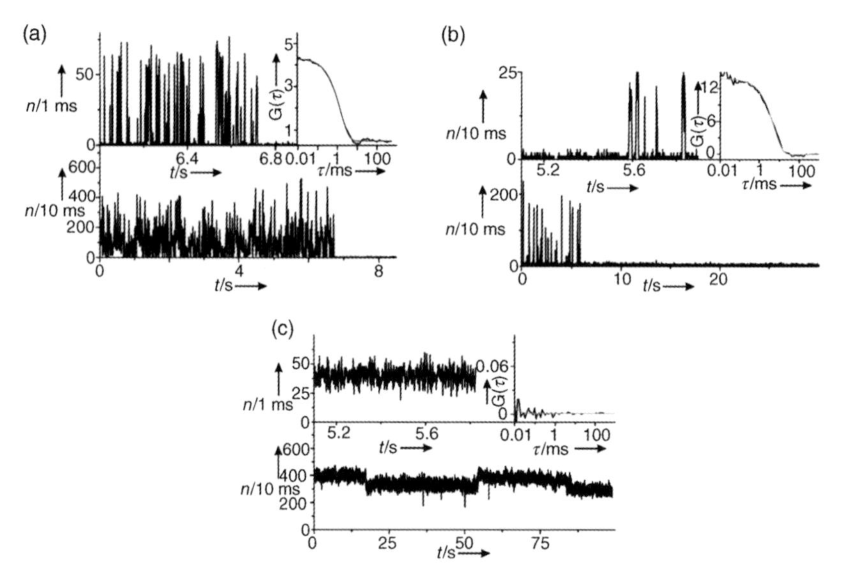

Figure 4.2 Fluorescence transients of ATTO647N labeled double-stranded DNA immobilized in an aqueous environment where the oxygen is removed using an oxygen scavenger [33]. (a) In the presence of 1 mM methylviologen serving as the oxidizing compound, (b) in the presence of 1 mM ascorbic acid serving as the reducing compound, and (c) in the presence of 1 mM of both the oxidizing and reducing compounds. The bottom traces are binned into 10 ms time bins, the left insets provide magnified views with 1 ms resolution, and the right insets show the second-order autocorrelation function $g(t)$ with monoexponential fit. Samples were excited on a confocal fluorescence microscopy at 635 nm with an average excitation intensity of approximately 2 kW cm^{-2} (Reproduced from Vogelsang *et al.* (2008) *Angew. Chem. Int. Ed.*, **47**, 5465 [38].)

The fast recovery of the singlet ground state is essential in order to compete successfully with the side reactions that lead to photobleached products. If applied appropriately, fluorophores show constant fluorescence, without any intermittence, and an intriguing photostability (Figure 4.2c).

Thermodynamic considerations of the underlying redox reactions support the model, providing a comprehensive picture of blinking and the photobleaching of organic fluorophores [38]. As ROXS takes into account several photobleaching pathways, the method is generally applicable and could find widespread implementation in modern fluorescence imaging and spectroscopy techniques. Very recently [39], it has been shown that Trolox, which is commonly known as a reducing compound, works according to the ROXS scheme. That is, the observed increased photostability and antiblinking effect of Trolox [37] is due to oxidizing quinoid derivatives of Trolox formed in aqueous buffer through reactions with molecular oxygen. Thus, Trolox and its oxidized quinoid form an efficient redox couple. Furthermore, when considering the photoswitching of an organic fluorophore and its application in various super-resolution or subdiffraction resolution imaging methods (see Chapter 8) [40], the controlled adjustability of the duration of on- and off-states is of critical importance [41]. Similarly, the addition of oxidizing and reducing compounds (usually applied at higher concentrations in the mM range to increase photostability and eliminate blinking) can be used advantageously at lower concentrations to generate off-states in standard organic fluorophores with a readily adjustable lifetime [42]. Interestingly, some reports demonstrated that thiol compounds, such as β-mercaptoethylamine, dithiothreitol, and glutathione can be used to efficiently reduce the triplet state of various organic fluorophores. In these instances, molecular oxygen, which is naturally present in aqueous solvents, at a concentration of 250 μM, acts as a reducer to repopulate the singlet ground state [41, 43, 44]. Thus, ROXS represents a universal method of choice to selectively control the lifetimes of excited states of small organic fluorophores.

References

1 Moerner, W.E. and Basche, T. (1993) *Angew. Chem., Int. Ed.*, **32**, 457–476.

2 Xie, X.S. (1996) *Acc. Chem. Res.*, **29**, 598–606.

3 Nie, S.M. and Zare, R.N. (1997) *Annu. Rev. Biophys. Biomol. Struct.*, **26**, 567–596.

4 Xie, X.S. and Trautman, J.K. (1998) *Annu. Rev. Phys. Chem.*, **49**, 441–480.

5 Ambrose, W.P., Goodwin, P.M., Jett, J.H., Van Orden, A., Werner, J.H., and Keller, R.A. (1999) *Chem. Rev.*, **99**, 2929–2956.

6 Deniz, A.A., Laurence, T.A., Dahan, M., Chemla, D.S., Schultz, P.G., and Weiss, S. (2001) *Annu. Rev. Phys. Chem.*, **52**, 233–253.

7 Moerner, W.E. (2002) *J. Phys. Chem. B*, **106**, 910–927.

8 Orrit, M. (2002) *J. Chem. Phys.*, **117**, 10938–10946.

9 Sunney Xie, X. (2002) *J. Chem. Phys.*, **117**, 11024–11032.

10 Keller, R.A., Ambrose, W.P., Arias, A.A., Cai, H., Emory, S.R., Goodwin, P.M., and Jett, J.H. (2002) *Anal. Chem.*, **74**, 316A–324A.

11 Kubitscheck, U. (2002) *Single Mol.*, **3**, 267–274.

12 Boehmer, M. and Enderlein, J. (2003) *ChemPhysChem*, **4**, 792–808.

13 Weiss, S. (2000) *Nat. Struct. Biol.*, **7**, 724–729.

14 Weiss, S. (1999) *Science*, **283**, 1670–1676.
15 Tinnefeld, P. and Sauer, M. (2005) *Angew. Chem. Int. Ed.*, **44**, 2642–2671.
16 Neuweiler, H. and Sauer, M. (2005) *Anal. Chem.*, **77**, 179A–185A.
17 Hischfeld, T. (1976) *Appl. Opt.*, **15**, 2965–2966.
18 Shera, E.B., Seitzinger, N.K., Davis, L.M., Keller, R.A., and Soper, S.A. (1990) *Chem. Phys. Lett.*, **174**, 553–557.
19 Kador, L., Horne, D.E., and E: Moerner, W. (1990) *J. Phys. Chem.*, **94**, 1237–1248.
20 Orrit, M. and Bernard, J. (1990) *Phys. Rev. Lett.*, **65**, 2716–2719.
21 Rigler, R., Mets, U., Widengren, J., and Kask, P. (1993) *Eur. Biophys. J. Biophys. Lett.*, **22**, 169–175.
22 Eigen, M. and Rigler, R. (1994) *Proc. Natl. Acad. Sci. USA*, **91**, 5740–5747.
23 Tsien, R. and Waggoner, A. (1995) in: *Handbook of Biological Confocal Microscopy* (ed. J.D. Pawley) Plenum Press, New York, pp. 267–279.
24 Ha, T. (2001) *Methods*, **25**, 78–86.
25 Kapanidis, A.N. and Weiss, S. (2002) *J. Chem. Phys.*, **117**, 10953–10964.
26 Song, L., Varna, C.A., Verhoeven, J.W., and Tanke, H.J. (1996) *Biophys. J.*, **70**, 2959–2968.
27 Widengren, J., Mets, Ü., and Rigler, R. (1995) *J. Phys. Chem.*, **99**, 13368–13379.
28 Eggeling, C., Widengren, J., Rigler, R., and Seidel, C.A.M. (1998) *Anal. Chem.*, **70**, 2651–2659.
29 Eggeling, C., Widengren, J., Brand, L., Schaffer, J., Felekyan, S., and Seidel, C.A.M. (2006) *J. Phys. Chem. A*, **110**, 2979–2995.
30 Kong, X., Nir, E., Hamadani, K., and Weiss, S. (2007) *J. Am. Chem. Soc.*, **129**, 4643–4654.
31 Widengren, J., Chmyrov, A., Eggeling, C., Löfdahl, P.A., and Seidel, C.A.M. (2007) *J. Phys. Chem. A*, **111**, 429–440.
32 Donnert, G., Eggeling, C., and Hell., S.W. (2007) *Nat. Methods*, **4**, 81–86.
33 Benesch, R.E. and Benesch, R. (1953) *Science*, **118**, 447–448.
34 Kishino, A. and Yanagida, T. (1988) *Nature*, **334**, 74–76.
35 Bates, M., Blosser, T.R., and Zhuang, X.W. (2005) *Phys. Rev. Lett.*, **94**, 108101.
36 Heilemann, M., Margeat, E., Kasper, R., Sauer, M., and Tinnefeld, P. (2005) *J. Am. Chem. Soc.*, **127**, 3801–3806.
37 Rasnik, I., McKinney, S.A., and Ha, T. (2006) *Nat. Methods*, **3**, 891–893.
38 Vogelsang, J., Kasper, R., Steinhauer, C., Person, B., Heilemann, M., Sauer, M., and Tinnefeld, P. (2008) *Angew. Chem. Int. Ed.*, **47**, 5465–5469.
39 Cordes, T., Vogelsang, J., and Tinnefeld, P. (2009) *J. Am. Chem. Soc.*, **131**, 15018–15019.
40 Hell, S.W. (2007) *Science*, **316**, 1153–1158.
41 Heilemann, M., Dedecker, P., Hofkens, J., and Sauer, M. (2009) *Laser Photon. Rev.*, **3**, 180–203.
42 Steinhauer, C., Forthmann, C., Vogelsang, J., and Tinnefeld, P. (2008) *J. Am. Chem. Soc.*, **130**, 16840–16841.
43 van de Linde, S., Endesfelder, U., Mukherjee, A., Schüttpelz, M., Wiebusch, G., Wolter, S., Heilemann, M., and Sauer, M. (2009) *Photochem. Photobiol. Sci.*, **8**, 465–469.
44 van de Linde, S., Kasper, R., Heilemann, M., and Sauer, M. (2008) *Appl. Phys. B*, **93**, 725–731.

5 Fluorescence Correlation Spectroscopy

5.1 Introduction

Fluorescence fluctuation spectroscopy (FFS) is the generalized name for a set of spectroscopic methods that are based on the measurement and evaluation of fluorescence intensity fluctuations originating from a small number of fluorescing molecules, usually contained within a sufficiently small detection range. Any process that influences the fluorescence intensity of these molecules (such as changes of their positions within the measurement systems, their photophysics, chemical reactions, conformational changes, etc.) will lead to a temporally changing fluorescence signal, most often in a stochastic way. For example, molecules that are free to diffuse in and out of the detection region will generate a stochastically changing fluorescence intensity signal. Similarly, molecules that, besides circling through the first excited singlet and ground states, can switch from time to time into a non-fluorescent triplet state, will generate a fluctuating fluorescence intensity signal. The important point is that the character of these fluorescence signal fluctuations is associated with the underlying physical processes and their parameters (i.e., diffusion coefficients, or photophysical rate constants). The core idea of FFS is to evaluate the observed intensity fluctuation in such a way that one can determine these parameters.

The classical example of FFS is fluorescence correlation spectroscopy (FCS), originally introduced by Elson, Magde and Webb in the early 1970s [1–3]. In its original form it was developed for measuring diffusion, concentration, and chemical/biochemical interactions/reactions of fluorescent or fluorescently labeled molecules at nanomolar concentrations in solution. Let us explain the basic principles of FFS through the example of measuring the diffusion of fluorescing molecules using FCS. Assume that one has a measurement system for exciting and measuring the fluorescence of solved molecules in solution within a sufficiently small detection range (for the moment, let us postpone all the technical details). The system will be characterized by a so-called volume of detection, that is, a region in the solution where efficient fluorescence excitation and detection take place. If the concentration of fluorescent molecules in solution is sufficiently small, so that only one of the very few molecules is within the detection volume at any moment in time, the resulting

Handbook of Fluorescence Spectroscopy and Imaging. M. Sauer, J. Hofkens, and J. Enderlein

ISBN: 978-3-527-31669-4

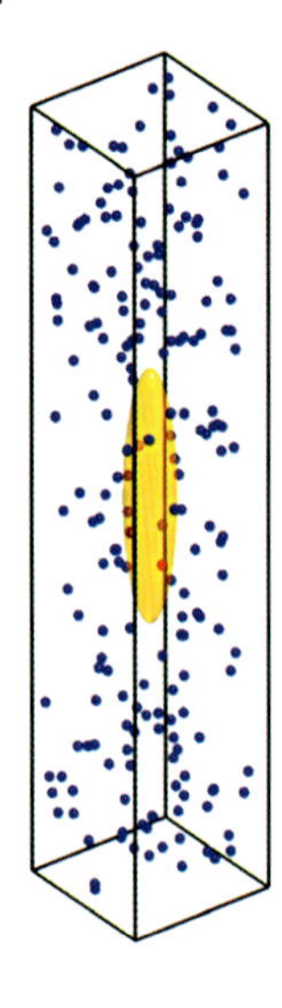

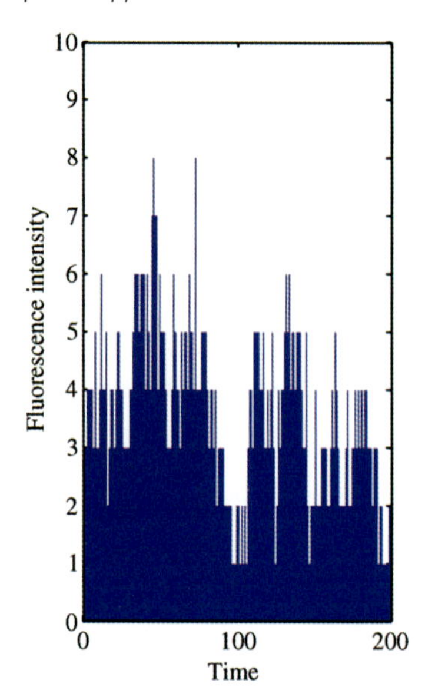

Figure 5.1 Typical FCS experiment. Fluorescing molecules (blue dots in left-hand box) diffuse in and out of a small detection volume (yellowish region in left box). Owing to the constantly changing number of molecules within the detection volume, the resulting fluorescence intensity out of that volume fluctuates significantly (right-hand graph).

measured fluorescence signal fluctuates significantly, in response to the entering and leaving of individual fluorescing molecules into or out of the detection volume (Figure 5.1).

In FCS, the detected fluorescence intensity is correlated with a time-shifted replica of itself for different values of the time shift (the lag time). The result is the so-called autocorrelation function (a second-order correlation function, abbreviated to ACF) which is calculated as

$$g(\tau) = \langle I(t) I(t+\tau) \rangle_t \tag{5.1}$$

where

$I(t)$ is the fluorescence intensity at time t and the triangular brackets denote averaging over all time values t.

The physical meaning of the autocorrelation is that it is directly proportional to the probability of detecting a photon at time τ if there was a photon detection event at time zero. This probability is composed of two basically different terms: The two photons detected at time zero and at time τ can originate from an uncorrelated background or from different fluorescing molecules, and therefore do not have any physical correlation (provided there is no interaction between the various fluorescing molecules). These events will contribute to a constant offset of $g(\tau)$, which is completely independent of τ (the joint probability of detecting two physically uncorrelated photons is completely independent on the length of time between their detection).

Alternatively, the two photons can originate from one and the same molecule and are then physically correlated. Let us start with some very simple qualitative considerations concerning the lag-time dependence of $g(\tau)$. Suppose that a molecule is close to the center of the detection volume. Then there will be a high probability of detecting a large number of consecutive fluorescence photons from this molecule, that is, the fluorescence signal will be highly correlated in time.

When the molecule, as a result of diffusion, begins to exit the detection volume, this correlation will decrease continuously, that is, the probability of seeing further fluorescence photons will decrease with time, until the molecule has diffused away completely and the correlation is totally lost. Of course, the temporal decay of the correlation, that is, the temporal decay of $g(\tau)$ with increasing lag time τ, will be proportional to the diffusion speed of the molecule: the larger the diffusion coefficient, the faster the fluorescence correlation decays. A typical example of this behavior is shown in Figure 5.2.

A second important property of the ACF is its dependence on the concentration of the fluorescing molecules. It is fairly obvious that the fluorescence intensity fluctuations will be larger for lower concentrations of the molecules, see Figure 5.3. Indeed, if one has, on average, only a signal molecule within the detection volume, then the diffusion of this molecule out of this volume or the diffusion of another molecule into this volume will cause a big change to the measured fluorescence intensity. In contrast, if the average number of fluorescing molecules within the detection volume is fairly large (e.g., several hundreds), then the leaving or entering of a molecule will

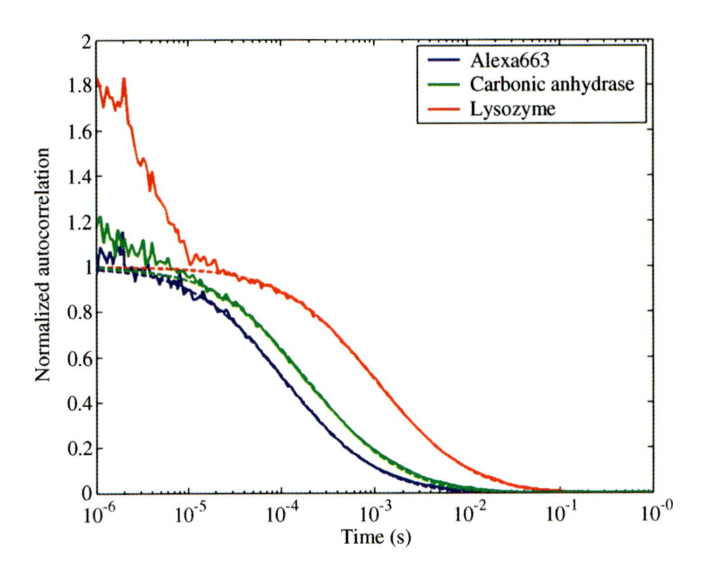

Figure 5.2 Typical FCS curves. Presented are the ACFs for three measurements on aqueous solutions of pure dye Alexa633, labeled carbonic anhydrase, and labeled lysozyme. The different long-time decay of the various ACFs reflects the different diffusion coefficients of the three samples, that is, their different molecular sizes. The short-time decay in the microsecond time range is due to photophysical transitions (in this instance singlet to triplet transitions) of the dye.

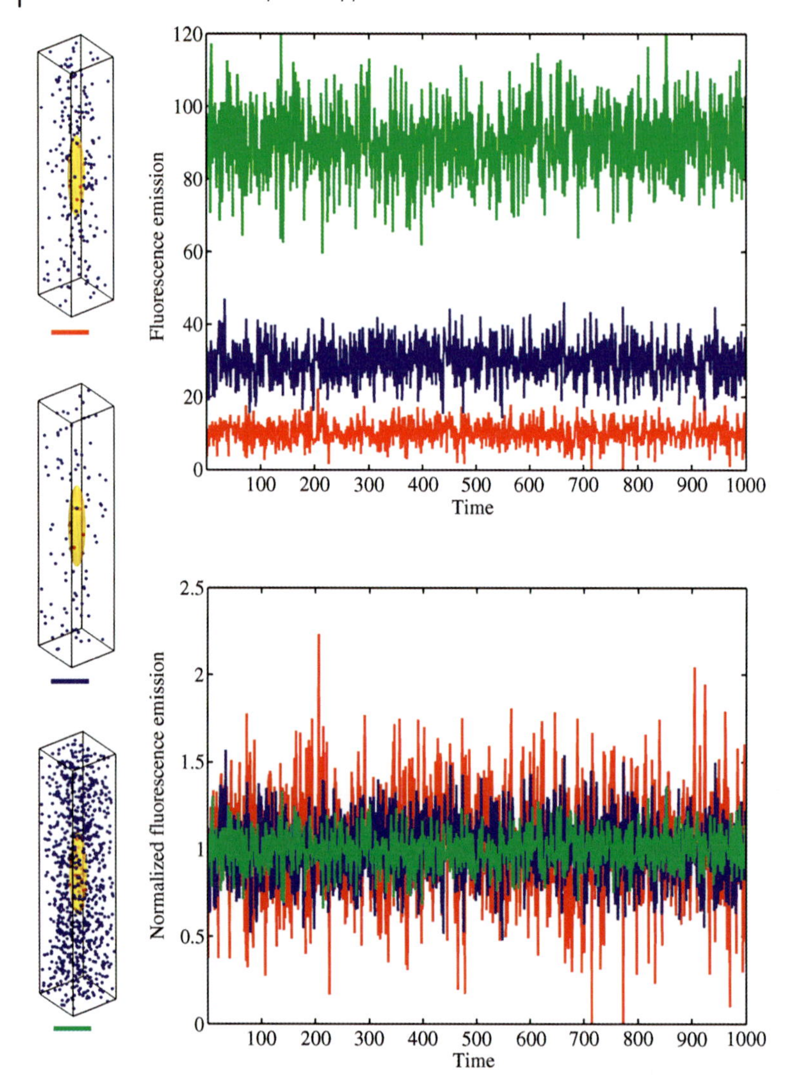

Figure 5.3 Concentration dependence of the fluorescence fluctuations. Shown are three fluorescence intensity traces for three concentrations of fluorescent molecules in the ratio of 1 : 3 : 9 (see left-hand panels). The top graph shows the absolute intensity traces, and the bottom graph the same curves normalized by their average value. As can be seen, the *relative* fluctuations of the fluorescence signal are larger the lower the concentration of the molecules.

cause only small signal variations. Intuitively one may expect a direct connection between the average number of molecules within the detection volume (i.e., concentration) and the amplitude of the fluorescence intensity fluctuations. Indeed, as will be seen below, there is a direct connection between the inverse concentration of the fluorescing molecules and the amplitude of the ACF.

Thus, FCS measurements can provide information about diffusion and concentration of fluorescing molecules. Any process that alters one (or both) of these quantities can also be measured by FCS. For example, consider the binding of two proteins in solution: by labeling one of the binding partners with a fluorescence label, and monitoring the changing value of the diffusion coefficient of the labeled molecules upon binding of their binding partner, using FCS, one can directly measure binding affinities and kinetics.

For various time scales, the temporal behavior of the autocorrelation function is determined by the different properties of the fluorescing molecules. On a nanosecond time scale, so-called photon antibunching can be observed, reflecting the fact that directly after the emission of a photon the molecule needs to be re-excited again to be able to emit the next photon, leading to a steep decrease in $g(\tau)$ towards shorter times. If excitation and/or detection are performed with polarization filters, the autocorrelation will also show contributions from the rotational diffusion dynamics of the molecules on a pico- to nanosecond time scale. On a microsecond time scale, $g(\tau)$ is dominated by fast photophysical processes, or fast intramolecular structural dynamics. As previously mentioned, on a millisecond to the second level, the autocorrelation function shows its typical decay, due to the lateral diffusion of the molecules out of the detection region, as shown in Figure 5.4.

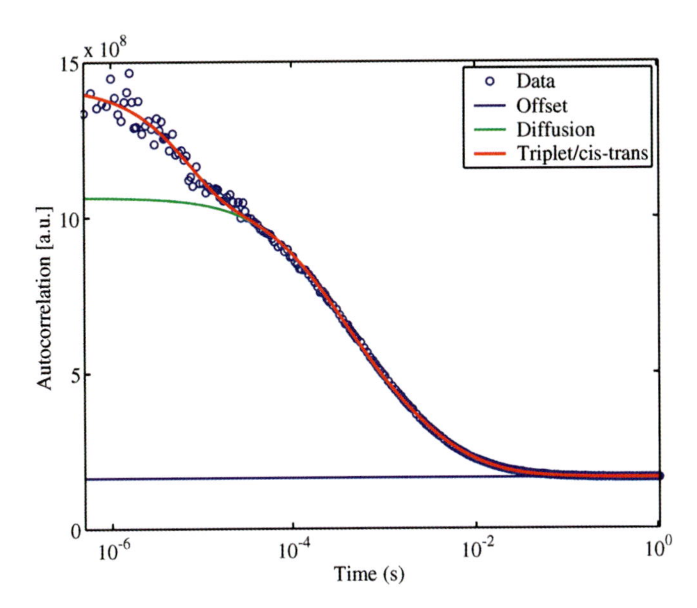

Figure 5.4 Typical anatomy of an ACF (measured on Alexa633 in aqueous solution). On a microsecond time scale, one observes the fast photophysical transition from the fluorescent singlet state into the non-fluorescent triplet state; on a millisecond time scale, the ACF decays due to diffusion of the dye out of the detection volume, and within the limit of a large lag time, the ACF decays towards an offset value, which is produced by physically non-correlated photon pairs coming from different molecules.

It took nearly two decades, until the development of new lasers with high beam quality and temporal stability, low-noise single-photon detectors, and high-quality microscope objectives with nearly perfect imaging quality at high numerical aperture, for the technique to see a new renaissance in SMS. Achieving values for the detection volume within the order of a few μm^3 made the technique applicable for samples at reasonably high concentrations and with short measurement times.

The advantage of FCS is its relative simplicity. Its drawback is that it works only within a very limited concentration range. If the concentration of fluorescing molecules becomes too large (typically $\gg 10^{-8}$ M), then the contribution from correlated photons from individual molecules, scaling with the number N of molecules within the detection volume, becomes very small compared with the contribution from uncorrelated photons from different molecules, scaling with N^2. If the concentration is too low (typically $<10^{-13}$ M), then the probability of finding a molecule within the detection region becomes extremely low. In both instances, the measurement time for obtaining a high-quality autocorrelation function becomes prohibitively large, although a remedy to this problem is to scan the laser focus rapidly through the solution [4, 5].

There are numerous excellent reviews and overviews of FCS available [6–11] and there is even a complete book devoted to the subject [12]. The present chapter gives a very general introduction into the philosophy of FCS, aiming to be self-contained and to develop the fundamental principles of FCS. However, it also describes recent methodological advances that are not well covered by previous reviews.

5.2 Optical Set-Up

A typical FCS measurement set-up is shown in Figure 5.5 [13]. Fluorescent molecules are dissolved in an aqueous solution that is placed on top of a chambered cover slide. A collimated laser beam with perfect Gaussian TEM_{00} mode [14] is coupled via a dichroic mirror into an objective with a high NA (numerical aperture), which focuses the laser into a diffraction-limited spot in the sample. The dichroic mirror is reflecting at the wavelength of the laser and transmitting at the wavelengths of the fluorescence emission. The use of a Gaussian TEM_{00} mode assures diffraction-limited focusing of light, thus achieving a minimum focus diameter in the sample. Fluorescence light generated in the sample is collected by the same objective (so-called epi-fluorescence set-up), transmitted through the dichroic mirror, and focused onto a circular confocal aperture. Behind the aperture, the fluorescence light is refocused onto a sensitive light detector, usually a single-photon avalanche diode (SPAD). The confocal aperture effectively rejects fluorescence light that is generated outside the focal plane, as shown in Figure 5.5. In combination, fluorescence generation (by diffraction-limited focusing of the excitation light) and fluorescence detection (by confocal detection) generate an effective detection volume of about 0.5 μm in diameter in the focal plane, and a few micrometers along the optical axis.

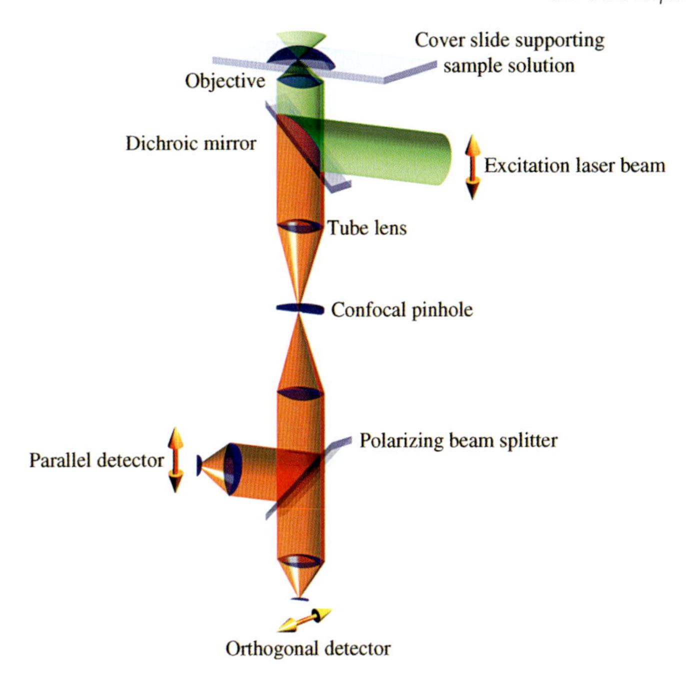

Figure 5.5 Schematic of the optical set-up used for FCS measurements. Shown is a set-up with a linearly polarized excitation laser and polarization-sensitive detection. The detectors are usually single-photon avalanche diodes or photomultiplier tubes.

The exact shape and size of the detection volume determines the shape and temporal decay of the autocorrelation function. For example, the smaller the detection volume, the faster molecules diffuse out of it and the faster the ACF decays, and vice versa. The actual quantity that defines the autocorrelation function is the so-called molecule detection function (MDF). The MDF describes the probability of seeing a fluorescence photon from a molecule at a given position $\mathbf{r}$ in the sample. Thus, the MDF, which we will denote by $U(\mathbf{r})$, is a function of position $\mathbf{r}$ and rapidly falls off to zero if one moves away from the optical axis and/or the focal plane. As we will see below, knowing the exact shape of the MDF allows one to calculate exactly the shape of the ACF, which can then subsequently be used to fit experimental data to obtain, for example, diffusion and/or concentration values of the fluorescent molecules. However, this is also the principal problem of FCS: a precise quantitative evaluation of an ACF critically depends on the exact knowledge of the MDF. We will discuss this topic in more detail in Section 5.4.

5.3
Data Acquisition and Evaluation

In the past, conventional FCS set-ups used to employ hardware correlators that calculated the ACF onboard, on the basis of the fed-in signal from the photodetectors.

Recently, most set-ups have been using fast photon counting electronics for asynchronously recording and storing the arrival times of the detected photons, and subsequently use software algorithms for calculating the ACF from the recorded photon data. This permits much more flexibility in data handling and evaluation, as will be seen for example in the case of fluorescence lifetime correlation spectroscopy (FLCS) (Section 5.7), and we will describe this approach here in more detail.

Asynchronously measured single-photon counting data consists of a linear file of detection times $\{t_1, t_2, \ldots, t_N\}$ of the detected photons, where N is the total number of detected photons during a given measurement. A special feature of these detection times is that they are integer multiples of some minimal time δt, determined by the temporal resolution of the detection electronics. Without the restriction of generality, it can be assumed that all times are measured in units of δt, so that all the numbers t_j take integer values. The value $g(\tau)$ of the autocorrelation function for a given lag time τ is defined in Equation 5.1. For a photon detection measurement with temporal resolution δt, the intensity values $I(t)$ within consecutive time intervals can only take the values $1/\delta t$ or 0, depending on whether or not there was a photon detection event during a time interval of width δt. The average in Equation 5.1 is then calculated as the sum over all consecutive time intervals of width δt, divided by the total number of summed intervals. In practice, one does not compute the autocorrelation function for all possible values of lag time τ, but at increasingly spaced lag time values. If the temporal resolution of the photon detection is, for example, 100 ns, and one desires to follow correlation processes up to a minute, possible values of lag time τ are any value between 100 ns and 60 s in intervals of 100 ns, resulting in 6×10^8 possible lag time values. Calculation of $g(\tau)$ for all of these values would be an enormous time-consuming numerical effort. Instead, the autocorrelation is calculated for only few, approximately logarithmically spaced values of τ.

A straightforward way of calculating the autocorrelation function is to divide the total measurement time, $t_N - t_1$, into intervals of unit length δt, and to sort the detected photons into these intervals corresponding to their arrival times t_j. The result is a synchronous photon detection intensity file I_j with j running from 1 through to $t_N - t_1$, where the I_j can only adopt the values one or zero. The fluorescence autocorrelation can then be calculated as given by Equation 5.1. In practice, such an approach is prohibitively memory demanding and computationally expensive. An alternative and much more efficient FCS algorithm works directly on the arrival times $\{t_1, t_2, \ldots, t_N\}$, without converting them into time-binned data. For a given lag time τ, the algorithm searches for all photon pairs in the data stream that are a temporal distance τ apart from each other. The number of photon pairs with a distance τ is directly proportional to the autocorrelation value at lag time τ. A convenient choice for the values of τ is

$$\tau_j = \begin{cases} 1 & \text{if } j = 1 \\ \tau_{j-1} + 2^{\lfloor (j-1)/B \rfloor} & \text{if } j > 1 \end{cases} \tag{5.2}$$

where

j takes integer values starting with one and running up to some maximum number $j_{max} = n_{casc} B$
B is some integer base number
the bracket $\lfloor\ \rfloor$ gives the integer part of the enclosed expression.

The resulting lag times are grouped into n_{casc} cascades with equal spacing of $2^{\lfloor j/B \rfloor}$. The advantage of such a choice of lag times is that all τ_j have integer values, so that fast integer arithmetic can be used in subsequent computations. For example, when using a base number value of $B = 10$ and $n_{casc} = 3$, one obtains the lag time sequence $\{\tau_j\} = \{1,2,\ldots,9,10,12,\ldots,28,30,34,\ldots,70\}$.

For a given lag time τ, a second vector of arrival times $\{t'_1, t'_2, \ldots, t'_N\}$ is generated, containing the time values $t'_j = t_j + \tau$. At the beginning, the value of the autocorrelation at lag time τ is set to zero. The algorithm starts with the time t_1 in the original vector and moves to consecutive time entries in that vector until it encounters a value t_j which is equal to or larger than t'_1. If $t_j = t'_1$, the value of the autocorrelation at lag time τ is increased by one. Next, the algorithm switches to the entries of the second vector and, starting with t'_1, moves to consecutive time entries in that vector until it encounters a value t'_{j_1} that is equal to or larger than t'_{j_1}. If $t'_{j_1} = t'_{j_1}$, the value of the autocorrelation at lag time τ is increased by one. The algorithm switches back to the first vector and, starting with t'_{j_1}, moves to consecutive time entries in that vector until it encounters a value t'_{j_2} which is equal to or larger than t'_{j_1}, and so on until the last entry in one of either vectors is reached. In its simplest form, the autocorrelation algorithm calculates, up to some constant factor, the probability of detecting a photon at some time $t + \tau$ if there was a photon detection event at time t. A visualization of the algorithm's working is shown in Figure 5.6.

When only applying the algorithm at the increasingly spaced lag times as given by Equation 5.2, it will completely miss, for example, the strong autocorrelation of any periodic signal with a repetition time not included within the vector of the lag times used. To avoid this, one usually applies an averaging procedure by coarsening the time resolution of the photon detection times t_j when approaching the calculation of the autocorrelation function at increasingly larger lag times. This is equivalent to the multiple-tau and multiple-sampling time correlation method employed in hardware correlators [15]. Such a procedure is easily incorporated into the present algorithm. Besides working just with the original and shifted vectors of the arrival times, $\{t_1, t_2, \ldots, t_N\}$ and $\{t'_1, t'_2, \ldots, t'_N\}$, all time entries t_j and t'_j are associated with weight values w_j and w'_j, which are all set to one at the start of the algorithm. For the case of an equality $t_j = t'_k$, the autocorrelation is increased by the weight product $w_j w'_k$ and not by one. A time coarsening step is inserted each time when finishing the calculations for one cascade of B lag times τ_j with equal spacing, and before starting with the next cascade of B lag times with double spacing. All values $\{t_1, t_2, \ldots, t_N\}$ used in the previous cascade are divided by two and rounded to the nearest lower integer value, which occasionally leads to the occurrence of consecutive entries with the same time value. Before continuing the autocorrelation computation, such double entries are reduced to one entry, and the corresponding weight of that remaining entry is increased by the weight of the eliminated one.

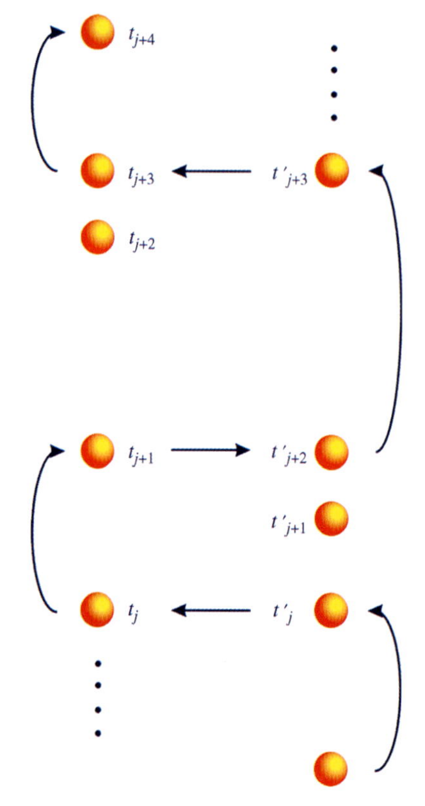

Figure 5.6 Visualization of the software correlation algorithm as described in the text.

Thus, with increasing lag time τ_j, the time scale underlying the autocorrelation calculation becomes increasingly coarser, and the total number of time entries to be processed increasingly smaller. To correct for the varying time scale of the autocorrelation calculation, one has finally to divide, at each lag time τ_j, the calculated autocorrelation value by the corresponding time scale factor $2^{\lfloor j/B \rfloor}$. As pointed out in reference [15], this time coarsening leads to a triangularly weighted average of the true value of the autocorrelation function.

The above algorithm can easily be generalized for more complex situations. In most photon detection experiments, it is desirable to detect the fluorescence within two detection channels, either to obtain more information about the fluorescence (e.g., detection of fluorescence polarization or different emission colors), or to eliminate the adverse effects of detector afterpulsing [16] on the short-time behavior of the FCS curve. In all these situations, cross-correlating the signals of both detection channels instead of autocorrelating the signals of each channel becomes an important task. This is easily realized with the above algorithm by using the time values of the first channel as the entries for the first vector $\{t_1, t_2, \ldots, t_N\}$ and the time values of the second channel for calculating the lag-time shifted vector $\{t'_1, t'_2, \ldots, t'_N\}$. In this case, one obtains the cross-correlation of the second channel with positive lag times relative to the first channel. By reversing the order, that is, assigning the time values of the second channel to $\{t_1, t_2, \ldots, t_N\}$ and using the time values of the first channel for $\{t'_1, t'_2, \ldots, t'_N\}$, one obtains the cross-correlation of the first channel with positive lag times relative to the second channel.

5.4 Milliseconds to Seconds: Diffusion and Concentration

Thermally induced translational diffusion is one of the fundamental properties exhibited by molecules within a solution. Via the Stokes–Einstein relationship,

$$D_{\text{trans}} = \frac{k_B T}{6\pi\eta R_{\text{trans}}} \tag{5.3}$$

it is directly coupled with the hydrodynamic radius R_{trans} of the molecules [17], where

D_{trans} is the translational diffusion coefficient
k_B is Boltzmann's constant
T the absolute temperature
η the viscosity of the solution.

Any change in this radius will change the associated diffusion coefficient of the molecules. Such changes occur for most biomolecules – in particular proteins, RNA and DNA – when interacting with their environment (e.g., binding of ions or other biomolecules) or performing biologically important functions (e.g., enzymatic catalysis) or reacting to changes in environmental parameters, such as pH, temperature, or chemical composition (e.g., protein unfolding). Therefore, the ability to precisely measure diffusion coefficients has a large range of potential applications, for monitoring, for example, conformational changes in proteins upon ion binding or unfolding. However, many biologically relevant conformational changes are connected with fairly small changes to the hydrodynamic radius, of the order of Ångstrøms (see e.g., [18]). To monitor these small changes, it is necessary to measure the diffusion coefficient with an accuracy of better than a small percentage. Standard methods for diffusion coefficient measurements achieving this accuracy are dynamic light scattering (DLS) [19], pulsed field gradient NMR (nuclear magnetic resonance) [20], size exclusion electrophoresis [21], or analytical ultracentrifugation [22]. However, all these methods operate at rather high sample concentrations, far removed from the limit of infinite dilution. To obtain the correct infinite-dilution limit and thus a correct estimate of the hydrodynamic radius, one frequently has to measure at different concentrations and to extrapolate the concentration/diffusion coefficient curve towards zero concentration (see e.g., [23]). Another problem is that proteins are often prone to aggregation [24] at the concentrations needed to obtain sufficient data quality. Thus, FCS is a relatively simple and attractive alternative for measuring diffusion coefficients, and the next sections will explain in detail how this is done.

5.4.1
Single-Focus FCS

Following Equation 5.1, the ACF is the correlation of the fluorescence intensity with a time-shifted replica of itself, calculated for all possible lag times τ. The measured signal $I(t)$ stems from the fluorescence of all the molecules within the sample plus the uncorrelated background I_{bg} (light scattering, electronic noise, etc.):

$$I(t) = I_{bg} + \sum_j I_j(t) \tag{5.4}$$

where

the index j refers to the jth molecule, and the summation runs over all molecules in the sample.

Thus, the ACF $g(\tau)$ is given by

$$\begin{aligned} g(\tau) &= \left\langle \left(I_{bg}(t) + \sum_j I_j(t) \right) \left(I_{bg}(t+\tau) + \sum_k I_k(t+\tau) \right) \right\rangle \\ &= \sum_j \langle I_j(t) I_j(t+\tau) \rangle + \sum_{j \neq k} \bar{I}_j \bar{I}_k + + \sum_j \bar{I}_j \bar{I}_{bg} + \bar{I}_{bg}^2 \end{aligned} \tag{5.5}$$

where the triangular brackets and bars denote averaging over all possible time values t. The last line takes into account that fluorescence photons coming from different molecules are completely uncorrelated (no intermolecular interaction provided). Because all molecules in solution are indistinguishable, the last equation can be simplified further to

$$g(\tau) = N\langle i(t)i(t+\tau)\rangle + N(N-1)\langle i(t)\rangle^2 \tag{5.6}$$

where

i is the measured fluorescence intensity of *any* molecule
N is the total number of molecules present in the sample.

Thus, the task of calculating the function $g(t)$ reduces to calculating $\langle i(t)i(t+\tau)\rangle$, the correlation of the fluorescence signal from one and the same molecule, and $\langle i \rangle$, the average detected fluorescence intensity of one molecule.

The correlation $\langle i(t)i(t+\tau)\rangle$ of the fluorescence signal from one and the same molecule can be easily derived if its physical meaning is remembered. It is proportional to the probability of seeing, from one and the same molecule, a photon at time $t+\tau$ if there was a photon detection at time t. The probability of finding a molecule within an infinitely small volume dV anywhere in the sample is equal to dV/V, where V is the total sample volume. Next, the probability of detecting a photon from a molecule at a given position $\mathbf{r}_0$ is directly proportional to the value of the MDF at this position, that is, to $U(\mathbf{r}_0)$. Furthermore, the chance that the molecule diffuses from position $\mathbf{r}_0$ to position $\mathbf{r}_1$ within time τ is given by the solution of the diffusion equation

$$\frac{\partial G}{\partial \tau} = D\Delta G \tag{5.7}$$

where

Δ is the three-dimensional Laplace operator in coordinate $\mathbf{r}_1$
D is the diffusion coefficient of the molecule
G approaches a three-dimensional Dirac function for $t \to 0$, $G(\mathbf{r}_1, \mathbf{r}_0, \tau = 0) = \delta(\mathbf{r}_1 - \mathbf{r}_0)$, that is, the molecule is exactly at position $\mathbf{r}_0$ at time zero.

For a sample with wide boundaries, this solution is explicitly given by

$$G(\mathbf{r}_1, \mathbf{r}_0, \tau) \equiv G(\mathbf{r}_1 - \mathbf{r}_0, \tau) = \frac{1}{(4\pi D\tau)^{3/2}} \exp\left(-\frac{|\mathbf{r}_1 - \mathbf{r}_0|^2}{4D\tau}\right) \tag{5.8}$$

Finally, the chance to detect a photon from the molecule at the new position $\mathbf{r}_1$ is again proportional to the value of the MDF at this position, that is, to $U(\mathbf{r}_1)$. Thus, the autocorrelation $\langle i(t)i(t+\tau)\rangle$ is calculated as the product of all these individual contributions and averaging over all possible initial and final positions of the molecule, that is, integrating over $\mathbf{r}_0$ and $\mathbf{r}_1$:

$$\langle i(t)i(t+\tau)\rangle = \frac{1}{V}\int_V d\mathbf{r}_1 \int_V d\mathbf{r}_0\, U(\mathbf{r}_1) G(\mathbf{r}_1, \mathbf{r}_0, t)\, U(\mathbf{r}_0) \tag{5.9}$$

Similarly, the average fluorescence intensity from a single molecule in the sample is given by

$$\langle i(t)\rangle = \frac{1}{V}\int_V d\mathbf{r}\, U(\mathbf{r}), \tag{5.10}$$

so that the full ACF, in its most general form, reads

$$g(\tau) = c\int_V d\mathbf{r}_1 \int_V d\mathbf{r}_0\, U(\mathbf{r}_1) G(\mathbf{r}_1, \mathbf{r}_0, t)\, U(\mathbf{r}_0) + \left[c\int_V d\mathbf{r}\, U(\mathbf{r})\right]^2 \tag{5.11}$$

where

c denotes the concentration of fluorescent molecules (numbers per volume)

and one has used the fact that in the limit of large sample volume $N/V \to c$ and $N(N-1)/V^2 \to c^2$

The above Equations 5.9 and 5.10 are of general validity, but before being able to apply them to the evaluation of real FCS experiments one has to specify the MDF $U(\mathbf{r})$. In the majority of publications on FCS, a very simple approximation of the MDF has been adopted, assuming that it is well described by a three-dimensional Gaussian distribution, that is,

$$U(\mathbf{r}) = \kappa \exp\left[-\frac{2}{a^2}(x^2 + y^2) - \frac{2}{b^2}z^2\right] \tag{5.12}$$

where

κ is some overall constant
(x, y, z) are Cartesian co-ordinates centered at the intersection of focal plane and optical axis and with $z = 0$ being the optical axis
a and b are the characteristic half axes of the cylindrically symmetrical, Gaussian-shaped detection volume.

This corresponds to the lowest-order polynomial expansion of ln $U(\mathbf{r})$ (due to both axial and mirror symmetry, terms linear in x, y, z are absent). The characteristic parameters a and b are not known a priori and are usually determined by reference measurements on a sample with know diffusion coefficient. Using the expression in Equation 5.12, the single-molecule autocorrelation $\langle i(t)i(t+\tau)\rangle$ can now be explicitly calculated as

$$\begin{aligned} \langle i(t)i(t+\tau)\rangle &= \frac{c\varepsilon^2}{(4\pi Dt)^{\frac{3}{2}}}\int_V d\mathbf{r}\int_V d\varrho\, U(\mathbf{r}+\varrho)\exp\left(-\frac{|\varrho|^2}{4Dt}\right)U(\mathbf{r}) \\ &= \left(\frac{\pi^{\frac{3}{2}}}{8}\right)\frac{c\varepsilon^2 a^2 b}{(1+4Dt/a^2)\sqrt{1+4Dt/b^2}} \end{aligned} \tag{5.13}$$

where

ε is a constant factor taking into account overall detection efficiency of the measurement system, absolute fluorescence brightness of the molecules (defined by absorption cross-section and fluorescence quantum yield) and so on.

In a similar way, the average fluorescence signal resulting from one molecule is given by

$$\langle i(t)\rangle = c\varepsilon\int_V d\mathbf{r}\, U(\mathbf{r}) = \left(\frac{\pi^3}{8}\right)^{\frac{1}{2}} c\varepsilon a^2 b \tag{5.14}$$

Thus, the final result for the total autocorrelation reads

$$g(\tau) = \left(\frac{\pi^{\frac{3}{2}}}{8}\right)\frac{c\varepsilon^2 a^2 b}{(1+4Dt/a^2)\sqrt{1+4Dt/b^2}} + \left[I_{bg} + \left(\frac{\pi^3}{8}\right)^{\frac{1}{2}} c\varepsilon a^2 b\right]^2 \tag{5.15}$$

where an additional background intensity I_{bg} has been included. An important property of the ACF is that the concentration of the fluorescent species can be derived from Equation 5.15 via

$$\frac{g(\infty)}{g(0)-g(\infty)} = c\frac{\left[\int d\mathbf{r}\, U(\mathbf{r}) + I_{bg}\right]^2}{\int d\mathbf{r}\, U^2(\mathbf{r})} \tag{5.16}$$

where we have taken into account that $G(\varrho, \tau)$ in Equation 5.8 approaches a δ-function for $\tau \to 0$. Using Equation 5.16 one can define the effective detection volume V_{eff} as

$$V_{\text{eff}} = \frac{\left[\int d\mathbf{r}\, U(\mathbf{r})\right]^2}{\int d\mathbf{r}\, U^2(\mathbf{r})} \tag{5.17}$$

so that, for negligible background, the left-hand side of Equation 5.16 equals cV_{eff}, that is, the mean particle number within V_{eff}. Thus, the ACF is often used for estimating concentrations of fluorescing molecules.

Although Equation 5.15 is remarkably successful at fitting measured autocorrelation curves, the physical meaning of the parameters a and b is rather obscure, because the actual MDF is usually much more complicated than as given by Equation 5.12. It is known that the real shape of the MDF is only poorly described as a three-dimensional Gaussian. A more serious problem is that the exact form of the MDF is extremely sensitive to several optical and photophysical artifacts and can easily change from one measurement to another. The most severe of these will be discussed here.

The first common problem is that state-of-the-art water immersion objectives used in FCS set-ups are designed to image through a cover slide of specific thickness. In this sense, the cover slide acts as the last optical element of an objective, and the optical quality of the imaging (and laser focusing) critically depends on the exact matching of the cover slide thickness the objective is adjusted to with its actual thickness. What happens when the cover slide thickness deviates from its design value by only a few micrometers is shown in Figure 5.7, where one can see the severe

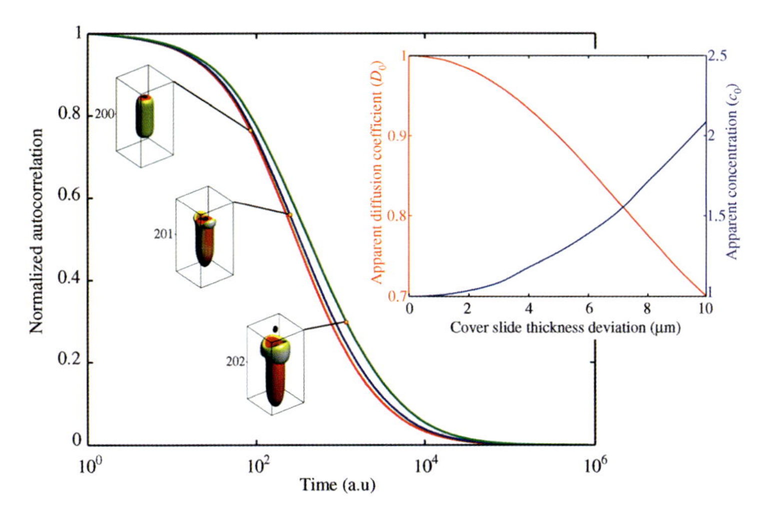

Figure 5.7 The large figure shows, from left to right, the MDF and ACF for three increasing values of cover slide thickness deviation, $\delta = 0$, $\delta = 5\,\mu\text{m}$, and $\delta = 10\,\mu\text{m}$. Box size of the MDF displays is $1 \times 1 \times 2\,\mu\text{m}^3$; the numbers next to each box give the center position along the optical axis in μm. Note the shift of the center of the MDF along the optical axis for increasing values of δ. The inset figure shows the dependence of apparent diffusion coefficient and the chemical concentration on thickness deviation value. These values would be obtained when performing a comparative FCS measurement using an ideal ACF ($\delta = 0$) as reference (the same applies to subsequent figures).

optical aberrations introduced by cover slide thickness mismatch and the resulting deformation of an ACF, and the shift of its decay towards longer lag times. The enlargement of the MDF results in increased diffusion times, that is, apparently lower diffusion coefficients, and in an apparently increased concentration (there are more molecules present in the detection volume because the latter has become larger). In general, any aberration results in an increased detection volume and thus leads to the same trend of an apparently lower diffusion coefficient and higher concentration with increasing aberration. The impact on the apparent concentration is much stronger than on the apparent diffusion, resulting, for example, for a cover slide thickness deviation of 10 μm, in an error of over 100% for the former and of roughly 30% for the latter. It should be noted that the errors shown do not change significantly when the focus position in the solution is changed.

This is in stark contrast to the effect of refractive index mismatch, which will be considered next. An optical microscope using a water immersion objective is optimally corrected for imaging in water. However, in many biophysical applications, one has to work in buffer solutions with slightly different refractive indexes. Also, when measuring in cells or tissues, one faces similarly slight refractive index variations. Typical values of interest are between 1.333 and 1.360. Figure 5.8 shows the impact of refractive index mismatch on the MDF and ACF and subsequently on the apparent diffusion coefficient and concentration. The impact of even a slight refractive index mismatch is much more dramatic than that of the cover slide thickness. This is mostly due to the large assumed distance of the focus position from the cover slide surface (200 μm, the default value of commercial instruments such as the Confocor I by Carl Zeiss). In contrast to cover slide thickness, the aberrations introduced by refractive index mismatch accumulate with increasing distance of the

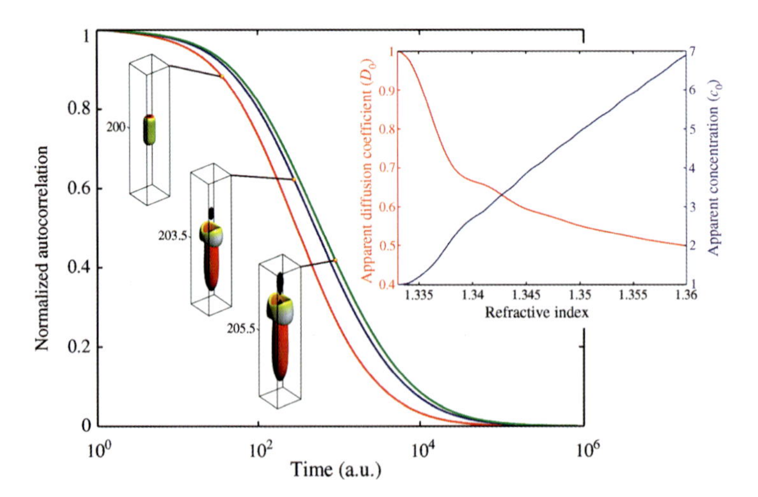

Figure 5.8 The large figure shows, from left to right, the MDF and ACF for three increasing values of refractive index of the sample solution, $n_m = 1.333$, $n_m = 1.346$, and $n_m = 1.360$. Box size of the MDF displays is $1 \times 1 \times 5\,\mu m^3$. Note again the shift of the center of the MDF along the optical axis for increasing values of n_m. The inset figure shows the dependence of apparent diffusion coefficient and the concentration on refractive index.

focus from the cover slide surface because an increasingly thicker layer of solution with the mismatched refractive index lies between the optics and the detection volume. The effect of refractive index mismatch can be much reduced when positioning the detection volume closer to the surface.

Another purely optical effect is laser beam astigmatism, that is, different focus positions within different axial planes of the laser beam. Astigmatism is easily introduced through slight curvatures of reflective elements in the optical set-up (such as the dichroic mirror), or by slight axial asymmetry of the optical fiber that is often used for guiding the excitation light towards the objective. The impact of astigmatism on the shape of the MDF and ACF in addition to the apparent diffusion coefficient and chemical concentration are shown Figure 5.9.

As can be seen, the effect of astigmatism on the measured diffusion and concentration is of similar magnitude to that of cover slide thickness deviation. With regard to cover slide thickness, the effect of astigmatism is fairly independent of the focus position in the sample.

A particularly intriguing effect in FCS measurements is the dependence of the ACF on excitation intensity due to optical saturation. Optical saturation occurs when the excitation intensity becomes so large that a molecule spends more and more time in a non-excitable state, so that increasing the excitation intensity does not lead to a proportional increase in the fluorescence intensity emitted. The most common sources of optical saturation are: (i) excited-state saturation, that is, the molecule is still in the excited state when the next photon arrives; (ii) triplet-state saturation, that is, the molecule undergoes intersystem-crossing from the excited to the triplet state

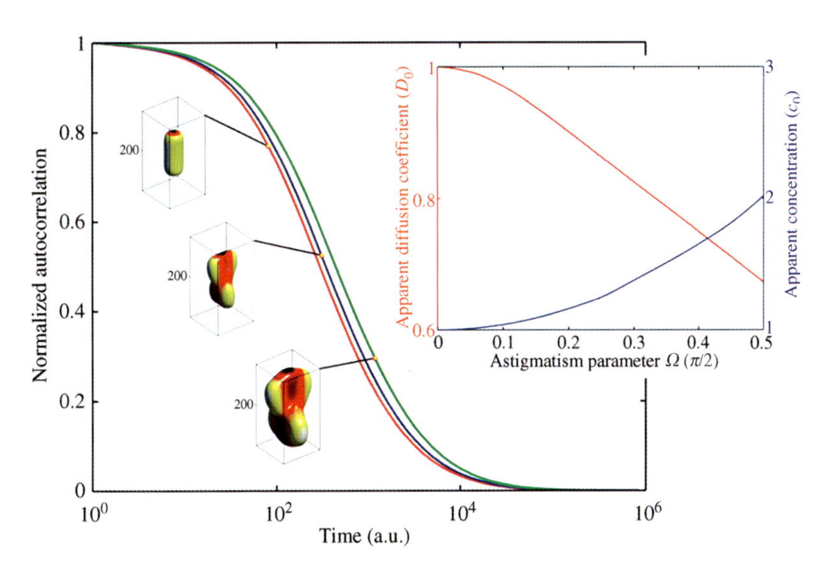

Figure 5.9 The large figure shows, from left to right, the MDF and ACF for three different values of laser beam astigmatism. Box size of the MDF displays is $1 \times 1 \times 2\,\mu m^3$. There is no shift of the center of the MDF along the optical axis for increasing astigmatism. The inset figure shows the dependence of apparent diffusion coefficient and the concentration on beam astigmatism.

so that it can no longer become excited until it returns back to the ground state; (iii) other photoinduced transitions into a non-fluorescing state, such as the photoinduced *cis–trans* isomerization in cyanine dyes, or the optically induced dark states in quantum dots. The exact relationship between fluorescence emission intensity and excitation intensity can be very complex [25] and can even depend on the excitation mode (pulsed or continuous wave) [26], but an adequate approximation of the dependence of fluorescence intensity on excitation intensity is given by the simple relationship

$$I_{\text{fluo}} \propto \frac{I_{\text{exc}}}{1 + I_{\text{exc}}/I_{\text{sat}}} \tag{5.18}$$

where

I_{fluo} and I_{exc} are the fluorescence and excitation intensity, respectively
I_{sat} is a parameter referred to as the saturation intensity, which describes the saturation behavior of a given dye.

Figure 5.10 shows how optical saturation changes the shape of the MDF and ACF and the apparent diffusion coefficient and concentration.

An important feature is the behavior of the curves for apparent diffusion and concentration at the limit of vanishing excitation intensity. Whereas for all optical effects studied so far, the slope of these curves tended to zero for vanishing aberration (or astigmatism), its value now is largest at zero intensity. To better understand the reason for this behavior, consider an ideal Gaussian excitation profile $I_0{\cdot}\exp(-x^2/2)$ with a mean square deviation of one. Figure 5.11 shows the widening of such a profile when transformed by a saturation to $I_0{\cdot}\exp(-x^2/2)/$

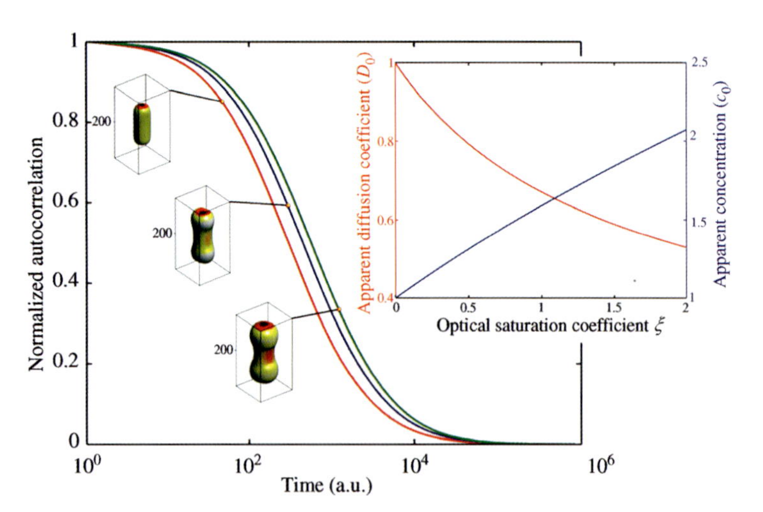

Figure 5.10 The large figure shows, from left to right, the MDF and ACF for three increasing values of optical saturation. Box size of the MDF displays is $1 \times 1 \times 2\,\mu\text{m}^3$. The inset figure shows the dependence of apparent diffusion coefficient and concentration on optical saturation, that is, excitation intensity.

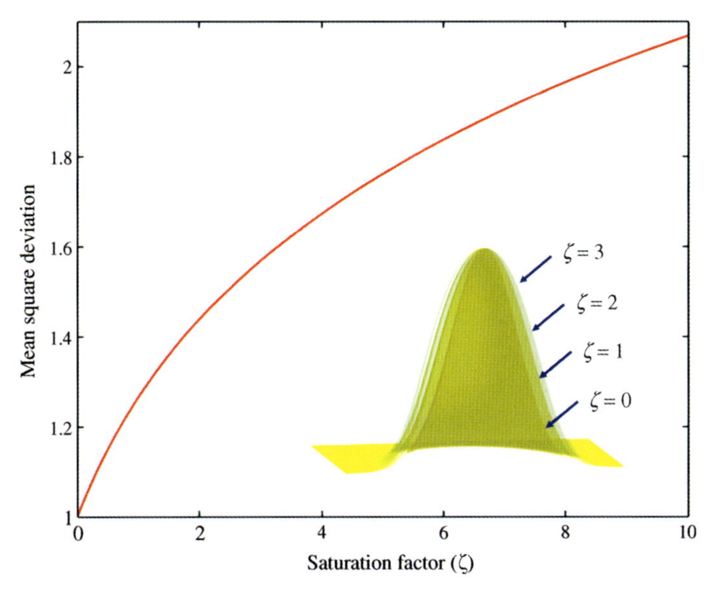

Figure 5.11 Change of the mean square deviation of the function $\exp(-x^2/2)/[1 + \varsigma\cdot\exp(-x^2/2)]$ with increasing value of ς.

$[1 + I_0\cdot\exp(-x^2/2)]$. As can be seen, the relative change in the profile width is fastest at the limit of zero intensity $I_0 \to 0$, explaining why one sees most of the changes in FCS at low saturation levels.

All of these effects make a quantitative evaluation of standard FCS measurements quantitatively unreliable. As pointed out previously, the core problem is the absence of an extrinsic and fixed length scale in the experiment. Even reference measurements, that is, using a dye with a known diffusion coefficient to determine the parameters a and b and then using them to measure the diffusion of a sample, can be problematic due to the strong dependence of an FCS result on optical saturation, which is itself determined, in a complex manner, by the photophysics of a particular dye. The photophysical parameters of one and the same dye can even change upon binding it to a protein or other target molecule! The next section describes a recent modification of the standard FCS measurement that seems to solve this long-standing problem and which allows for reproducible, quantitative, and absolute measurements of diffusion coefficients.

5.4.2
Dual-Focus FCS

Several modified concepts of FCS have been proposed to obtain a better defined detection volume, that is, better defined MDF by introducing an external ruler into the measurement, which is lacking in standard FCS. Among these attempts were excitation in front of dielectric mirrors [27], standing wave excitation [28], or spatial correlation FCS between two detection volumes generated by detecting the fluorescence through two laterally shifted pinholes [29]. The external ruler was provided either by the known modulation length of a standing light wave, or the estimated distance between the detection volumes. However, all the proposed methods suffer from a similar problem, in that for a precise quantification of the diffusion coefficient one still needs precise knowledge of the overall shape of the MDF, evoking the same problems as in conventional FCS.

Recently, a new and straightforward modification of FCS was developed, namely dual-focus FCS or 2fFCS, which fulfils two requirements: (i) it introduces an external ruler into the measurement by generating two overlapping laser foci of precisely known and fixed distance, (ii) it generates the two foci and corresponding detection regions in such a way that the corresponding MDFs are sufficiently well described by a simple two-parameter model, yielding accurate diffusion coefficients when applied to 2fFCS data analysis. Both of these properties allow for the measuring of absolute values of the diffusion coefficient with an accuracy of a small percentage. Moreover, the new technique is robust against refractive index mismatch and optical saturation effects, which are problems in standard FCS measurements.

The 2fFCS set-up, as shown in Figure 5.12, is based on a standard confocal epifluorescence microscope as was shown in Figure 5.5. However, instead of using a single excitation laser, the light from two identical, linearly polarized pulsed diode lasers is combined by a polarizing beam splitter. Both lasers are pulsed *alternately* with a high repetition rate (about 40–80 MHz), and by an excitation scheme that is termed pulsed interleaved excitation or PIE [30]. Both beams are then coupled into a polarization-maintaining single-mode fiber. At the output, the light is again collimated. Thus, the combined light consists of a train of laser pulses with alternating orthogonal polarization. The beam is then reflected by a dichroic mirror towards the water-immersion objective of the microscope, but before entering the objective, the light beam is passed through a Nomarski prism, which is normally exploited for differential interference contrast (DIC) microscopy. The Nomarski prism is an optical element that deflects the laser pulses into two different directions according to their corresponding polarizations. Thus, after focusing the light through the objective, two overlapping excitation foci are generated, with a small lateral shift between them. The distance between the beams is uniquely defined by the chosen Nomarski prism and is independent of the sample's refractive index, cover slide thickness, or laser beam astigmatism, because all these properties may introduce severe aberrations but will not change the main distance between the axes of propagation of both focused laser beams.

As in one-focus FCS, the generated fluorescence is collected by the same objective, passed through the Nomarski prism and the dichroic mirror, and focused onto a single circular aperture (diameter 200 μm), which is positioned symmetrically with respect to both focus positions and chosen to be large enough to allow the light from both foci to easily pass. After the pinhole, the light is collimated, split by a non-polarizing beam splitter cube, and focused onto two SPADs. Electronics for single-photon counting are used to record the detected photons from both SPADs with picosecond temporal resolution. The picosecond temporal resolution is used to

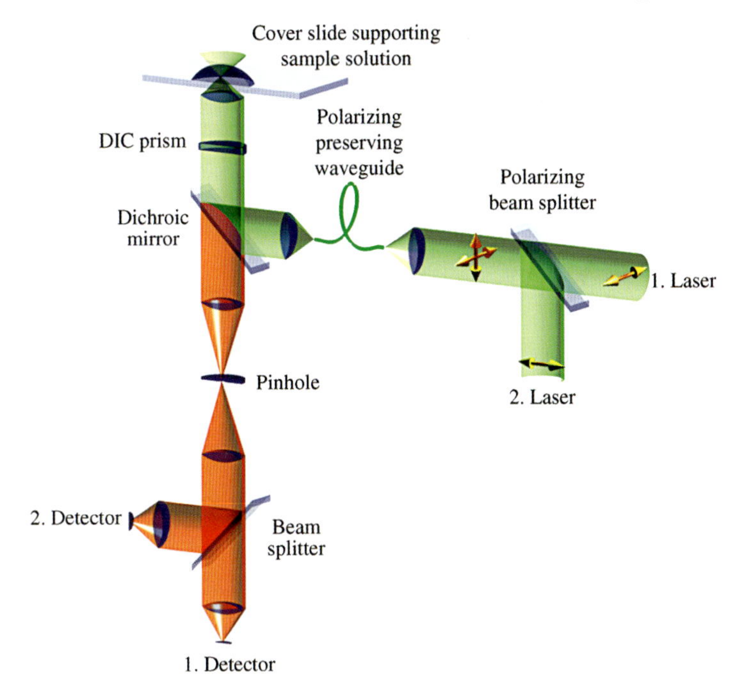

Figure 5.12 Schematic of the 2fFCS set-up. Excitation is achieved by two interleaved pulsed lasers of the same wavelength. The polarization of each laser is linear but orthogonal to each other. Light is then combined by a polarizing beam splitter and coupled into a polarization maintaining optical single-mode fiber. After exiting the fiber, the laser light is collimated by an appropriate lens and reflected by a dichroic beam splitter through a DIC prism. The DIC prism separates the laser light into two beams according to the polarization of the incoming laser pulses. The microscope objective focuses the two beams into two laterally shifted foci. Fluorescence is collected by the same objective. The tube lens focuses the detected fluorescence from both excitation foci onto a single pinhole. Subsequently, the fluorescence light is split by a 50/50 beam splitter and detected by two single photon avalanche diodes.

decide which laser has excited which fluorescence photon, that is, within which laser focus/detection volume the light was generated. This is done by correlating the detection time of each photon with the time of the last preceding laser pulse. A typical histogram of these time correlations, a so-called time-correlated single-photon counting or TCSPC histogram [31], is shown in Figure 5.13.

In this figure, two different fluorescence decay curves that correspond to the two alternately pulsing lasers can be seen. In the data evaluation, all photons that fall into the first time window are associated with the first laser, and all photons that fall into the second time window with the second laser. For this method to work successfully, it is, of course, necessary that the time between the laser pulses is significantly larger than the fluorescence lifetime of the fluorescent molecules. Knowing which photon is generated in which detection volume, the ACFs for each detection volume in addition to the cross-correlation function (CCF) between the two detection volumes can be calculated. The CCF is calculated in a similar way to

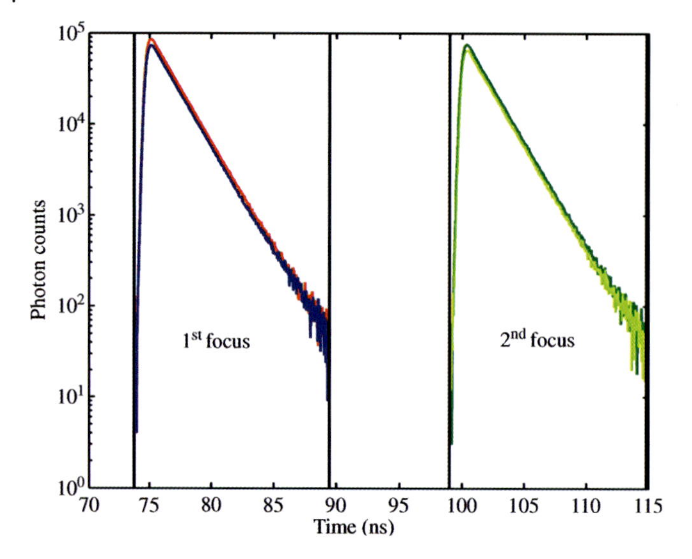

Figure 5.13 TCSPC histograms measured on an aqueous solution of Atto655. The photon counts in the left time window (73 ns $\leq t \leq$ 89 ns) are generated by the first laser, that is, the first focus, the photon counts in the second time window (99 ns $\leq t \leq$ 115 ns) are generated by the second laser, that is, the second focus. In both time windows (limited by vertical lines in the figure), there are two curves corresponding to the two SPAD detectors, respectively.

the ACF but the correlating photons are from different detection volumes. The CCF at lag time τ is thus proportional to the probability of seeing a photon from the second detection volume at any time $t + \tau$ if there was a detection event from the first detection volume at time t, and vice versa. A typical 2fFCS measurement is presented in Figure 5.14.

A crucial point for a successful 2fFCS data analysis is to have a sufficiently appropriate model function for the MDF. It was found that a suitable expression is given by the following

$$U(\mathbf{r}) = \frac{\kappa(z)}{w^2(z)} \exp\left[-2\frac{x^2+y^2}{w^2(z)}\right] \tag{5.19}$$

where $\kappa(z)$ and $w(z)$ are functions of the axial coordinate z (optical axis) defined by

$$w(z) = w_0 \left[1 + \left(\frac{\lambda_{\text{ex}} z}{\pi w_0^2 n}\right)^2\right]^{\frac{1}{2}} \tag{5.20}$$

and

$$\kappa(z) = 2\int_0^a \frac{d\varrho \varrho}{R^2(z)} \exp\left[-\frac{2\varrho^2}{R^2(z)}\right] = 1 - \exp\left[-\frac{2a^2}{R^2(z)}\right] \tag{5.21}$$

where the $R(z)$ itself is defined by an expression similar to Equation 5.20:

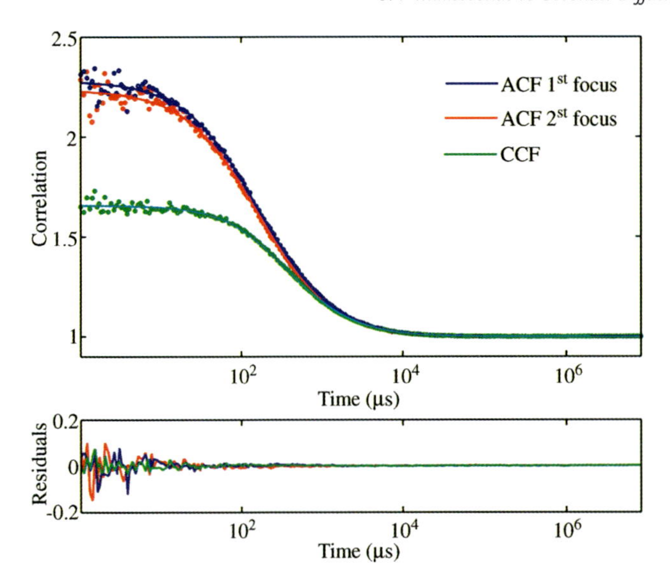

Figure 5.14 2fFCS measurement on a nanomolar aqueous solution of Atto655. Shown are the autocorrelation functions for the first focus (ACF 1st focus), second focus (ACF 1st focus), and the cross-correlation between both foci (CCF). The shapes of both ACF curves are virtually identical. Dots are experimental values, solid lines represent global fits. The bottom panel shows the difference between experimental and fitted values.

$$R(z) = R_0 \left[1 + \left(\frac{\lambda_{em} z}{\pi R_0^2 n} \right)^2 \right]^{\frac{1}{2}} \tag{5.22}$$

where, in the above equations

λ_{ex} is the excitation wavelength
λ_{em} is the center emission wavelength
n is the refractive index of the immersion medium (water)
a is the radius of the confocal aperture divided by magnification
w_0 and R_0 are two (generally unknown) model parameters.

Equation 5.20 is nothing other than the scalar approximation for the radius of a diverging laser beam with beam waist radius w_0. Equation 5.19 is a modification of the three-dimensional Gaussian we have already discusssed with respect to one-focus FCS, and indicates that in each plane perpendicular to the optical axis, the MDF is approximated by a Gaussian distribution having width $w(z)$ and amplitude $\kappa(z)/w^2(z)$. How well such a simple model fits a real MDF of a well-adjusted confocal epi-fluorescence set-up is shown in Figures 5.15 and 5.16, where the model function is compared with three-dimensional scans of the actual MDF using fluorescent beads of 100 nm diameter.

All that remains is to calculate the auto- and cross-correlation curves of the two-focus set-up. One derives these expressions by following a similar philosophy to the

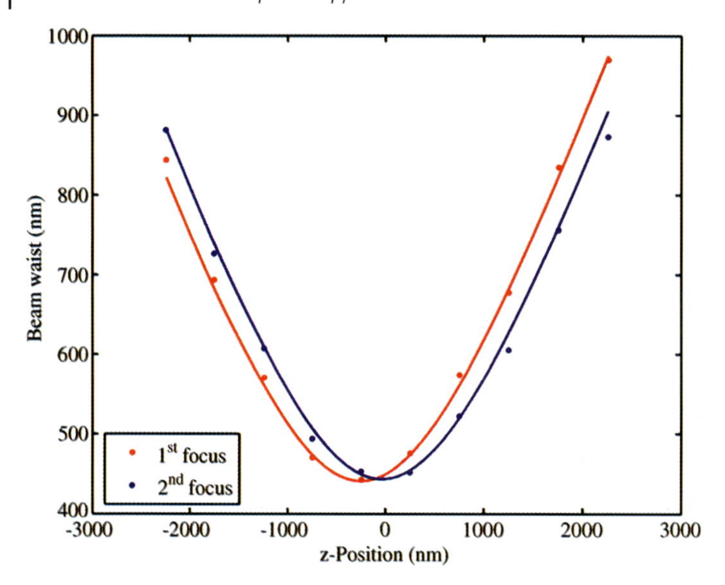

Figure 5.15 Dependence of the effective beam radius of the two PSFs on the vertical scan position. Solid lines are fits of Equation 5.20 to the measured values (circles). Fitted effective beam radius is 440 nm for the first and 445 nm for the second focus.

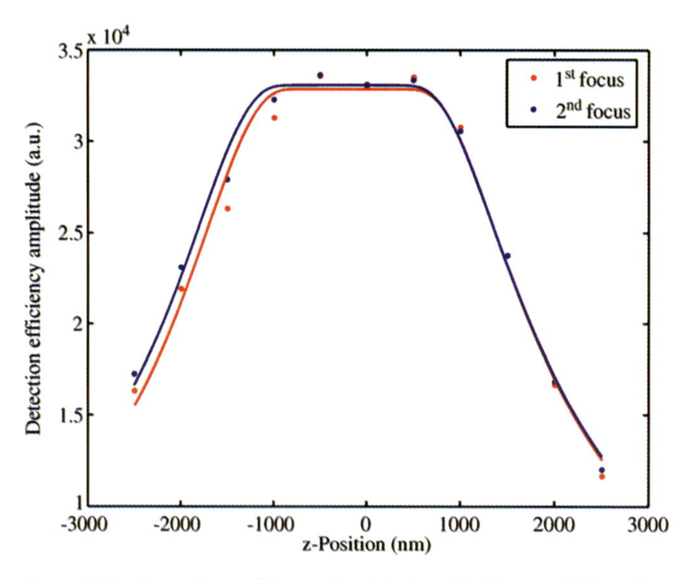

Figure 5.16 Dependence of the amplitude factor $\kappa(z)$ of the two PSFs on the vertical scan position. Solid lines are fits of Equations 5.21 and 5.22 to the measured values (circles). Fitted value of R_0 is 130 nm for the both foci.

calculation of the photon detection and diffusion probabilities, as described in the previous section. For example, the cross-correlation function between the fluorescence signals from the two different detection volumes is given by a similar integral to that in Equation 5.15:

$$g(\tau,\delta) = \frac{c\varepsilon_1\varepsilon_2}{(4\pi Dt)^{3/2}}\int_V d\mathbf{r}_1\int_V d\mathbf{r}_0\, U(\mathbf{r}_1)\exp\left(-\frac{|\mathbf{r}_1-\mathbf{r}_0+\delta\hat{\mathbf{x}}|^2}{4Dt}\right)U(\mathbf{r}_0) + \varepsilon_1\varepsilon_2\left[c\int_V d\mathbf{r}\, U(\mathbf{r})\right]^2 \tag{5.23}$$

Here, we have taken into account that the MDFs of both detection volumes are identical but shifted by a distance δ along the x-axis (along unit vector $\hat{\mathbf{x}}$) and having potentially two different overall detection efficiencies ε_1 and ε_2. Inserting Equations 5.19–5.23 into Equation 5.23 yields

$$g(t,\delta) = g_\infty(\delta) + \frac{\varepsilon_1\varepsilon_2 c}{4}\sqrt{\frac{\pi}{Dt}}\int_{-\infty}^{\infty} dz_1\int_{-\infty}^{\infty} dz_2 \frac{\kappa(z_1)\kappa(z_2)}{8Dt+w^2(z_1)+w^2(z_2)}\exp\left[-\frac{(z_2-z_1)^2}{4Dt}-\frac{2\delta^2}{8Dt+w^2(z_1)+w^2(z_2)}\right] \tag{5.24}$$

which is certainly more complicated than the simple expression of the second line in Equation 5.15, but it is not much harder to handle numerically in the age of powerful PCs. For numerical purposes, it is useful to slightly modify this result by changing the variables to

$$\xi = \frac{z_2-z_1}{2\sqrt{Dt}},\quad \eta = \frac{z_2+z_1}{2} \tag{5.25}$$

leading to the expression

$$g(t,\delta) = g_\infty(\delta) + 2\varepsilon_1\varepsilon_2 c\sqrt{\pi}\int_0^{\infty} d\xi\int_0^{\infty} d\eta \frac{\kappa(\eta-\sqrt{Dt}\xi)\kappa(\eta+\sqrt{Dt}\xi)}{8Dt+w^2(\eta-\sqrt{Dt}\xi)+w^2(\eta+\sqrt{Dt}\xi)} \exp\left[-\xi^2-\frac{2\delta^2}{8Dt+w^2(\eta-\sqrt{Dt}\xi)+w^2(\eta+\sqrt{Dt}\xi)}\right] \tag{5.26}$$

Because w and κ are rapidly decaying functions for large argument values, the infinite integrations over η and ς can be approximated by numerically evaluating the integrals within a finite two-dimensional strip defined by $|\eta \pm (Dt)^{1/2}\varsigma| < M$, where M is a truncation value chosen in such a way that the numerical integration result does not change on increasing M further. Numerical integration can be performed, for example, using a finite element scheme [32], and convergence is checked by testing whether the numerical result remains the same upon refining the finite element size and when increasing the threshold value M.

Data fitting is usually performed with least-square fitting of the model curve, Equation 5.26, against the measured ACF ($\delta = 0$, $\varepsilon_1\varepsilon_2$ replaced by either ε_1^2 or ε_2^2) and cross-correlation CCF *simultaneously* in a global fit. As fit parameters one has $\varepsilon_1 c^{1/2}$, $\varepsilon_2 c^{1/2}$, D, w_0 and R_0, in addition to three offset values g_∞. The distance δ between the detection regions is determined by the properties of the Nomarski prism and has to be exactly known a priori, thus introducing an external length scale into the data evaluation. It is important to notice that a crucial criterion of fit quality is not only to simultaneously reproduce the temporal shape of both ACFs and the cross-correlation function, but also to reproduce their three amplitudes $g_{t \to 0} - g_\infty$ using only the two parameters $\varepsilon_1 c^{1/2}$ and $\varepsilon_2 c^{1/2}$. The relationship between the amplitudes of the CCF and the amplitudes of the ACFs is determined by the overlap between the two MDFs, and thus by the shape parameters w_0 and R_0. Hence, achieving a good quality fit for the relative amplitudes of the ACFs and the CCF helps considerably in finding the correct values of these parameters.

Owing to the presence of an external length scale determined by the distance δ between the detection volumes and the reasonably accurate model of the MDF, 2fFCS is indeed a method of superior accuracy and stability for measuring diffusion. An optimal distance between foci is equal to their radius in the focal plane, giving a sufficiently large overlap between detection volumes, so that the amplitude of the cross-correlation function between both detection volumes is roughly one half of the amplitude of each autocorrelation function. Larger distances will lead to significantly longer measurement times for accumulating a sufficiently good cross-correlation, whereas smaller distances will lead to a cross-correlation function too similar to the autocorrelation functions so that data fitting becomes unreliable.

A crucial aspect of 2fFCS is the precise knowledge of the interfocal distance, as generated by the DIC prism. Any error in determining this distance will lead to twice as large a relative error of the determined value of a diffusion coefficient. An elegant way to determine this distance is to perform a comparative measurement of the diffusion of fluorescently labeled beads by both dynamic light scattering (DLS) and 2fFCS, as described in reference [33].

The insensitivity of 2fFCS with respect to optical saturation effects, which cause so much trouble for conventional single-focus FCS, is demonstrated in Figure 5.17, where we measured the diffusion of the cyanine dye Cy5 in aqueous solution at different values of laser excitation power. To evaluate the ACFs of conventional FCS, the standard model approach assuming a three-dimensional Gaussian distribution for the MDF would also have sufficed, because conventional FCS is not an absolute method for determining diffusion coefficients: it must be calibrated against a reference standard with known diffusion coefficient. In Figure 5.17, we used as reference the diffusion coefficient of Cy5, as determined by 2fFCS. The figure shows the dependence of the determined values of the diffusion coefficient as a function of total excitation power. As one can see, the values as determined with 2fFCS are insensitive to the excitation power.

We used 2fFCS to systemically determine the absolute values of diffusion coefficients for several dyes across the visible spectrum. The values obtained are summarized in Table 5.1. These values could serve as reference values for calibrating

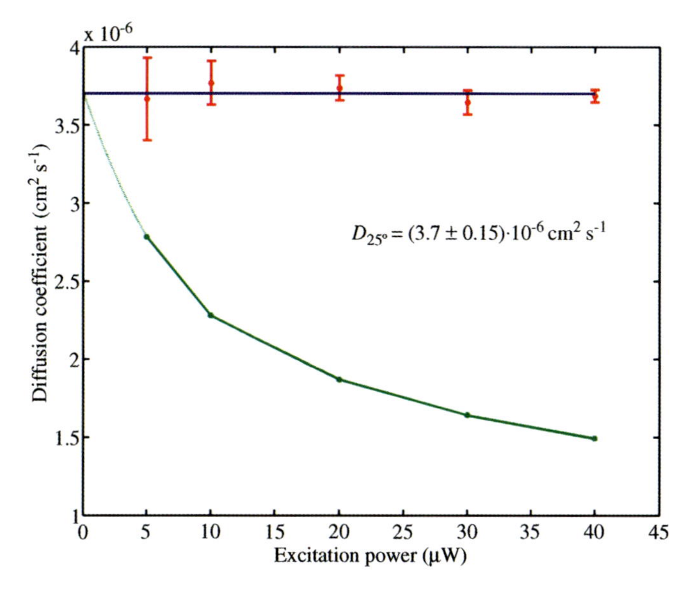

Figure 5.17 Determined diffusion coefficient as a function of total laser excitation power per focus. Points with error bars are the results of 2fFCS, using ten measurements for each point to determine a standard deviation of the diffusion coefficient. Solid horizontal line shows the average value of all 2fFCS measurements. Lower intensity-dependent curve refers to the results of conventional FCS, using the extrapolated zero-intensity value as reference. Dotted line is an extrapolation of the determined power dependence toward zero power.

FCS measurements in the future. It should be pointed out that the value found for Rhodamine 6G is 37% larger than the value that was used for nearly 30 years as the reference value for conventional single-focus FCS measurements [3]. However, the new 2fFCS value is in perfect agreement with a recent measurement using peak broadening in capillary flow [34].

Table 5.1 Absolute values of diffusion coefficients for several dyes across the visible spectrum using 2fFCS.

Dye	Diffusion coefficient ($10^6\ cm^2\ s^{-1}$) at 25 °C
Atto655-COOH (free acid) (AttoTec, Siegen, Germany)	4.26 ± 0.08
Atto655-maleimide (AttoTec, Siegen, Germany)	4.07 ± 0.1
Cy5-maleimide (Amersham)	3.7 ± 0.15
Rhodamin6G (No. R634, Invitrogen, Karlsruhe, Germany)	4.15 ± 0.05
Oregon Green 488 (No. D6145, Invitrogen, Karlsruhe, Germany)	4.11 ± 0.06

5.5
Nanoseconds to Microseconds: Photophysics, Conformational Fluctuations, Binding Dynamics

Thus far we have considered the temporal behavior of an ACF on the millisecond to second time scale, where it is governed mainly by diffusion. On a time scale of dozens of nanosconds up to several microseconds, the positions of the molecules are virtually frozen, and diffusion can be neglected. Any temporal changes of the ACF on this time scale are due to processes that are much faster than molecular diffusion on the length scale of a typical detection volume. The most prominent process showing up on a microsecond time scale in almost all FCS experiments is the triplet-state dynamics of the fluorescent molecules. For a better understanding of its impact on an ACF, consider the following rate equations (in matrix notation) for the transitions between the singlet ground, first excited, and first triplet states:

$$\frac{d}{dt}\begin{pmatrix} s_0 \\ s_1 \\ t_1 \end{pmatrix} = M\begin{pmatrix} s_0 \\ s_1 \\ t_1 \end{pmatrix} \tag{5.27}$$

where

s_0, s_1, and t_1 are the time-dependent probabilities of finding a molecule in the singlet ground, first excited, and the triplet state, respectively
M is the rate matrix given by

$$M = \begin{pmatrix} -k_e & k_{10} & k_t \\ k_e & -k_{10}-k_{\text{isc}} & 0 \\ 0 & k_{\text{isc}} & -k_t \end{pmatrix} \tag{5.28}$$

where

k_e is the excitation rate
k_{10} is the transition rate between the singlet first excited and the ground state
k_t is the transition rate from the triplet to the ground state
k_{isc} is the intersystem crossing rate from the first excited singlet to the triplet state.

Usually k_{isc} and k_t are smaller than the k_{10}, by orders of magnitude, so that one may assume, on the time scale of the triplet-state transitions, that there is a stationary equilibrium between s_0 and s_1 leading to $k_{10}s_1 \approx k_e s_0$ and the simplified rate equations

$$\frac{d}{dt}\begin{pmatrix} s_0 \\ t_1 \end{pmatrix} = \begin{pmatrix} -k_e k_{\text{isc}}/k_{10} & k_t \\ k_e k_{\text{isc}}/k_{10} & -k_t \end{pmatrix}\begin{pmatrix} s_0 \\ t_1 \end{pmatrix} \tag{5.29}$$

Now, an FCS experiment always measures the probability of detecting a photon at lag time τ if there was a photon detected at time zero. Consider for a moment the fluorescence from a single molecule. If there was a photon at time zero, this indicates that the molecule has just jumped back from state S_1 to S_0. The probability of seeing another photon at lag time τ is proportional to $k_{10}s_1(\tau) \approx k_e s_0(\tau)$, which can be found as the solution to Equation 5.29 with the initial conditions $s_0(\tau=0)=1$ and $t_1(\tau=0)=0$.

Thus, one finds that the single molecule related part of the ACF on the microsecond time scale will be proportional to

$$g_{\text{triplet}}(\tau) = \frac{k_t}{k_t + (k_e k_{\text{isc}}/k_{10})} + \frac{k_e k_{\text{isc}}/k_{10}}{k_t + (k_e k_{\text{isc}}/k_{10})} \exp[-(k_t + k_e k_{\text{isc}}/k_{10})\tau] \quad (5.30)$$

that is, via a fast exponential decay towards a stationary value with characteristic decay time τ_{triplet},

$$\tau_{\text{triplet}} = \left(k_t + \frac{k_e k_{\text{isc}}}{k_{10}}\right)^{-1} \quad (5.31)$$

The constant part of Equation 5.30 is usually denoted by T and gives the average probability of finding a molecule in the triplet state after stationary equilibrium has been reached between the three photophysical states. Then, Equation 5.30 can be rewritten as

$$g_{\text{triplet}}(\tau) = T + (1-T)\exp\left(-\frac{\tau}{\tau_{\text{triplet}}}\right) \quad (5.32)$$

Finally, the full ACF from microseconds to infinity is given by multiplying g_{triplet} by the time dependent diffusion-related part of Equation 5.11. This is, of course, only justified if the time scale of the triplet-state dynamics and that of the diffusion out of the detection volume are clearly distinct.

While deriving Equation 5.32 we made one significant simplification. It was assumed that the excitation rate k_e of a molecule is constant and independent of position. Of course, in reality this is a position-dependent function, determined for example by the laser intensity distribution in the detection volume, but also by the relative orientation of a molecule's absorption dipole with respect to the local polarization of the exciting laser light. For rapidly rotating molecules, the orientation dependence usually averages out over the time scale of the triplet-state dynamics; however, it remains position dependent because of the non-uniform laser intensity distribution. In general, the excitation rate can depend on excitation intensity in a fairly complicated non-linear way (as has already been mentioned when discussing optical saturation effects in Section 5.2.1), and a rigorous treatment would require a complicated averaging of Equation 5.32 over all possible molecule positions within the detection volume. As it happens, Equation 5.32 fits measured ACFs remarkably well, yielding some type of effective triplet time τ_t. However, it must be clear that this effective triplet time τ_t bears a somewhat empirical character, and it is not advisable to confuse it with absolute quantitative values for the underlying photophysical rate constants (such as intersystem crossing rate or triplet-state lifetime). This is especially true in the light of all the uncertainties in knowing the true light intensity distribution in the detection region (and fluorescence excitation peculiarities), as was discussed extensively in Section 5.2.1.

Some molecules exhibit other light-driven processes that influence their fluorescence rate in a manner similar to triplet-state dynamics. The most prominent example is the light-induced *cis-trans* isomerization of cyanine dyes, such as the popular NIR dye Cy5. The mathematical description of these dynamics is fully equivalent to the triplet-state dynamics, the only difference is in a reinterpretation of the involved states. Now, the triplet state corresponds to a dark *cis*-conformation of the molecule, and the intersystem crossing rate, in addition to the triplet-to-ground state rates, corresponds to the *trans*-to-*cis* and *cis*-to-*trans*-transition rates, respectively. Again, although fitting of the experimental ACFs usually looks adequate when using Equation 5.32, one has to keep in mind that behind it is a somewhat semi-empirical model, which does not allow for obtaining precise quantitative numbers for the transition rate constants, although some attempts in this direction have been made in the literature [35, 36].

The situation becomes much better when considering fast dynamic processes that are not light driven. The most prominent example is the photoelectron transfer (PET) dynamics of the oxazine dye Atto655 in the presence of tryptophan or guanosine (see Chapter 7). Attto655 has the tendency to form conjugates with tryptophan or guanosine, and when in the conjugated form, any photoexcitation of the dye is quickly transferred to the conjugated molecule via a fast electron transfer. This results in complete quenching of the fluorescence. The PET of Atto655 with, for example, tryptophan, can be used for monitoring fast conformational dynamics in peptides by binding Atto655 to a position in the peptide at a well-defined distance from a tryptophan site. For the fast behavior of the ACF, we will again find a relationship similar to Equation 5.32, but now the intersystem crossing rate and the triplet-to-ground state rate correspond to the transition rates between an open (Atto655 far away from the tryptophan) and a closed (Atto655 in conjugation with the tryptophan) peptide conformation. However, in contrast to light-induced triplet-state or *cis-trans* conformational dynamics, Equation 5.32 is now an exact description and can be used for determining precise quantitative transition rates for the conformational dynamics.

A similar argument holds for any other fast fluorescence-influencing dynamics that are not light-driven, for example, fast binding or reaction kinetics. For all such processes, FCS can be a very powerful tool for obtaining rate constants on the nano- to microsecond time scale.

5.6 Picoseconds to Nanoseconds: Rotational Diffusion and Fluorescence Antibunching

When approaching the picosecond to nanosecond time scale, one enters the realm of the fast photophysical transition between singlet first excited and ground state and of fast molecular rotation. We will first consider fluorescence antibunching neglecting any polarization effects, and then turn our attention to the impact of rotational diffusion.

5.6.1 Antibunching

Let us consider an experiment where the sample is excited by two consecutive pulses of negligible pulse width. What is the probability of detecting, from one and the same molecule, two photons with lag time t between them? If we assume that the fluorescence decay is mono-exponential with decay time τ, and if we further take into account that a molecule can emit, after one excitation pulse, only one photon, this probability will be proportional to

$$\int_{\min(\delta-t,0)}^{\delta} dt_1 \frac{\kappa_1}{\tau} e^{-t_1/\tau} \frac{\kappa_2}{\tau} e^{-(t_1+t-\delta)/\tau} = \kappa_1 \kappa_2 F_1(t,\tau,\delta) \tag{5.33}$$

where we have introduced the function

$$F_1(t,\tau,\delta) = \frac{1}{\tau} \begin{cases} \sinh(t/\tau) e^{-\delta/\tau} & t \leq \delta \\ \sinh(\delta/\tau) e^{-t/\tau} & t > \delta \end{cases} \tag{5.34}$$

where

δ is the time delay between the two pulses
κ_1 and κ_2 represent the chances of the first or the second pulse leading to a photon detection event, respectively.

For a time distance between the two pulses that is much longer than the fluorescence decay time, $\delta \gg \tau$, and lag time values much larger than the fluorescence decay time, $t \gg \tau$, this function approaches the simple form

$$F_1(t,\tau,\delta) \rightarrow \frac{1}{\tau} \exp\left(-\frac{|t-\delta|}{\tau}\right) \tag{5.35}$$

The value of $F_1(t,\tau,\delta)$ tends to zero when the pulse delay δ goes to zero, an effect termed fluorescence antibunching, which reflects the fact that a single molecule cannot emit more than a single photon per excitation.

The chance of detecting two photons with lag time t from *two different* molecules is similar to the expression in Equation 5.34, with the difference being that the upper integration limit is now extended to infinity, leading to

$$\int_{\min(\delta-t,0)}^{\delta} dt_1 \frac{\kappa_1}{\tau} e^{-t_1/\tau} \frac{\kappa_2}{\tau} e^{-(t_1+t-\delta)/\tau} = \kappa_1 \kappa_2 \left[F_1(t,\tau,\delta) + \frac{1}{2\tau} \exp\left(-\frac{t+\delta}{\tau}\right) \right] \equiv \kappa_1 \kappa_2 F_2(t,\tau,\delta) \tag{5.36}$$

If we neglect that excitation and detection efficiency will, in general, be a function of a molecule's orientation, which is constantly changing due to rotational diffusion, then the coefficients κ_1 and κ_2 in the above equations are lag-time independent constants, so that the full ACF is simply proportional to

$$g(t) \propto \gamma(\gamma-1) \sum_{k=0}^{\infty} F_2(t,\tau,\delta+kT) + \gamma \sum_{k=0}^{\infty} F_1(t,\tau,\delta+kT) \tag{5.37}$$

where γ is the probability of finding a molecule within the detection volume, so that the first sum in Equation 5.37 accounts for photon pairs arising from two different molecules, and the second sum for photon pairs originating from one and the same molecule.

Equation 5.37 shows nicely that one can use a measured ACF for estimating the average number γ of independent emitters within the detection volume. By fitting Equation 5.37 to a measured ACF, one extracts the coefficients $\gamma(\gamma-1)$ and γ in front of both sums and the value of γ can then be calculated. Combining this information with the average number of diffusing molecules within the detection volume as extracted from the long-time behavior of the ACF, compare with Equation 5.16, one thus has the possibility of determining the number of independent emitters per diffusing molecule, which does not have to be the same, for example in the case of multiply labeled complexes. A typical example for such a fluorescence antibunching measurement in solution is shown in Figure 5.18.

Most reported measurements of fluorescence antibunching have been carried out on immobilized molecules [37–41], because for the detection of a sufficient number of photon pairs with only nanosecond temporal distance, one needs large numbers of emitted photons, which usually makes antibunching measurements fairly noisy, due to the limited number of fluorescence photons that can be extracted from an individual molecule until photobleaching occurs. In solution measurements, however, one averages over many transits of different molecules through the detection volume, which makes antibunching measurements feasible even for weak or photolabile emitters [42–44].

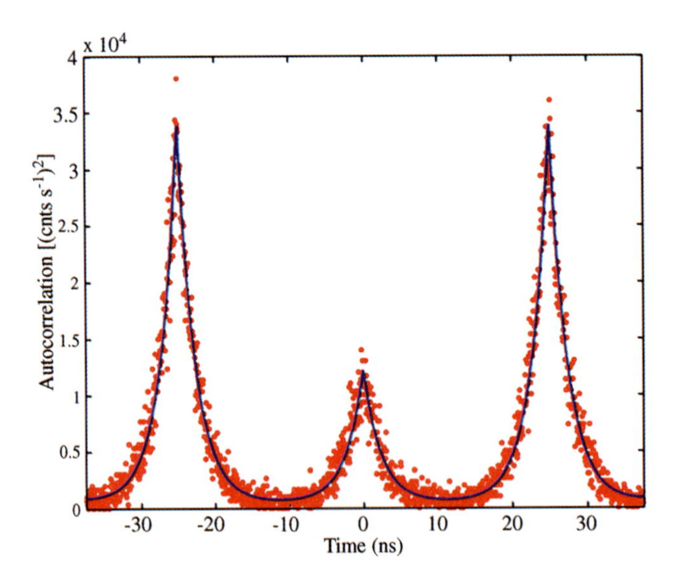

Figure 5.18 Normalized measured ACF (circles) of a nanomolar solution of Atto655 in water at nanosecond lag-time values. Minimum temporal resolution is 50 ps. Positive lag-time values correspond to pair correlations of photons detected by the first detector against photons detected by the second detector, and negative lag-time values correspond to pair correlations of photons detected by the second detector against photons detected by the first detector. Curve fitting was done using Equation 5.37.

5.6.2
Rotational Diffusion

If fluorescence detection is done in a polarization-sensitive manner, that is, by placing a polarization filter into the detection channel, the ACF will be affected by rotational diffusion of the fluorescing molecules. Nearly all fluorescing molecules of interest behave as electric dipole emitters (and absorbers – exceptions are some rare-earth luminophores such as certain europium complexes), so that their fluorescence is anisotropic and polarized. In the far field, the electric field amplitude **E** of a dipole emitter is described by the simple relationship

$$\mathbf{E}(\mathbf{r}) = \frac{\exp(ikr)}{\varepsilon r}[\mathbf{p} - \hat{\mathbf{e}}_r(\hat{\mathbf{e}}_r \cdot \mathbf{p})] \tag{5.38}$$

where

$k = 2\pi/\lambda$ is the absolute wave vector of the emitted light
ε is the dielectric constant of the embedding medium
$\mathbf{p}$ is the electric dipole amplitude
r the distance from the dipole to position $\mathbf{r}$
$\hat{\mathbf{e}}_r$ is a unit vector pointing from the dipole towards $\mathbf{r}$.

Owing to the rotation of a molecule between photon emission and detection and thus rotation of the molecule's dipole axis in or out of the polarization plane of the detector, the correlation of the recorded fluorescence signal will show a temporal component that is related to the rotational diffusion of the molecule [45–49].

Measuring the rotational diffusion of a molecule is an interesting alternative for obtaining values for its hydrodynamic radius. This connection between rotational diffusion coefficient and hydrodynamic radius is provided by the Stokes–Einstein–Debye equation [50]:

$$D_{\text{rot}} = \frac{k_B T}{8\pi\eta R_{\text{rot}}^3} \tag{5.39}$$

where k_B, T, and η have the same meaning as in Equation 5.3.

In the previous section, we considered the antibunching-induced shape of the ACF. The contributions of rotational diffusion are hidden in the pre-factors κ_1 and κ_2 in Equations 5.33 and 5.36. To gain more insight into these contributions, one has to resort to the theory of rotational diffusion of a generally anisotropic rotor, as first applied to correlation spectroscopy and light scattering by Aragón and Pecora [46], see also [19]. Let us start with the rotational diffusion equation

$$\frac{\partial P}{\partial t} = -\left(D_a \hat{J}_a^2 + D_b \hat{J}_b^2 + D_c \hat{J}_c^2\right) P \tag{5.40}$$

where

a, b, and c denote the principal axes of rotation of the molecule
$P = P(\psi,\theta,\phi)$ is the probability of finding the molecule's principal axes rotated by Euler angles ψ, θ, and ϕ with respect to the lab frame

$D_{a,b,c}$ are the generally different rotational diffusion coefficients around the molecule's principal axes
$\hat{J}_{a,b,c}$ are the three angular momentum operators around these axes.

Equation 5.40 is derived analogously to the more familiar translational diffusion equation. The difficulty with Equation 5.40 is that the angular momentum operators relate to the intrinsic frame of the molecule's principal axis, which is rotating in time with respect to the fixed lab frame. To simplify matters, one can first rotate the molecule back to the lab's frame so that its axes align with the fixed Cartesian coordinate axes of the lab frame, then apply the operator, and finally rotate the molecule back, that is,

$$\frac{\partial P}{\partial t} = -R\left(D_a\hat{J}_x^2 + D_b\hat{J}_y^2 + D_c\hat{J}_z^2\right)R^{-1}P \tag{5.41}$$

where

R denotes the operation of rotating the molecule's frame from an orientation aligned with the lab's Cartesian x,y,z-coordinates to its actual orientation as specified by the Euler angles ϕ, θ, and ψ, see Figure 5.19.

The rotation operator R can be decomposed into

$$R = R_z(\phi)R_y(\theta)R_z(\psi) \tag{5.42}$$

where

$R_{y,z}(\beta)$ denotes a rotation by angle β around axis y or z, respectively.

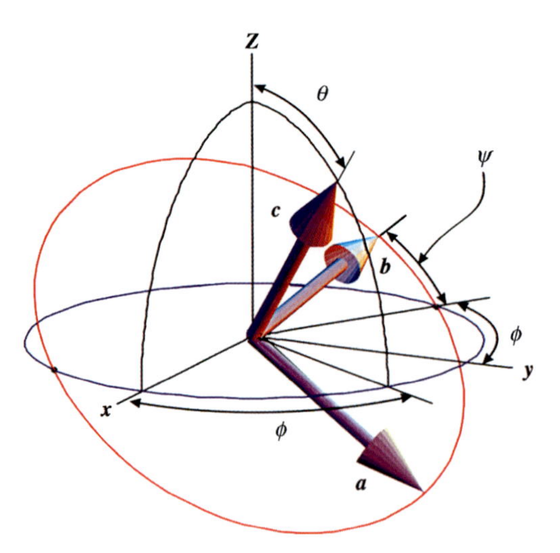

Figure 5.19 Geometric meaning of the three Euler angles ϕ, θ, and ψ. Shown are the molecule's three principal (and orthogonal) axes of rotation a, b, and c, and the three Cartesian axes x, y, and z of the lab frame.

The advantage of Equation 5.41 is that the angular momentum operators are now referring to the fixed lab frame. To further analyze Equation 5.41, let us consider the special case that the function P is replaced by

$$P = R|l, m\rangle \tag{5.43}$$

where

$|l,m\rangle$ is an eigenfunction of the angular momentum operator obeying the two relationships

$$\hat{\mathbf{J}}^2|l, m\rangle = \left(\hat{J}_x^2 + \hat{J}_y^2 + \hat{J}_z^2\right)|l, m\rangle = l(l+1)|l, m\rangle \tag{5.44}$$

and

$$\hat{J}_z|l, m\rangle = m|l, m\rangle. \tag{5.45}$$

Inserting Equation 5.43 into Equation 5.41 yields

$$\frac{\partial(R|l, m\rangle)}{\partial t} = -R\left(D_a\hat{J}_x^2 + D_b\hat{J}_y^2 + D_c\hat{J}_z^2\right)|l, m\rangle. \tag{5.46}$$

Next, one has to clarify how the rotation operator R acts on $|l,m\rangle$. One of the most lucid derivations of this action has been given by Feynman in reference. [51] using the possibility of representing any state $|l,m\rangle$ through a combination of spin-½ states, for which the transformation relationships under the action of R are known, see for example, Chapter 3.3 in reference [52]. Here, we will give only the final result,

$$R(\phi, \theta, \psi)|l, m\rangle = e^{im\psi} \sum_{k=-l}^{l} e^{ik\phi} S_{mk}^{l}(\theta)|l, k\rangle \tag{5.47}$$

The functions S_{mk}^{l} are Wigner's rotation matrices defined by

$$S_{mk}^{l}(\theta) = \langle l, k|R_y(\theta)|l, m\rangle = \left[\frac{(j+k)!(j-k)!}{(j+m)!(j-m)!}\right]^{\frac{1}{2}} \tag{5.48}$$

$$\cdot \sum_{n} \frac{(-1)^{j+k-n}(j+m)!(j-m)!}{k!(j+m-n)!(j+k-n)!(n-m-k)!} C^{2n-m-k} S^{2j+m+k-2n}$$

Here we have introduced the abbreviations $C = \cos(\theta/2)$ and $S = \sin(\theta/2)$. For the sake of simplicity, we will further consider the special case of a symmetric top rotor where one has $D_a = D_b = D_\perp$ and $D_{||} = D_c$. The general case of the fully asymmetric rotor will be briefly discussed later. For the symmetric top rotor, one finds, by multiplying Equation 5.46 with $\langle l,k|$, that the functions

$$\exp\{-[D_\perp l(l+1) + (D_{||} - D_\perp)m^2]t\} C_{mk}^{l}(\phi, \theta, \psi) \tag{5.49}$$

with

$$C^l_{mk}(\phi,\theta,\psi) = c^l_{mk}\exp(ik\phi + im\psi)S^l_{mk}(\theta) \tag{5.50}$$

are eigenfunctions of the rotational diffusion equation. In Equation 5.50 we introduced a normalizing factor c^l_{mk} so that the $C^l_{mk}(\phi,\theta,\psi)$ represents a complete orthonormal system of eigenfunctions obeying the relationships

$$\int_0^{\pi} d\theta\sin\theta \int_0^{2\pi} d\phi \int_0^{2\pi} d\psi\, C^l_{mk}(\phi,\theta,\psi)C^{l'^*}_{m'k'}(\phi,\theta,\psi) = \delta_{l,l'}\delta_{k,k'}\delta_{m,m'}. \tag{5.51}$$

The $\delta_{l,l'}$ are Kronecker symbols taking the value one for $l = l'$ and zero otherwise. The orthogonality of the functions $C^l_{mk}(\phi,\theta,\psi)$ with respect to the variables ϕ and ψ is obvious from their definition in Equation 5.50. The orthogonality with respect to θ is less obvious, but is a consequence of the fundamental orthogonality theorem of group theory (see e.g., [53]), which is applied here to the functional representation of the three-dimensional rotation group, as given by the functions $C^l_{mk}(\phi,\theta,\psi)$. With this complete orthonormal system of eigenfunctions, the probability that a molecule has rotated, within time t, from an initial orientation Ω' described by the Euler angles ϕ', θ' and ψ' into a final orientation Ω described by Euler angles ϕ, θ, and ψ is given by Green's function in the standard way [54] as

$$\begin{aligned} G(\Omega,\Omega',t) = &\sum_{l=0}^{\infty}\sum_{m,k=-l}^{l} \exp\{-[D_\perp l(l+1) + (D_\parallel - D_\perp)m^2]t\} \\ &C^l_{mk}(\phi,\theta,\psi)C^{l^*}_{mk}(\phi',\theta',\psi') \end{aligned} \tag{5.52}$$

where the asterisk superscript denotes complex conjugation.

For the sake of completeness, we will briefly discuss the most general case of a completely asymmetric rotor. Now, it is not possible to obtain simple eigenfunctions in the form of Equation 5.50 However, it is helpful to introduce the operators

$$\hat{J}_\pm = \hat{J}_x \pm i\hat{J}_y \tag{5.53}$$

so that the $\hat{J}_x$ and $\hat{J}_y$ operators on the right -hand side of Equation 5.41 can be written as

$$\hat{J}_x^2 = \frac{1}{4}\left(\hat{J}_+^2 + \hat{J}_-^2 + \hat{J}^2 - \hat{J}_z^2 + \hat{J}_z\right) \tag{5.54}$$

and

$$\hat{J}_y^2 = \frac{1}{4}\left(-\hat{J}_+^2 - \hat{J}_-^2 + \hat{J}^2 - \hat{J}_z^2 + \hat{J}_z\right) \tag{5.55}$$

where the commutation property of the angular momentum operators

$$\left[\hat{J}_x, \hat{J}_y\right] \equiv \hat{J}_x\hat{J}_y - \hat{J}_y\hat{J}_x = i\hat{J}_z \tag{5.56}$$

has been used. When taking into account how the operators $\hat{J}_{\pm}$ act on the eigenstates $|l,m\rangle$ (see e.g., Chapter 3.4 in reference [52]):

$$\hat{J}_{\pm}|l,m\rangle = \sqrt{l(l+1)-m(m\pm 1)}|l,m\pm 1\rangle \tag{5.57}$$

it is straightforward to see that Equation 5.41 separates, for each value of l, into a set of $2l+1$ coupled ordinary and linear differential equations on the basis of the $2l+1$ state vectors $|l,m\rangle$ (more correctly: into two sets of equations with $l+1$ equations coupling the values of m with $m \in [-l, -l+2, \ldots, l]$ and l equations coupling the values of m with $m \in [-l+1, -l+3, \ldots, l-1]$), which can be solved in a standard way [55]. This yields $2l+1$ orthonormal eigenfunctions as superpositions of the states $|l,m\rangle$ with the corresponding eigenvalues as characteristic temporal exponents, from which Green's function can be constructed as previously. Because the case of a fully asymmetric rotor is of rather little interest for almost all fluorescence-based measurements of molecular rotation, we will not pursue this topic further.

After having found Green's function for the rotation diffusion equation, one has to specify the fluorescence excitation and detection conditions of the measurement, in particular its polarization properties. Let us assume that the fluorescence lifetime is considerably shorter than the rotational diffusion time, which is mostly the case when studying rotational diffusion of large proteins when using short-lifetime dyes. Thus, one needs only to consider the so-called molecule detection function (MDF), which describes the probability of exciting and detecting a photon of a dye molecule as a function its orientation and position. The calculation of this function can be done using a wave-optics approach as described in reference [56]. For our subsequent considerations it is important that the MDF can be expanded into a series of spherical harmonics in the angles α and β that describe the angular orientation ω of the excitation/emission dipole (which are assumed to be collinear) as depicted in Figure 5.20. The coefficients of this series expansion are functions of the molecule's position $\mathbf{r}$, and the MDF, which will be denoted by $U(\alpha,\beta,\mathbf{r})$, and is thus represented through

$$U(\omega,\mathbf{r}) \equiv U(\alpha,\beta,\mathbf{r}) = \sum_{l=0}^{\infty}\sum_{m=-l}^{l} u_{lm}(\mathbf{r})Y_{lm}(\beta,\alpha) \tag{5.58}$$

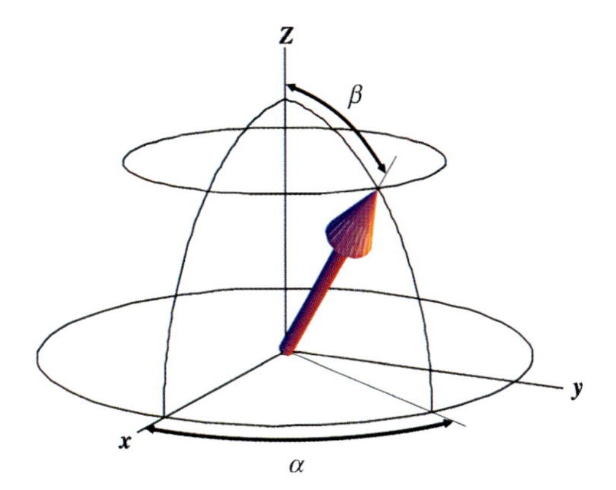

Figure 5.20 Geometric meaning of the orientation angles β and α with respect to the lab frame.

where the spherical harmonics $Y_{lm}(\beta,\alpha)$ are defined by

$$Y_{lm}(\beta,\alpha) = P_l^{|m|}(\cos\beta)\exp(im\alpha). \tag{5.59}$$

Here the functions $P_l^m(\cos\beta)$ are associated Legendre polynomials [57]. Using the orthogonality of spherical harmonics, the coefficients $u_{lm}(\mathbf{r})$ can be found from the full MDF via the backward transformation

$$u_{lm}(\mathbf{r}) = \int_0^{\pi} d\beta \sin\beta \int_0^{2\pi} d\alpha\, U(\beta,\alpha,\mathbf{r}) Y_{lm}^*(\beta,\alpha) \tag{5.60}$$

The importance of Equation 5.60 lies in the fact that the spherical harmonics themselves are representations of the three-dimensional rotation group and transform under rotation according to Equation 5.47. The MDF depends, of course, on the peculiarities of the excitation, and can be different for different excitation pulses (for example, when exciting the sample with a train of pulses with alternating polarization). A first laser pulse with corresponding MDF $U_1(\omega,\mathbf{r})$ thus "prepares" the sample in such a way that $U_1(\omega,\mathbf{r})$ describes the probability of detecting a photon from an excitation/emission dipole at position $\mathbf{r}$ having orientation ω. The next important consideration is that we are interested in measurements where the protein is tagged with a dye molecule in such a way that the relative orientation of the dye with respect to the protein's principal axes is *random* but fixed (co-rotation of the dye with the protein). Thus, rotating the distribution $U_1(\omega,\mathbf{r})$ back into the protein's frame of principal axes, which has orientation Ω' with respect to the lab frame, gives the average probability of exciting and detecting a photon from the protein–dye complex. Next, Green's function $G(\Omega,\Omega',t)$, Equation 5.52 provides the probability that the protein–dye complex rotates from orientation Ω' into orientation Ω within time t, and through a similar argument as before, the probability of exciting and detecting a photon by a second laser pulse with MDF $U_2(\omega,\mathbf{r})$ is given by a back-rotation Ω of $U_2(\omega,\mathbf{r})$ into the protein's frame. Finally, by integrating over all possible positions and orientations, one obtains the average of the product $\kappa_1\kappa_2$ (averaged over many repeats of the double-pulse excitation and many different relative protein–dye orientations):

$$\begin{aligned}\langle\kappa_1\kappa_2\rangle_t = &\int d\mathbf{r} \int d\omega \int d\Omega \int d\Omega' [R^{-1}(\Omega) U_2(\omega,\mathbf{r})] \\ &\cdot G(\Omega,\Omega',t)[R^{-1}(\Omega') U_1(\omega,\mathbf{r})]\end{aligned} \tag{5.61}$$

where R^{-1} is the back-rotation operator. The integrations run over all possible initial and final orientations Ω' and Ω of the protein, all possible dye–label orientations ω, and all possible positions $\mathbf{r}$. It should be emphasized that the above expression is fairly general, allowing for different excitation and detection geometries/polarizations for the first and second laser pulse. Now, using the transformation relationship in Equation (5.47), and the orthonormality of the eigenfunctions $C_{mk}^l(\phi,\theta,\psi)$ and of spherical harmonics Y_{lm}, the integrations over

Ω, Ω' and ω can be performed analytically, resulting in

$$\langle \kappa_1 \kappa_2 \rangle_t = \sum_{l=0}^{\infty} \sum_{m=-l}^{l} \left[\int d\mathbf{r}\, u_{2,lm}^*(\mathbf{r}) u_{1,lm}(\mathbf{r}) \right] \cdot \exp\left[-l(l+1) D_\perp t - (D_\| - D_\perp) m^2 \right] \tag{5.62}$$

For a spherically symmetric molecule with $D_\perp = D_\| \equiv D$ this expression simplifies to

$$\langle \kappa_1 \kappa_2 \rangle_t = \sum_{l=0}^{\infty} \left[\sum_{m=-l}^{l} \int d\mathbf{r}\, u_{2,lm}^*(\mathbf{r}) u_{1,lm}(\mathbf{r}) \right] \exp[-l(l+1)Dt] \tag{5.63}$$

The explicit calculation of the coefficients $u_{\alpha,lm}(\mathbf{r})$ is a formidable task, and for the details the reader is referred to reference [56] and citations therein. Remarkably, when neglecting optical saturation (i.e., the excitation rate is directly proportional to the absolute square of the scalar product of the excitation light electric field amplitude multiplied by the molecule's absorption dipole vector), only coefficients with $\ell = (0,2,4)$ will differ from zero. Even taking into account depolarization in excitation and detection caused by objectives with high numerical aperture [58, 59] does not change the computation noticeably. As an example, Figure 5.21 shows the result of a numerical calculation for a 1.2 NA (numeric aperture) water immersion objective as a function of the laser beam diameter coupled into the objective's back focal plane. In these calculations, it was assumed that detection is done by two detectors looking at orthogonal emission polarizations.

Without loss of generality, we will denote the detection polarization for the first photon by the symbol $\|$, and that for the second photon by $\perp$ ($\| \times \perp$ detection polarization mode). Then, there are three principally different excitation modes: (i) polarizations of excitation for the first and second photons are both parallel to the respective detection polarization ($\| \times \perp$ excitation polarization mode), (ii) polarizations of excitation for the first and second photons are both orthogonal to the respective detection polarization ($\perp \times \|$ excitation polarization mode), and (iii) excitation polarizations for both photons are the same ($\| \times \|$ or $\perp \times \perp$ excitation polarizations mode), so that the first (second) photon is excited with an excitation polarization parallel to its detection polarization, and the second (first) orthogonally to its detection polarization. Figure 5.21 shows several remarkable features: Firstly, the amplitude ratios in the $\| \times \perp$, the $\perp \times \|$ and the $\| \times \|$ excitation modes are close to $9:1:3$ for $l=0$, $(-18):1:3$ for $l=2$, and $(-6):8:9$ for $l=4$, which are the values at the limit of zero numerical aperture, the situation considered by Aragón and Pecora [46]. Secondly, one always has non-zero contributions with $l=4$. However, the relative weight of these contributions when compared with the $l=2$ term is least for the $\| \times \perp$ excitation mode, where it is about 1/15th of the amplitude for $l=2$.

Thirdly, when getting closer to diffraction-limited focusing (values at the right in Figure 5.3), depolarization effects have a non-negligible impact on the different pre-exponential amplitudes in Equation 5.63. The lowest impact is observed for the $\| \times \perp$ excitation mode, which makes this mode of excitation/detection the most favorable

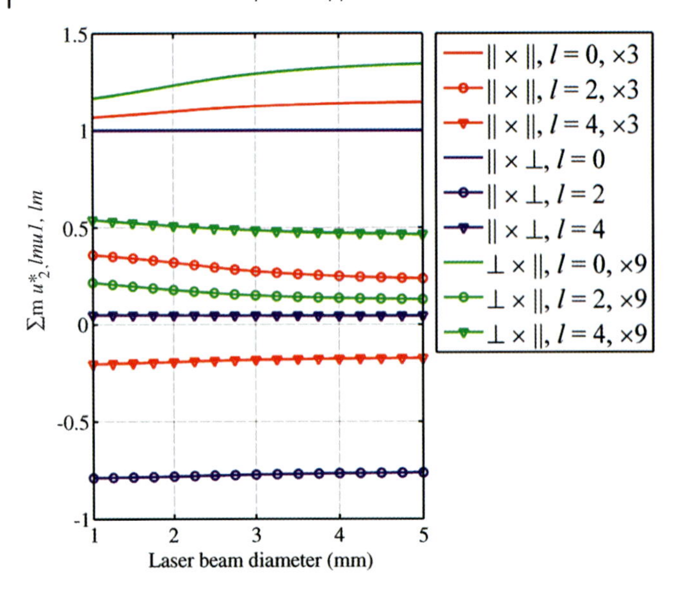

Figure 5.21 Dependence of the (normalized) coefficients $[\sum_m \int d\,\mathbf{r}\, u^*_{2,lm} \cdot u_{1,lm}]$ in Equation 5.63 for $l = 0$ (solid lines), $l = 2$ (solid lines with circles), $l = 4$ (solid lines with triangles) and for different excitation/detection polarizations as a function of laser beam diameter (measured at the objective's back focal plane). It is assumed that detection is done through two polarizers with orthogonally aligned polarization axes for the first and the second photon. The red curves show the case when the first and second laser pulse are both polarized along the same direction as the first or the second detector polarizer; the blue curves show the case when both laser pulses have the same polarization as the corresponding detector polarizers; and the green curves show the case when both laser pulses are polarized perpendicular to the corresponding detector polarizers. The calculations were done for a perfectly aplanatic 1.2 NA water immersion objective.

one for measuring rotational diffusion via fluorescence correlation spectroscopy in a confocal microscope with high NA. It yields the maximum amplitude for the lag-time dependent part of the correlation function with smallest contribution from the $l = 4$ mode, and the smallest impact from depolarization effects. As an example, the modeled correlation functions for a globular protein (isotropic rotor) with 20 ns rotational diffusion time $\tau_{\text{rot}} = 1/6D_{\text{rot}}$ are shown in Figure 5.22.

Often, fluorescent molecules exhibit a non-negligible angle between the absorption and emission dipoles. This will change the amplitudes of the different exponential terms in the autocorrelation function, but not the exponents themselves. Thus, any data analysis of autocorrelation curves that relies solely on these exponents will be independent of these peculiarities.

As a typical example of a rotational diffusion measurement, Figure 5.23 shows the ACF for a double-pulse experiment on the fluorescently labeled protein aldolase. An important aspect when devising such experiments is to ensure that the fluorescent label used co-rotates with the labeled protein, and that its fluorescence lifetime is sufficiently short. For the measurements on aldolase, the protein was non-specifically

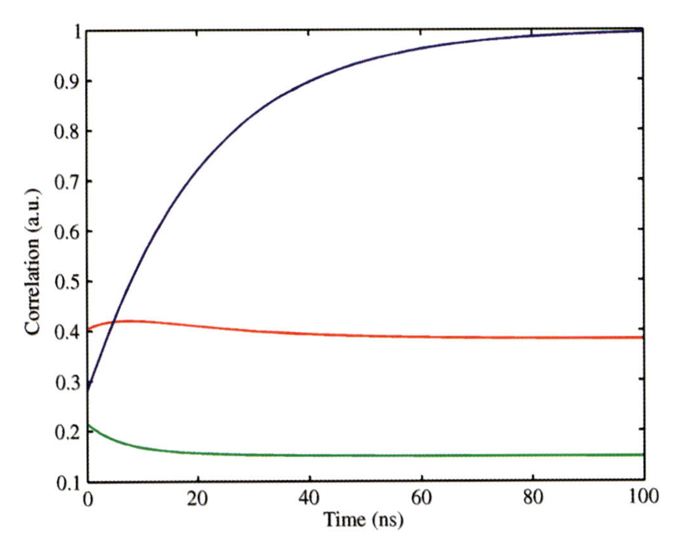

Figure 5.22 Correlation functions for $\| \times \perp$ (blue), $\perp \times \|$ (green) and $\| \times \|$ (red) excitation modes for a spherical globular protein with 20 ns rotational diffusion time.

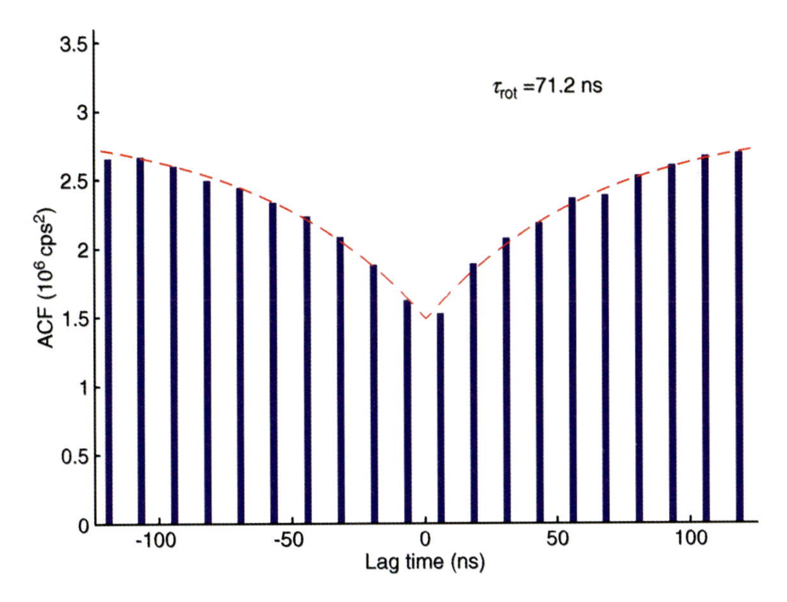

Figure 5.23 Measured $\| \times \perp$ correlation function (blue bars, compare with blue curve in Figure 5.22) and fitted mono-exponential lag-time dependence (dashed red line) for aldolase. The fitted exponential time was 71.2 ns. Similar to the antibunching measurement of Figure 5.18, positive lag-time values correspond to pair correlations of photons detected by the first detector versus photons detected by the second detector, and negative lag-time values correspond to pair correlations of photons detected by the second detector versus photons detected by the first detector.

labeled with the bis-functional fluorescence label Cy5 bis-succinimidyl (fluorescence lifetime <1 ns), which binds at two ends to lysin residues on the surface of the protein. The obtained hydrodynamic radius of 4.1 ± 0.1 nm is in excellent agreement with that obtained by dual-focus FCS.

At the end of this section, we will briefly discuss when it is necessary to take into account the non-spherical shape of a molecule, and when the assumption of a rotationally symmetric shape is still sufficient. As already noted, any molecule can be modeled by an object with three orthogonal axes of rotation (principal axes) with, in the most general case, three different rotational diffusion constants around each of these axes. In almost all cases of practical interest, it is sufficient to approximate a molecule by a symmetrical top, that is, an object that has two identical rotational diffusion constants around two of its principal axes and one different one around the third. This corresponds to approximating the shape of a molecule by a prolate or oblate ellipsoid of rotation. The question arises as to how large the axis ratio between the axes of the ellipsoid has to be in order to be clearly discernible in a rotational diffusion measurement. Following Perrin [60, 61] and Koenig [62], the rotational diffusion coefficients for an oblate ellipsoid of rotation with aspect ratio $\varepsilon = R_\perp / R_{||} < 1$ are given by

$$\frac{D_{||}}{D_0} = \frac{3\varepsilon^2}{2(1-\varepsilon^4)} \left\{ \frac{2-\varepsilon^2}{\sqrt{1-\varepsilon^2}} \ln\left[\frac{1+\sqrt{1-\varepsilon^2}}{\varepsilon}\right] - 1 \right\} \tag{5.64}$$

and

$$\frac{D_\perp}{D_0} = \frac{3}{2(1-\varepsilon^2)} \left\{ 1 - \frac{\varepsilon^2}{\sqrt{1-\varepsilon^2}} \ln\left[\frac{1+\sqrt{1-\varepsilon^2}}{\varepsilon}\right] \right\} \tag{5.65}$$

whereas for a prolate ellipsoid of rotation ($\varepsilon > 1$) they read

$$\frac{D_{||}}{D_0} = \frac{3\varepsilon^2}{2(\varepsilon^4-1)} \left\{ \frac{\varepsilon^2-2}{\sqrt{\varepsilon^2-1}} \arctan\left(\sqrt{\varepsilon^2-1}\right) + 1 \right\} \tag{5.66}$$

and

$$\frac{D_\perp}{D_0} = \frac{3}{2(\varepsilon^2-1)} \left\{ \frac{\varepsilon^2}{\sqrt{\varepsilon^2-1}} \arctan\left(\sqrt{\varepsilon^2-1}\right) - 1 \right\} \tag{5.67}$$

Here, D_0 is the diffusion coefficient of a sphere of radius R_0 with the same volume as the ellipsoid, that is,

$$R_0^3 = R_{||} R_\perp^2 \tag{5.68}$$

In all the above expressions, the subscript $||$ refers to the symmetry axis, and the subscript $\perp$ to the two transversal axes of the ellipsoid. Figure 5.24 shows the dependence of the two rotational diffusion coefficients on the eccentricity ε of the ellipsoid.

As can be seen, the values of rotational diffusion coefficients change fairly quickly with changing eccentricity. Theoretically, it should be possible to observe the

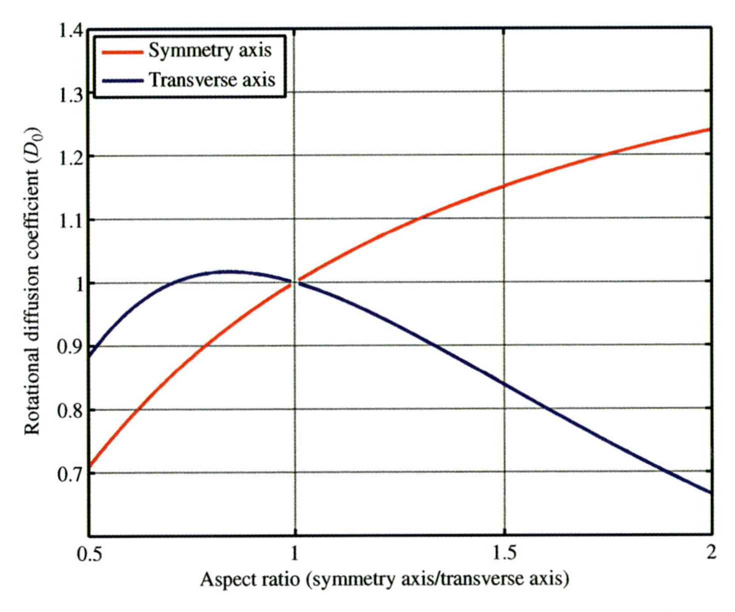

Figure 5.24 Dependence of the rotational diffusion coefficients on ellipsoid eccentricity.

difference in the rotational diffusion coefficients around the symmetrical and the transverse axes by the emergence of a more complex multi-exponential behavior of the correlation function compared with the correlation function produced by an ideal spherical rotor. However, in practice the measured correlation curves are usually too noisy to extract this information except the axes ratio becomes exceedingly large. Usually one fits the correlation function assuming a spherically shaped molecule and obtains a *mean* rotational diffusion coefficient and a *mean* hydrodynamic radius. This corresponds to taking the mean of the diffusion coefficients, $\langle D \rangle = (2 \cdot D_{\perp} + D_{||})/3$, and to use Equation 5.39 to obtain the hydrodynamic radius. owing to the cubic relationship between radius and diffusion coefficient, the dependence of the thus defined mean value of hydrodynamic radius changes much less with eccentricity than the individual rotational diffusion coefficients. This is shown in Figure 5.25, where one can see that the mean value of the hydrodynamic radius changes only slightly in the range $0.75 < \varepsilon < 1.5$, at maximum by only 2%. Thus, assuming a spherical shapc is a fairly reasonable approach for moderate values of eccentricity, which applies to most globular proteins.

5.7 Fluorescence Lifetime Correlation Spectroscopy

A powerful extension of FCS is to use multi-color excitation and detection [63], or to use two-photon excitation at a single wavelength together with multi-color detection [64]. Two-photon excitation uses the fact that most fluorescing molecules show

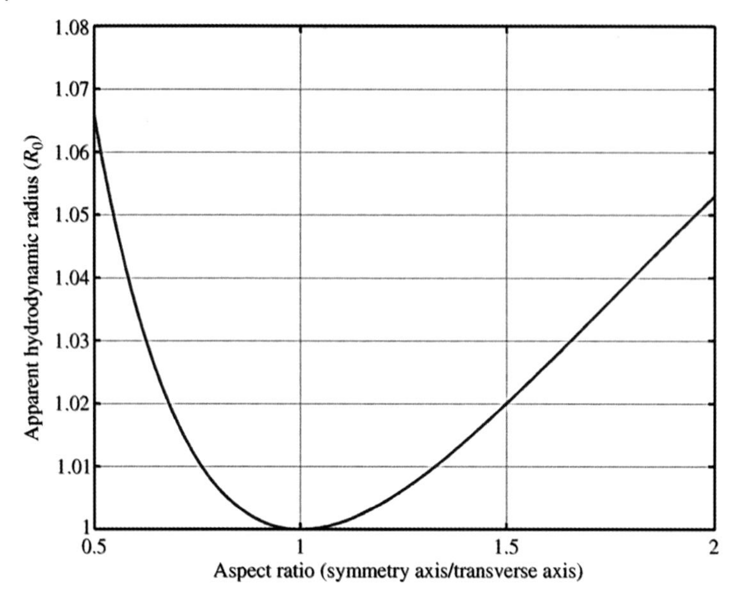

Figure 5.25 Dependence of the mean hydrodynamic radius on ellipsoid eccentricity.

extremely broad excitation spectra, allowing for the simultaneous excitation of molecules with significantly different emission spectra. Detecting the fluorescence in different spectral regions offers the possibility to perform, besides an autocorrelation, a cross-correlation analysis on the signals from different detection channels (fluorescence cross correlation spectroscopy or FCCS). Again, the cross-correlation will show a nontrivial temporal behavior only if there is a physical correlation between photons detected within the various detection channels. Thus, cross-correlation is an ideal tool to follow binding/unbinding or dissociation processes. FCCS has been successfully applied in studies of DNA [65], prion proteins [66], vesicle fusion [67], gene expression [68], DNA–protein interactions [69], protein–protein interactions [70], enzyme kinetics [71], and high-throughput screening [72]; for more details and literature citations see reference [73]. However, dual-color FCCS is technically challenging due to the necessity of simultaneously exciting two spectrally different fluorescent labels byeither two different excitation sources, or by employing a single-wavelength, femtosecond-pulsed high-repetition and high-power laser by using the broad absorption bands of many fluorescent dyes upon two-photon excitation [74]. Recently, the emergence of quantum dots with broad overlapping absorption bands, with distinct narrow emission bands, have also shown promise for simplifying dual-color FCCS [75]. Another problem is always the imperfect overlap of the detection volumes at the two emission wavelengths, due to chromatic aberrations of the optics used. An alternative to dual-color FCCS is fluorescence lifetime correlation spectroscopy or FLCS [76]. The core advantage of FLCS is that one has only a single excitation and a single emission channel, so that optical pathways and detection volumes are identical for all fluorescent labels involved, whereas the distinction between the different labels is done solely on the basis of their fluorescence lifetime.

In an FLCS measurement set-up, a high-repetition pulsed laser is used for fluorescence excitation, and detection is done using time-correlated single-photon counting (TCSPC) [31] electronics in so-called time-tagged time-resolved (TTTR) mode [23], so that both the macroscopic detection time of the photons, with about 100 ns temporal resolution, and the time delay between the last laser pulse and the detected photon, on a picosecond time scale, are recorded.

Let us consider a sample emitting fluorescence with two different lifetime signatures, so that the measured intensity signal I_j has the form

$$I_j(t) = w^{(1)}(t)p_j^{(1)} + w^{(2)}(t)p_j^{(2)} \tag{5.69}$$

where

index j refers to the jth discrete TCSPC time channel used for timing the photon detection events with respect to the exciting laser pulses

$p_j^{(1,2)}$ are the *normalized* fluorescence decay distributions over these channels for the two different fluorescence decay signatures of the sample (e.g., two mono-exponential decays with different decay constants)

$w^{(1,2)}(t)$ are the total intensities of both fluorescence contributions measured at a given time t of the macroscopic time scale.

When inspecting Equation 5.69, it should be emphasized that two completely different times scales are involved: the macroscopic time scale of t, for which the ACF is calculated, and the (discrete) TCSCP time scale labeled by the numbers j of the corresponding TCSPC time channel. Fluorescence decay-specific auto- (ACF) and cross-correlation (CCF) functions can now be defined by

$$g_{\alpha\beta}(t) = \left\langle w^{(\alpha)}(t_0)w^{(\beta)}(t_0+t)\right\rangle_{t_0} \tag{5.70}$$

where the α, β can take either the values 1 or 2, and the angular brackets denote averaging over time t_0. Please take into account that reference to the TCSPC time scale is no longer present. The question now is how to extract the weights $w^{(\alpha)}(t)$ from the measured photon count data. Let us rewrite Equation 5.69 in matrix notation as

$$\mathbf{I} = \mathbf{M}\cdot\mathbf{w} \tag{5.71}$$

where

$\mathbf{I}$ and $\mathbf{w}$ are column vectors with elements I_j and $w^{(\alpha)}$, respectively

the elements of matrix $\mathbf{M}$ are given by $M_{j\alpha} = p_j^{(\alpha)}$.

The most likely values of $w^{(\alpha)}(t)$ at every moment t are found by minimizing the quadratic form [77, 78]

$$\left(\mathbf{I}-\overline{\mathbf{M}}\mathbf{w}\right)^{\mathrm{T}}\cdot\mathbf{V}^{-1}\cdot\left(\mathbf{I}-\overline{\mathbf{M}}\mathbf{w}\right) \tag{5.72}$$

where

$\overline{\mathbf{M}}$ is the average of $\mathbf{M}$ over many excitation cycles
$\mathbf{V}$ is the covariance matrix given by

$$\mathbf{V} = \left\langle (\mathbf{I}-\overline{\mathbf{M}}\mathbf{w})\cdot(\mathbf{I}-\overline{\mathbf{M}}\mathbf{w})^{\mathrm{T}}\right\rangle - \langle(\mathbf{I}-\overline{\mathbf{M}}\mathbf{w})\rangle\cdot\langle(\mathbf{I}-\overline{\mathbf{M}}\mathbf{w})\rangle^{\mathrm{T}} = \mathrm{diag}\langle\mathbf{I}\rangle \tag{5.73}$$

Here, triangular brackets denote averaging over an infinite measurement time interval of t. In the last equation, it was assumed that the photon detection obeys Poisson statistics so that $\langle I_j I_k\rangle - \langle I_j\rangle\langle I_k\rangle = \delta_{jk}\langle I_k\rangle$. The solution of the above minimization task is given by using a weighted quasi-inverse matrix operation and has the explicit form

$$\mathbf{w} = \left[\overline{\mathbf{M}}^{\mathrm{T}}\cdot\mathrm{diag}\langle\mathbf{I}\rangle^{-1}\cdot\overline{\mathbf{M}}\right]^{-1}\cdot\overline{\mathbf{M}}^{\mathrm{T}}\cdot\mathrm{diag}\langle\mathbf{I}\rangle^{-1}\cdot\mathbf{I} = \mathbf{F}\cdot\mathbf{I} \tag{5.74}$$

Thus,

$$\mathbf{u} = \left[\overline{\mathbf{M}}^{\mathrm{T}}\cdot\mathrm{diag}\langle\mathbf{I}\rangle^{-1}\cdot\overline{\mathbf{M}}\right]^{-1}\cdot\overline{\mathbf{M}}^{\mathrm{T}}\cdot\mathrm{diag}\langle\mathbf{I}\rangle^{-1} \tag{5.75}$$

is the desired filter function that recovers $w^{(\alpha)}(t)$ from the measured $I_j(t)$,

$$w^{(\alpha)}(t) = \sum_{j=1}^{N} u_j^{(\alpha)} I_j(t) \tag{5.76}$$

Notice that $\mathbf{u}$ is a $2\times N$ matrix, with elements $u_j^{(1,2)}$, $1 \leq j \leq N$, and a visualization of the meaning of the filter functions $u_j^{(\alpha)}$ is depicted in Figure 5.26. Finally, the auto- and cross-correlations are calculated as

$$g_{\alpha\beta}(t) = \sum_{j=1}^{N}\sum_{k=1}^{N} u_j^{(\alpha)} u_k^{(\beta)} \langle I_j(t_0+t) I_k(t_0)\rangle_{t_0} \tag{5.77}$$

The concept just described can be expanded to an arbitrary number of different fluorescence components in a straightforward way. It is often advisable to include, besides the distinct fluorescence contributions of the sample, an additional component with a uniform distribution among the TCSPC diagram, corresponding to a uniform background (e.g., dark counts, electronic noise, detector afterpulsing). This automatically eliminates background contributions from the fluorescence ACFs and CCFs finally calculated, see for example, [79].

A potential application of FLCS is the study of molecular conformational changes of a biomolecule (protein, DNA, RNA, etc.) that are reflected as lifetime changes of a fluorescence label. In the experimental example described in the next section, a fluorescing molecule is covalently attached to a protein, and one observes two distinct fluorescence decay times that supposedly reflect two different states of the protein–dye. In many instances, one also has to take into account fast photophysical processes of the fluorescent label itself, such as triplet state dynamics (intersystem crossing from the excited singlet state to the first triplet state with subsequent phosphorescence back to the singlet ground state) or conformational transitions

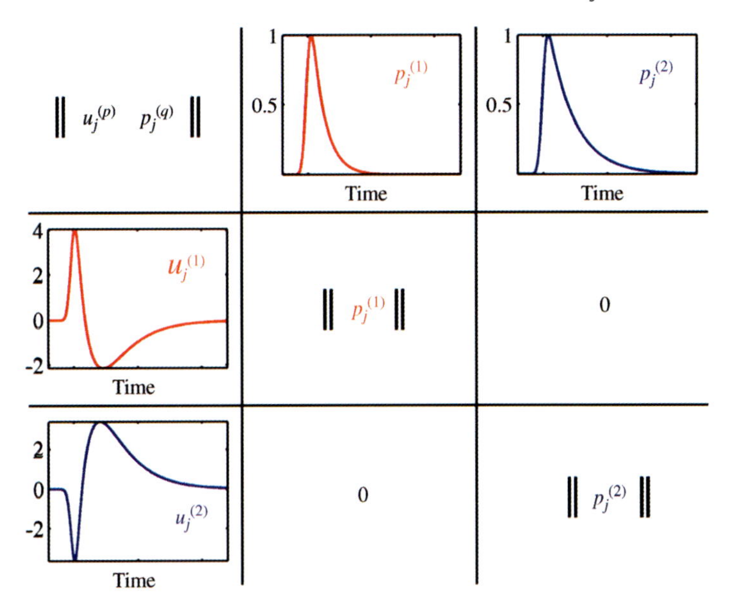

Figure 5.26 Visualization of the meaning of the filter functions $u_j^{(1,2)}$. At the top of the table, the fluorescence decay curves of two states 1 and 2 are depicted. The filter functions, shown at the left side of the table, are designed in such a way that element-wise multiplication and summation of these functions with the fluorescence decay curves yields the identity matrix. In the table, we used the abbreviation $\|x_j\| = \Sigma_j x_j$.

between a fluorescent *cis*- and a non-fluorescent *trans*-state (as happens for many cyanine dyes).

Thus, we will consider the general case of a system depicted in Figure 5.27: A dye–molecule complex undergoes major transitions between states A and B (from left to right and back in Figure 5.27), whereas the dye itself makes transitions between a fluorescent (S) and a non-fluorescent (N) state (from top to bottom and back in

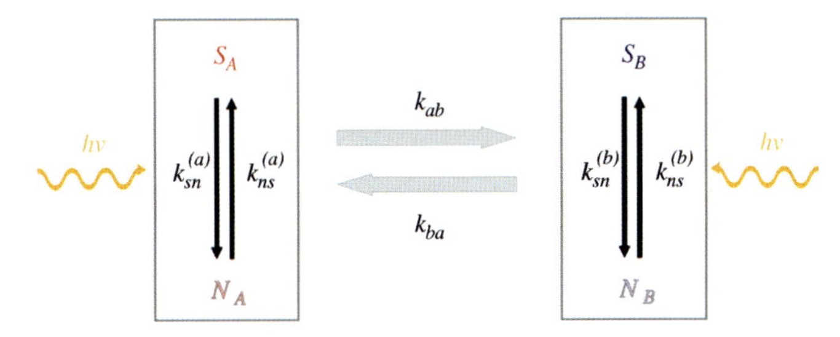

Figure 5.27 Schematic of the four-state model for a dye–protein complex. The whole complex toggles back and forth between states A and B (left to right and back). In both states, the fluorescent dye can reside either in a fluorescent state S or a non-fluorescent sate N. Fluorescent decay in states SA and SB is distinct and is used for FLCS. Transitions between fluorescent and non-fluorescent states may be light-driven (indicated by the waving lines with $h\nu$ on top).

Figure 5.27). In the general case, the photophysical transition rate constants k_{sn} and k_{ns} may themselves depend on whether the complex is in state A or B. Thus, the essential model parameters are the two transition rate constants k_{ab} and k_{ab} for the transition from A to B and from B to A, respectively, and the transition rate constants $k_{sn}^{(a)}$, $k_{sn}^{(b)}$, $k_{ns}^{(a)}$, and $k_{ns}^{(b)}$, describing transitions of the label between a fluorescent and a non-fluorescent state. The rate constants of most interest are k_{ab} and k_{ab}, which may describe, for example, conformational changes of a protein or a DNA complex.

Let us denote the four probabilities of finding the molecular complex in states S_A, S_B, N_A, and N_B by s_a, s_b, n_a, n_b, respectively, which all have to take values between zero and one, all adding up to one. By introducing the column vector $\mathbf{v} = (s_a, s_b, n_a, n_b)^T$, where the superscript T denotes transposition, the rate equations for the temporal evolution of these states are given by

$$\frac{d\mathbf{v}}{dt} = \hat{\mathbf{M}} \cdot \mathbf{v} \tag{5.78}$$

where the matrix $\hat{\mathbf{M}}$ has the explicit form

$$\hat{\mathbf{M}} = \begin{pmatrix} -k_{ab}-k_{sn}^{(a)} & k_{ba} & k_{ns}^{(a)} & 0 \\ k_{ab} & -k_{ba}-k_{sn}^{(b)} & 0 & k_{ns}^{(b)} \\ k_{sn}^{(a)} & 0 & -k_{ab}-k_{ns}^{(a)} & k_{ba} \\ 0 & k_{sn}^{(b)} & k_{ab} & -k_{ba}-k_{ns}^{(b)} \end{pmatrix} \tag{5.79}$$

This linear system of differential equations can be solved in a standard way by finding the eigenvalues λ_j and eigenvectors $\hat{\mathbf{e}}_j$ of matrix $\hat{\mathbf{M}}$ obeying the equation $\hat{\mathbf{M}} \cdot \hat{\mathbf{e}}_j = \lambda_j \hat{\mathbf{e}}_j$. Then, the general solution for $\mathbf{v}(t)$ takes the form

$$\mathbf{v}(t) = \sum_{j=1}^{4} [\mathbf{v}_0 \cdot \hat{\mathbf{d}}_j]\, \hat{\mathbf{e}}_j \exp(-\lambda_j t) \tag{5.80}$$

where the vectors $\hat{\mathbf{d}}_j$ form a conjugate basis to the eigenvectors $\hat{\mathbf{e}}_j$, that is, obey the relationship $\hat{\mathbf{d}}_j \cdot \hat{\mathbf{e}}_k = \delta_{jk}$, and $\mathbf{v}_0$ is the initial value of $\mathbf{v}$ at $t = 0$. Knowing this general solution to the rate equations, the *fast part* of the ACFs and CCFs of the fluorescence emerging from states A and B are explicitly given by

$$g_{\alpha\beta}^{\text{fast}}(t) = \kappa_\alpha \kappa_\beta v_\beta \left(t \middle| v_{0,\gamma} = \delta_{\alpha\gamma}\right) = \sum_{j=1}^{4} g_{\alpha\beta,j} \exp(-\lambda_j t) \tag{5.81}$$

where the coefficients $g_{\alpha\beta,j}$ have the form

$$g_{\alpha\beta,j} = \kappa_\alpha \kappa_\beta \hat{d}_{j,\alpha} \hat{e}_{j,\beta} \tag{5.82}$$

and the κ_α are coefficients accounting for the relative brightness of the different fluorescent states. The "fast part" of the ACFs and CCFs means that one considers correlation lag times much shorter than the typical diffusion time, so that one may assume that a molecule is not moving significantly within the spatially inhomogeneous molecule detection function, and the temporal dynamics of the ACFs and CCFs is dominated by the fast photophysical and molecular transitions. For the matrix $\hat{\mathbf{M}}$ in Equation 5.79, the eigenvalues are given by

$$\lambda_1 = 0,\ \lambda_2 = k_{ab} + k_{ba} \text{ and } \lambda_{3,4} = \frac{\sigma \pm \delta}{2} \tag{5.83}$$

where the abbreviations

$$\sigma = k_{ab} + k_{ba} + k_{sn}^{(a)} + k_{ns}^{(a)} + k_{sn}^{(b)} + k_{ns}^{(b)} \tag{5.84}$$

and

$$\delta = \left\{\sigma^2 - 4\left[k_{ba}\left(k_{sn}^{(a)} + k_{ns}^{(a)}\right) + \left(k_{sn}^{(b)} + k_{ns}^{(b)}\right)\left(k_{ab} + k_{sn}^{(a)} + k_{ns}^{(a)}\right)\right]\right\}^{1/2} \tag{5.85}$$

where introduced. The zero value of the first eigenvalue, λ_1, reflects the conservation of the sum of all state occupancies (neglecting photobleaching for the short time scale considered). The second eigenvalue, λ_2, is solely dominated by the transition dynamics between A and B, whereas $\lambda_{3,4}$ are determined also by the transitions between the fluorescent and non-fluorescent states of the label. The quantities of interest are the transition rate constants k_{ab} and k_{ba}. The second eigenvalue λ_2 yields their sum; to separate this sum one can use the amplitude coefficients $g_{\alpha\beta,j}$ from Equation 5.82. After some tedious algebraic calculations one finds the two relationships

$$\frac{g_{ab,2}g_{aa,2}}{g_{ba,2}g_{bb,2}} = \left(\frac{k_{ab}}{k_{ba}}\frac{\kappa_a}{\kappa_b}\right)^2 \tag{5.86}$$

and

$$\frac{g_{ab,1}g_{aa,1}}{g_{ba,1}g_{bb,1}} = \left(\frac{\kappa_a}{\kappa_b}\right)^2 \tag{5.87}$$

allowing determination of the brightness ratio κ_a/κ_b and, together with Equation 5.83, the separate values of k_{ab} and k_{ba}.

Of course, for fitting experimentally obtained ACFs and CCFs one also has to consider the long-time behavior of these functions that is determined by the diffusion of the molecules out of the detection volume. This can be done in a standard way as described in Section 5.4. For fitting purposes, it is also important to notice that all amplitude coefficients $g_{\alpha\beta,j}$ are non-negative except for $g_{ab,2}$ and $g_{ba,2}$, which are connected with the transition from A to B and back (and thus with λ_2) and generate higher terms in the CCFs. It should also be noted that the particular model used for describing the diffusion part of the ACFs and CCFs is unimportant, because we are not interested in determining diffusion coefficients by extracting rate constants acting on a much faster time scale.

As a typical example of an FLCS, measurements on the Cy5-labeled protein streptavidin are presented. The fluorescence lifetime decay of this complex is shown in Figure 5.28.

Using the short and long lifetime components as the patterns for the two "pure" states of the system, we subsequently measured and calculated the FLCS auto- and cross-correlations, which are shown in Figure 5.29.

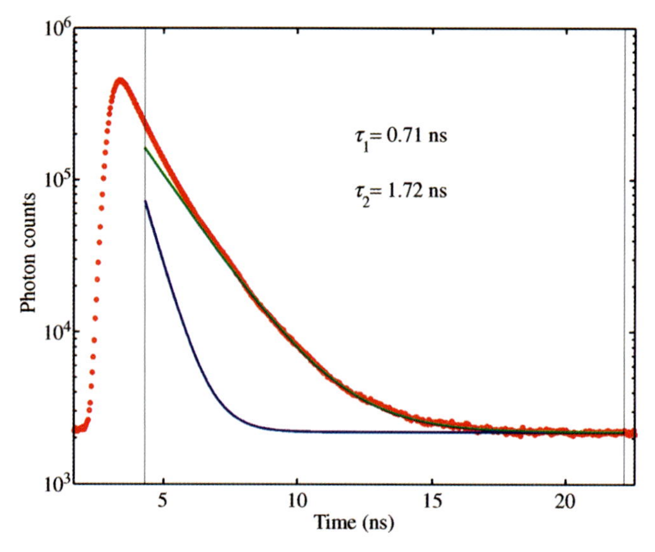

Figure 5.28 Measured TCSPC curve for the Cy5–streptavidin complex (red dots). Blue and green lines are the fitted short and long lifetime components used as fluorescence decay patterns in the FLCS. Thin vertical lines delimit the time window used for the FLCS calculations.

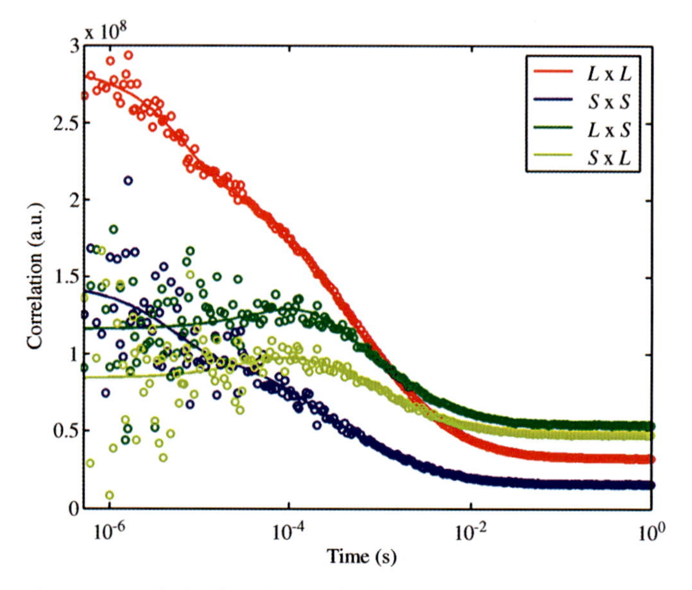

Figure 5.29 Calculated (circles) and fitted (lines) ACFs and CCFs for the measurements at 400 μW excitation power. *L* denotes the long lifetime state, *S* the short lifetime state. The amplitude of the $L \times L$ autocorrelation is divided by a factor of ten for better comparability with the other curves.

The observable short-time dynamics of the auto- and cross-correlation functions is due to a *cis–trans*-isomerization dynamics of the fluorescent label Cy5. In contrast to free Cy5, one observes two fluorescing states, both having a fluorescence lifetime different from the mono-exponential lifetime observed for free Cy5 ($\sim$1 ns). Also, the FLCS curves cannot be fitted satisfactorily without taking into account the two additional exponential terms with $\lambda_{2,3}$. This shows that, in contrast to free Cy5, the Cy5–streptavidin is a more complex multi-state system, with two fluorescing states and two non-fluorescing states.

5.8 Conclusion

In this chapter, we have tried to give both an introduction to FCS, without demanding any prior knowledge, but also to present several recent and, what we consider, important advances of the technique. However, in no way have we covered the whole subject exhaustively, as this would certainly go far beyond the space available in this book. For the many exciting and diverse applications of FCS and its variants, the reader is referred to the many excellent reviews already cited in the introduction [6–12]. It was our aim, beyond presenting the basics of the method, to highlight several aspects of FCS that are fairly difficult to find in other reviews or books on the subject.

References

1 Magde, D., Elson, E., and Webb, W.W. (1972) Thermodynamic fluctuations in a reacting system - measurement by fluorescence correlation spectroscopy. *Phys. Rev. Lett.*, **29**, 705–708.

2 Elson, E.L. and Magde, D. (1974) Fluorescence correlation spectroscopy. I. Conceptual basis and theory. *Biopolymers*, **13**, 1–27.

3 Magde, D., Elson, E., and Webb, W.W. (1974) Fluorescence correlation spectroscopy. II. An experimental realization. *Biopolymers*, **13**, 29–61.

4 Petersen, N.O. (1986) Scanning fluorescence correlation spectroscopy. I. Theory and simulation of aggregation measurements. *Biophys. J.*, **49**, 809–815.

5 Petersen, N.O., Johnson, D.C., and Schlesinger, M.J. (1986) Scanning fluorescence correlation spectroscopy. II. Application to virus glycoprotein aggregation. *Biophys. J.*, **49**, 817–820.

6 Widengren, J. and Mets, Ü. (2002) Conceptual basis of Fluorescence Correlation Spectroscopy and related techniques as tools in bioscience, in *Single-Molecule Detection in Solution - Methods and Applications* (eds. C. Zander, J. Enderlein, and R.A. Keller), Wiley-VCH Verlag GmbH, Wenheim, pp. 69–95.

7 Schwille, P. (2001) Fluorescence correlation spectroscopy and its potential for intracellular applications. *Cell. Biochem. Biophys.*, **34**, 383–408.

8 Hess, S.T., Huang, S., Heikal, A.A., and Webb, W.W. (2002) Biological and chemical applications of fluorescence correlation spectroscopy: a review. *Biochemistry*, **41**, 697–705.

9 Krichevsky, O. and Bonnet, G. (2002) Fluorescence correlation spectroscopy: the technique and its applications. *Rep. Prog. Phys.*, **65**, 251–297.

10 Thompson, N.L., Lieto, A.M., and Allen, N.W. (2002) Recent advances in fluorescence correlation spectroscopy. *Curr. Opin. Struc. Biol.*, **12**, 634–641.

11 Müller, J.D., Chen, Y., and Gratton, E. (2003) Fluorescence correlation spectroscopy. *Biophotonics Pt. B*, **361**, 69–92.

12 Rigler, R. and Elson, E. (eds.) (2001) *Fluorescence Correlation Spectroscopy*, Springer, Berlin.

13 Böhmer, M., Pampaloni, F., Wahl, M., Rahn, H.J., Erdmann, R., and Enderlein, J. (2001) Advanced time-resolved confocal scanning device for ultrasensitive fluorescence detection. *Rev. Sci. Instrum.*, **72**, 4145–4152.

14 Enderlein, J. and Pampaloni, F. (2004) Unified operator approach for deriving Hermite-Gaussian and Laguerre-Gaussian laser modes. *J. Opt. Soc. Am. A*, **21**, 1553–1558.

15 Schätzel, K. (1987) Correlation techniques in dynamic light scattering. *Appl. Phys. B*, **42**, 193–213.

16 Enderlein, J. and Gregor, I. (2005) Using fluorescence lifetime for discriminating detector afterpulsing in fluorescence-correlation spectroscopy. *Rev. Sci. Instrum.*, **76**, 033102.

17 Einstein, A. (1985) *Investigations on the Theory of the Brownian Movement*, Dover, New York.

18 Weljie, A.M., Yamniuk, A.P., Yoshino, H., Izumi, Y., and Vogel, H.J. (2003) Protein conformational changes studied by diffusion NMR spectroscopy: Application to helix-loop-helix calcium binding proteins. *Protein Sci.*, **12**, 228–236.

19 Berne, B.J. and Pecora, R. (2000) *Dynamic Light Scattering*, Dover, New York.

20 Callaghan, P.T. (1991) *Principles of Nuclear Magnetic Resonance Microscopy*, Clarendon Press, Oxford.

21 Harvey, D. (2000) *Modern Analytical Chemistry*, McGraw-Hill, Boston, pp. 593–595.

22 Cole, J.L. and Hansen, J.C. (1999) Analytical ultracentrifugation as a contemporary biomolecular research tool. *J. Biomol. Technol.*, **10**, 163–176.

23 Liu, W., Cellmer, T., Keerl, D., Prausnitz, J.M., and Blanch, H.W. (2005) Interactions of lysozyme in guanidinium chloride solutions from static and dynamic light-scattering measurements. *Biotechnol. Bioeng.*, **90**, 482–490.

24 Kiefhaber, T., Rudolph, R., Kohler, H.H., and Buchner, J. (1991) Protein aggregation in vitro and in vivo: a quantitative model of the kinetic competition between folding and aggregation. *Nat. Biotechnol.*, **9**, 825–829.

25 Enderlein, J. (2005) Dependence of the optical saturation of fluorescence on rotational diffusion. *Chem. Phys. Lett.*, **410**, 452–456.

26 Gregor, I., Patra, D., and Enderlein, J. (2005) Optical saturation in fluorescence correlation spectroscopy under continuous-wave and pulsed excitation. *ChemPhysChem*, **6**, 164–170.

27 Rigneault, H. and Lenne, P.F. (2003) Fluorescence correlation spectroscopy on a mirror. *J. Opt. Soc. Am. B*, **20**, 2203–2214.

28 Davis, S.K. and Bardeen, C.J. (2002) Using two-photon standing waves and patterned photobleaching to measure diffusion from nanometers to microns in biological systems. *Rev. Sci. Instrum.*, **73**, 2128–2135.

29 Jaffiol, R., Blancquaert, Y., Delon, A., and Derouard, J. (2006) Spatial fluorescence cross-correlation spectroscopy. *Appl. Opt.*, **45**, 1225–1235.

30 Müller, B.K., Zaychikov, E., Bräuchle, C., and Lamb, D. (2005) Cross talk free fluorescence cross correlation spectroscopy in live cells. *Biophys. J.*, **89**, 3508–3522.

31 O'Connor, D.V. and Phillips, D. (1984) *Time-Correlated Single Photon Counting*, Academic Press, New York.

32 Vetterling, W.T., Press, W.H., Teukolsky, S.A., and Flannery, B.P. (2003) *Numerical Recipes in C+ +*, 2nd edn, Cambridge University Press, Cambridge.

33 Müller, C.B., Weiß, K., Richtering, W., Loman, A., and Enderlein, J. (2008) Calibrating differential interference contrast microscopy with dual-focus fluorescence correlation spectroscopy. *Opt. Express*, **16**, 4322–4329.

34 Culbertson, C.T., Jacobson, S.C., and Ramsey, J.M. (2002) Diffusion coefficient measurements in microfluidic devices. *Talanta*, **56**, 365–373.

35 Widengren, J., Rigler, R., and Mets, Ü. (1994) Triplet-state monitoring by fluorescence correlation spectroscopy. *J. Fluoresc.*, **4**, 255–258.

36 Widengren, J., Mets, Ü., and Rigler, R. (1995) Fluorescence correlation spectroscopy of triplet states in solution: A theoretical and experimental study. *J. Phys. Chem.*, **99**, 13368–13379.

37 Ambrose, W.P., Goodwin, P.M., Enderlein, J., Semin, D.J., Martin, J.C., and Keller, R.A. (1997) Fluorescence photon antibunching from single molecules on a surface. *Chem. Phys. Lett.*, **269**, 365–370.

38 Lounis, B., Bechtel, H.A., Gerion, D., Alivisatos, P., and Moerner, W.E. (2000) Photon antibunching in single CdSe/ZnS quantum dot fluorescence. *Chem. Phys. Lett.*, **329**, 399–404.

39 Treussart, F., Clouqueur, A., Grossmann, C., and Roch, J.F. (2001) Photon antibunching in the fluorescence of a single dye molecule embedded in a thin polymer film. *Opt. Lett.*, **26**, 1504–1506.

40 Tinnefeld, P., Weston, K.D. *et al.* (2002) Antibunching in the emission of a single tetrachromophoric dendritic system. *J. Am. Chem. Soc.*, **124**, 14310–14311.

41 Weston, K.D., Dyck, M. *et al.* (2002) Measuring the number of independent emitters in single-molecule fluorescence images and trajectories using coincident photons. *Anal. Chem.*, **74**, 5342–5349.

42 Kask, P., Piksarv, P., and Mets, Ü. (1985) Fluorescence correlation spectroscopy in the nanosecond time range: Photon antibunching in dye fluorescence. *Eur. Biophys. J.*, **12**, 163–166.

43 Mets, Ü., Widengren, J., and Rigler, R. (1997) Application of the antibunching in dye fluorescence: measuring the excitation rates in solution. *Chem. Phys.*, **218**, 191–198.

44 Sýkora, J., Kaiser, K., Gregor, I., Bönigk, W., Schmalzing, G., and Enderlein, J. (2007) Exploring fluorescence antibunching in solution for determining the stoichiometry of molecular complexes. *Anal. Chem.*, **79**, 4040–4049.

45 Ehrenberg, M. and Rigler, R. (1974) Rotational Brownian motion and fluorescence intensity fluctuations. *Chem. Phys.*, **4**, 390–401.

46 Aragón, S.R. and Pecora, R. (1975) Fluorescence correlation spectroscopy and Brownian rotational diffusion. *Biopolymers*, **14**, 119–138.

47 Kask, P., Piksarv, P., Mets, Ü., Pooga, M., and Lippmaa, E. (1986) Fluorescence correlation spectroscopy in the nanosecond time range: rotational diffusion of bovine carbonic anhydrase B. *Eur. Biophys. J.*, **14**, 257–261.

48 Tsay, J.M., Dose, S., and Weiss, S. (2006) Rotational and translational diffusion of peptide-coated CdSe/CdS/ZnS nanorods studied by fluorescence correlation spectroscopy. *J. Am. Chem. Soc.*, **128**, 1639–1647.

49 Felekyan, S., Kühnemuth, R., Kudryavtsev, V., Sandhagen, C., Becker, W., and Seidel, C.A.M. (2005) Full correlation from picoseconds to seconds by time-resolved and time-correlated single photon detection. *Rev. Sci. Instrum.*, **76**, 083104.

50 Debye, P. (1929) *Polar Molecules*, The Chemical Catalogue Company, New York, pp. 77–108.

51 Feynman, R. (1964) *Feynman's Lectures on Physics*, Addison-Wesley, London, Ch 18.4.

52 Thompson, W.J. (1994) *Angular Momentum*, John Wiley & Sons, Ltd, Chichester.

53 Jones, H.F. (1998) *Groups, Representations and Physics*, IOP Publishing, Bristol and Philadelphia, Ch 4.

54 Morse, P.M. and Feshbach, H. (1953) *Methods of Theoretical Physics*, McGraw-Hill, New York, Ch 7.

55 Zwillinger, D. (1990) *Handbook of Differential Equations*, 3rd edn, Academic Press, New York, London, Ch 96.

56 Enderlein, J., Gregor, I., Patra, D., Dertinger, T., and Kaupp, B. (2005) Performance of fluorescence correlation spectroscopy for measuring diffusion and concentration. *ChemPhysChem*, **6**, 2324–2336.

57 Abramowitz, M. and Stegun, I.A. (1965) *Handbook of Mathematical Functions*, Dover, New York.

58 Bahlmann, K. and Hell, S.W. (2000) Electric field depolarization in high

aperture focusing with emphasis on annular apertures. *J. Microsc.*, **200**, 59–67.

59 Bahlmann, K. and Hell, S.W. (2000) Depolarization by high aperture focusing. *Appl. Phys. Lett.*, 77, 612–614.

60 Perrin, F. (1934) Mouvement brownien d'un ellipsoide (I). Dispersion dielectrique pour des molecules ellipsoidales. *J. Phys. Radium*, **Ser. VII 5**, 497–511.

61 Perrin, F. (1936) Mouvement brownien d'un ellipsoide (II). Rotation libre et depolarisation des fluorescences. Translation et diffusion de molecules ellipsoidales. *J. Phys. Radium*, **Ser. VII 7**, 1–11.

62 Koenig, S.H. (1975) Brownian motion of an ellipsoid. A correction to Perrin's results. *Biopolymers*, **14**, 2421–2423.

63 Schwille, P., Meyer-Almes, F.J., and Rigler, R. (1997) Dual-color fluorescence cross-correlation spectroscopy for multicomponent diffusional analysis in solution. *Biophys. J.*, **72**, 1878–1886.

64 Heinze, K., Koltermann, A., and Schwille, P. (2000) Simultaneous two-photon excitation of distinct labels for dual-color fluorescence crosscorrelation analysis. *Proc. Nat. Acad. Sci. USA*, **97**, 10377–10382.

65 Jahnz, M. and Schwille, P. (2005) An ultrasensitive site-specific DNA recombination assay based on dual-color fluorescence cross-correlation spectroscopy. *Nucleic Acid Res.*, **33**, e60.

66 Bieschke, J., Giese, A., Schulz-Schaeffer, W., Zerr, I., Poser, S., Eigen, M., and Kretzschmar, H. (2000) Ultrasensitive detection of pathological prion protein aggregates by dual-color scanning for intensely fluorescent targets. *Proc. Nat. Acad. Sci. USA*, **97**, 5468–5473.

67 Swift, J.L., Carnini, A., Dahms, T.E.S., and Cramb, D.T. (2004) Anesthetic-enhanced membrane fusion examined using two-photon fluorescence correlation spectroscopy. *J. Phys. Chem. B*, **108**, 11133–11138.

68 Camacho, A., Korn, K., Damond, M., Cajot, J.F., Litborn, E., Liao, B.H., Thyberg, P., Winter, H., Honegger, A., Gardellin, P., and Rigler, R. (2004) Direct quantification of mRNA expression levels using single molecule detection. *J. Biotechnol.*, **107**, 107–114.

69 Rippe, K. (2000) Simultaneous binding of two DNA duplexes to the NtrC-enhancer complex studied by two-color fluorescence cross-correlation spectroscopy. *Biochemistry*, **39**, 2131–2139.

70 Baudendistel, N., Müller, G., Waldeck, W., Angel, P., and Langowski, J. (2005) Two-hybrid fluorescence cross-correlation spectroscopy detects protein-protein interactions in vivo. *ChemPhysChem*, **6**, 984–990.

71 Kettling, U., Koltermann, A., Schwille, P., and Eigen, M. (1998) Real-time enzyme kinetics monitored by dual-color fluorescence cross-correlation spectroscopy. *Proc. Nat. Acad. Sci. USA*, **95**, 1416–1420.

72 Koltermann, A., Kettling, U., Bieschke, J., Winkler, T., and Eigen, M. (1998) Rapid assay processing by integration of dual-color fluorescence cross-correlation spectroscopy: High throughput screening for enzyme activity. *Proc. Nat. Acad. Sci. USA*, **95**, 1421–1426.

73 Bacia, K., Kim, S.A., and Schwille, P. (2006) Fluorescence cross-correlation spectroscopy in living cells. *Nat. Methods*, **3**, 83–89.

74 Xu, C., and Webb, W.W. (1996) Measurement of two-photon excitation cross sections of molecular fluorophores with data from 690 to 1050nm. *J. Opt. Soc. Am. B*, **13**, 481–491.

75 Han, M., Gao, X., Su, J.Z., and Nie, S. (2001) Quantum-dot-tagged microbeads for multiplexed optical coding of biomolecules. *Nat. Biotechnol.*, **19**, 631–635.

76 Böhmer, M., Wahl, M., Rahn, H.J., Erdmann, R., and Enderlein, J. (2002) Time-resolved fluorescence correlation spectroscopy. *Chem. Phys. Lett.*, **353**, 439–445.

6
Excited State Energy Transfer

6.1
Introduction

Excited fluorophores dissipate their excess energy mainly via fluorescence, through intersystem crossing (ISC) and other non-radiative pathways, such as internal conversion as the competitive pathways. In the presence of quenchers, additional new pathways are available; for instance, in quenching, the loss of fluorescence can occur through electron transfer or energy transfer. This chapter will take a closer look at the mechanisms of the energy transfer phenomenon and a number of examples will be elucidated on how energy transfer can be applied to gain insights into chemical and biochemical processes.

In its most simplified form, energy transfer can be explained as the transfer of the energy of an excited donor molecule (D) to an acceptor molecule (A). This transfer causes the donor to relax to its ground state, while the acceptor molecule ends up in a (higher) excited state. As a result, the donor's excited state lifetime (decay time) decreases, while the fluorescence of the acceptor is sensitized. There are some important conditions that apply for excited state energy transfer. For instance, transfer to a higher energy level is forbidden by the laws of energy conservation. Moreover, sufficient spectral overlap between the emission of the donor and the absorption of the acceptor is required.

In general two types of energy transfer can be distinguished: radiative and non-radiative transfer. In radiative transfer an emitted photon from a donor molecule is re-absorbed by an acceptor. Non-radiative energy transfer implies that a secondary photon is not involved. In this case energy is transferred directly, for example, through long-range (1–10 nm) dipolar interactions between donor and acceptor (Förster transfer [1]), or through short-range (<1 nm) interactions between molecular orbitals leading to a concerted electron exchange (Dexter mechanism). The latter is an important process in, for example, quenching by molecular oxygen. This chapter will focus on the Förster mechanism, also known as fluorescence resonance energy transfer (FRET) or Förster type energy transfer, because it is broadly applicable in various (bio)chemical systems. However, when studying FRET processes, one should keep in mind that at short distances (<2 nm), interference with Dexter transfer may occur [2].

Handbook of Fluorescence Spectroscopy and Imaging. M. Sauer, J. Hofkens, and J. Enderlein

ISBN: 978-3-527-31669-4

6.2 Theory of (Förster) Energy Transfer

In recent years, FRET has become a popular tool in biological and chemical investigations. Although the first reported FRET experiments date back to 1922 [3–6], Förster described and modeled the phenomenon of FRET correctly during the 1940s [1, 7, 8]. The widespread use of FRET is due to the fact that it is highly sensitive to the distance between donor and acceptor, usually in a range between 2 and 10 nm, and hence can be considered as a "molecular ruler" [9].

6.2.1 Mechanism and Mathematical Formalism of FRET

Owing to long-range dipole–dipole coulomb interactions, FRET can occur between an excited donor and an acceptor (Figure 6.1) provided that there is a spectral overlap between the normalized emission of the donor and the absorption of the acceptor. Although this transfer mechanism is in fact based on the interaction between transition densities, a fairly good mathematical approximation can be obtained by considering transition dipoles instead of densities. Thus, the rate of FRET (k_{FRET}) is given by [10–12]:

$$k_{\mathrm{FRET}} = 8.79 \times 10^{-5} \cdot \frac{\varphi_{D,0} \cdot \kappa^2 \cdot J}{\tau_{D,0} \cdot r^6 \cdot n^4} \tag{6.1}$$

$$\kappa^2 = (\cos\theta_T - 3\cos\theta_D \cos\theta_A)^2 \tag{6.2}$$

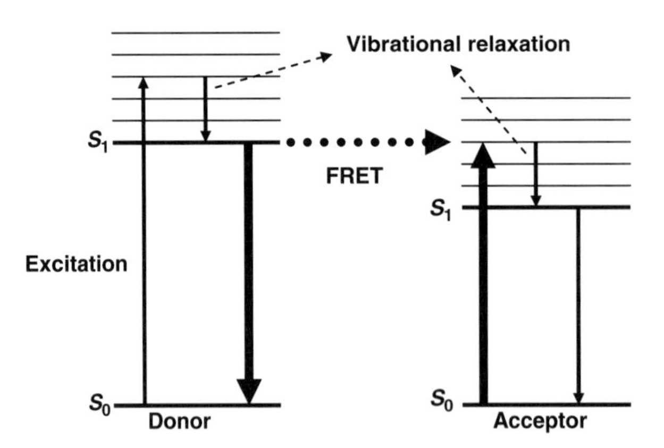

Figure 6.1 Schematic representation of fluorescence energy transfer (FRET). After excitation and internal conversion, the donor molecule reaches the vibrational ground level of the first excited state S_1. Through FRET it can pass on its excitation energy to an acceptor that in turn is excited. Conservation of energy implies a spectral overlap between the donor emission and the acceptor absorbance.

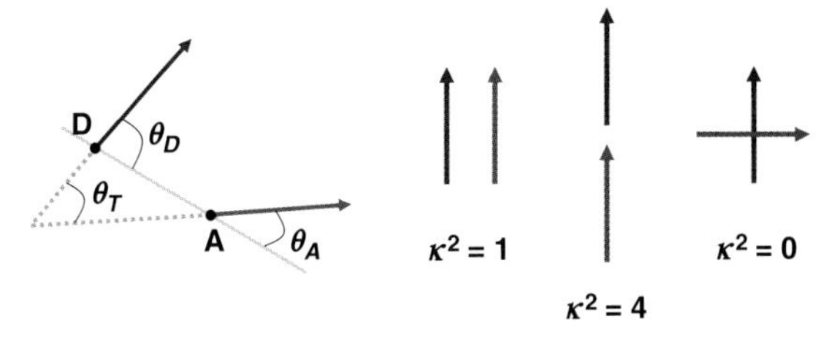

Figure 6.2 The orientation factor κ^2. The grey and black arrows represent, respectively, the donor (D) and acceptor (A) transition dipoles. Left: definition of the angles for calculating the orientation factor κ^2. Right: some examples of orientations with their corresponding κ^2 values.

$$J = \int_0^\infty F_D(\lambda) \cdot \varepsilon_A(\lambda) \cdot \lambda^4 \, d\lambda \qquad (6.3)$$

where

$\varphi_{D,0}$ and $\tau_{D,0}$ are, respectively, the quantum yield and the excited-state lifetime of the donor in the absence of the acceptor
n is the refractive index of the surrounding medium

κ^2 is an orientation factor that is dependent on the angle between the donor and acceptor transition dipoles (θ_T) and the angles θ_D and θ_A, as defined in Figure 6.2, this orientation factor ranges from 0 for perpendicularly oriented dipoles to 4 for head-to-tail arranged dipoles.

For randomly oriented dipoles with fast rotational diffusion the ensemble averaged κ^2 yields 2/3. In the case of randomly oriented, but static orientations, 0.476 is used for κ^2 [13]. When the orientations are restricted in plane, κ^2 is 4/5. J is the overlap integral between the donor emission F_D, normalized such that $\int_0^\infty F_D(\lambda)d\lambda = 1$, and the acceptor absorption spectrum expressed in exctinction coefficients ε (Figure 6.3). Finally, r is the center-to-center distance between donor and acceptor. For correct use of these formulae, J should be expressed in $M^{-1}\,cm^{-1}\,nm^4$, while r should be expressed in Å. The Förster radius R_0 is defined as

$$R_0^6 = 8.79 \times 10^{-5} \cdot \frac{\varphi_{D,0} \cdot \kappa^2 \cdot J}{n^4} \qquad (6.4)$$

Equation 6.1 then simplifies to

$$k_{FRET} = \left(\frac{R_0}{r}\right)^6 \cdot \frac{1}{\tau_{D,0}} \qquad (6.5)$$

It is clear from this equation that there is a very strong dependency of the energy transfer rate on the distance (inverse sixth power).

The FRET efficiency E is defined as the fraction of excited molecules that undergo energy transfer from the donor to the acceptor and can be expressed in terms of the rate constants of the processes involved:

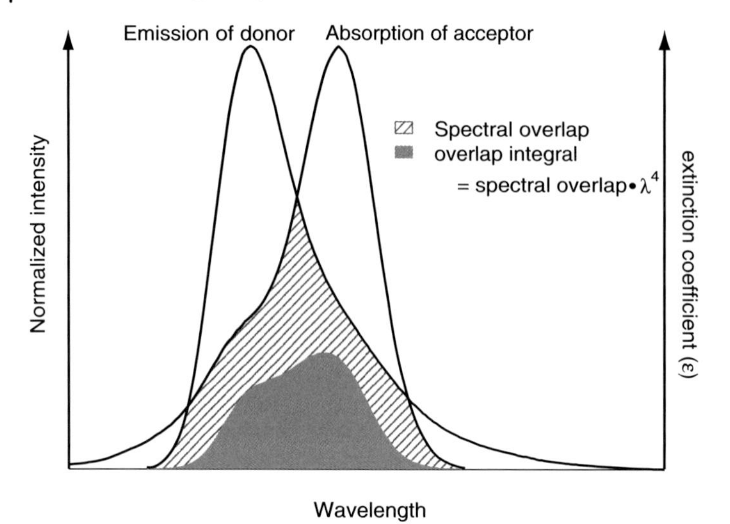

Figure 6.3 Spectral overlap between emission of donor and absorption of acceptor. The dashed area represents the spectral overlap of the donor emission and acceptor excitation, while the grey area represents the overlap corrected to the fourth power of the wavelength, as defined by Equation 6.3. For correct calculation, the donor emission spectrum must be normalized such that the area under the curve is 1, while the absorption spectrum should be expressed in extinction coefficients ε. The bigger the overlap integral J, the more efficient the energy transfer will be. Note: intensity scaling is relative.

$$E = \frac{k_{\mathrm{FRET}}}{k_{\mathrm{FRET}} + k_{f,\mathrm{donor}}} \tag{6.6}$$

Combining this expression with Equation 6.5 results in a different way of expressing E:

$$E = \frac{R_0^6}{R_0^6 + r^6} = 1 - \frac{I_{D,\mathrm{FRET}}}{I_{D,0}} = \frac{I_{A,\mathrm{FRET}}}{I_{D,\mathrm{FRET}} + \gamma I_{A,\mathrm{FRET}}} = 1 - \frac{\tau_{D,\mathrm{FRET}}}{\tau_{D,0}} \tag{6.7}$$

$$\gamma = \frac{\eta_A \varphi_A}{\eta_D \varphi_D} \tag{6.8}$$

Experimentally the amount of energy transferred can be calculated from the observed quenching of the donor intensity ($I_{D,\mathrm{FRET}}$ versus $I_{D,0}$), by the enhanced acceptor fluorescence ($I_{A,\mathrm{FRET}}$) in the presence of the donor, or from the ratio of the intensities from the donor and acceptor channels (or by the corresponding lifetime ratios $\tau_{D,\mathrm{FRET}}$ versus $\tau_{D,0}$). In Equation 6.7 it is assumed that only the donor molecule is directly excited by the light source. The factor γ is a detector correction factor, which compensates for differences in the sensitivity of the detector system to the donor and acceptor fluorescences (φ_D and φ_A are the fluorescence quantum yields and η_D and η_A are the detector efficiencies for both channels) [14].

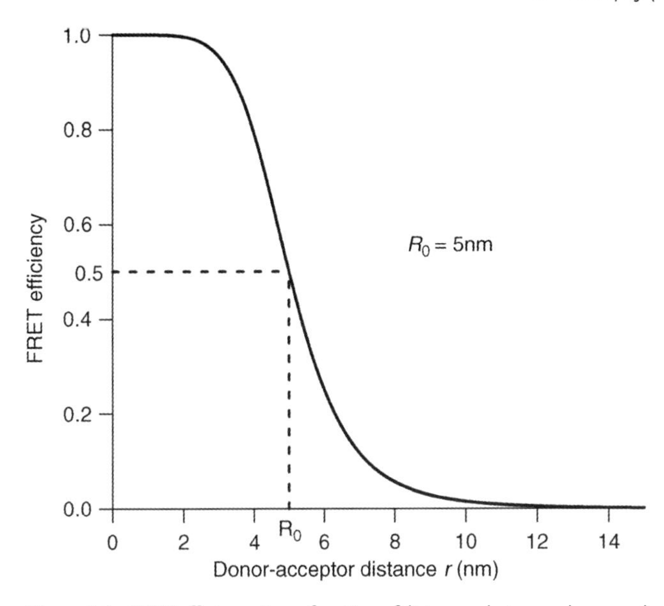

Figure 6.4 FRET efficiency E as a function of distance r between donor and acceptor. When r equals the characteristic FRET distance R_0, the FRET efficiency is 50%.

R_0 can be interpreted as the characteristic distance for the Förster interaction between the donor and acceptor: when r equals R_0 the FRET efficiency equals 50%. R_0 typically has a value in the range of 1–10 nm, meaning that distances within this range can be accurately probed by use of FRET experiments (Figure 6.4). Förster distances of some commonly used donor–acceptor pairs in FRET experiments can be found in Table 6.1. The structures of the dyes used are shown in Figure 6.5. References [15, 16] contain some more R_0 values of frequently used FRET pairs.

The above mentioned formulae are based on the assumption that the chromophores behave as perfect point transition dipoles. This means that the shape of the wave function of the donor and of the acceptor is neglected and thus these formulae only yield an accurate value when the size of the chromophores is much smaller than the intermolecular distance. If this was not the case, the interchromophoric interaction would be more complex as the total transition densities around the dyes have to be taken into account. This can be done by expressing the total electronic coupling as

Table 6.1 Förster radii of some widely used donor–acceptor pairs in FRET experiments. The structures of the dyes are shown in Figure 6.5. The values are based on references [17–21].

Donor **Acceptor**	**Cy3** **Cy5**	**Atto 495** **Atto 590**	**Alexa Fluor 488** **Alexa Fluor 546**	**PDI** **TDI**	**ECFP** **EYFP**
R_0	3.6 nm	5.6 nm	5.5 nm	7.3 nm	4.9 nm

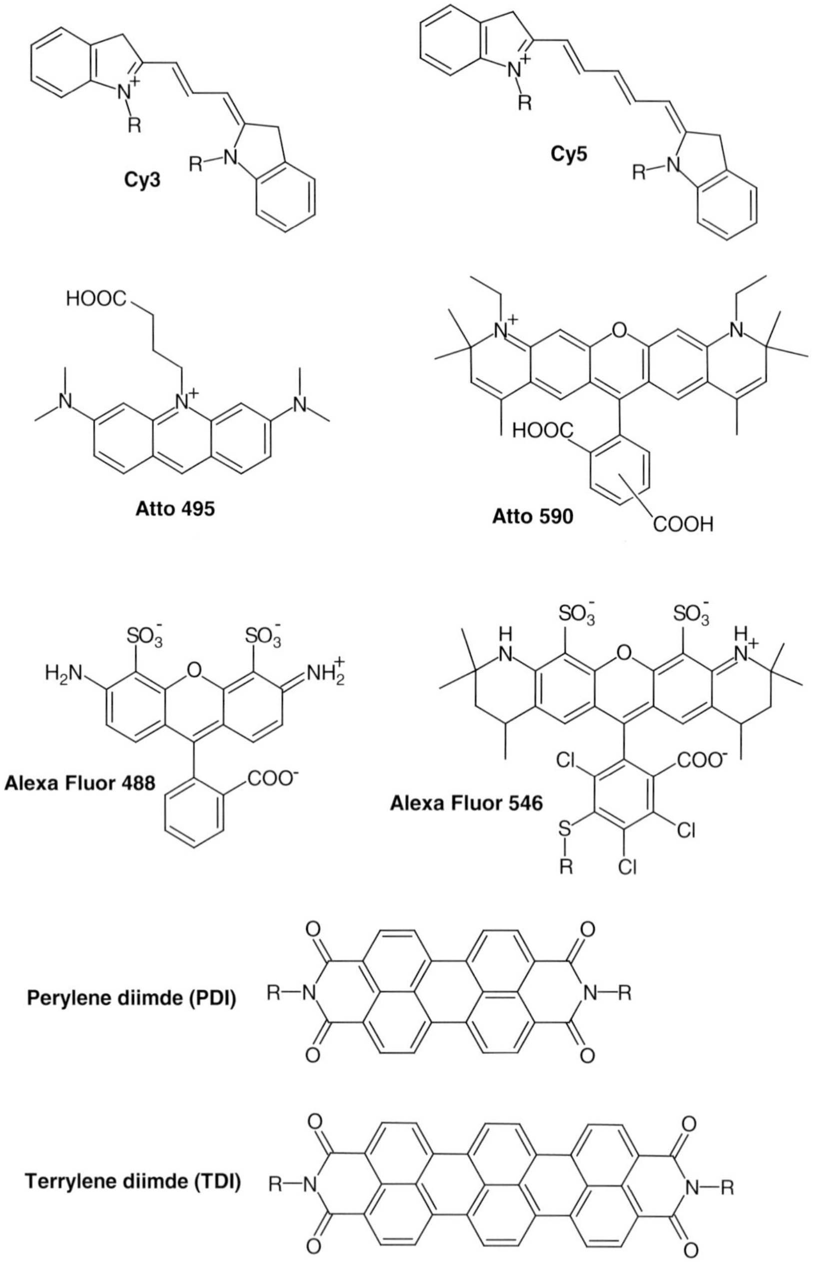

Cy3

Cy5

Atto 495

Atto 590

Alexa Fluor 488

Alexa Fluor 546

Perylene diimde (PDI)

Terrylene diimde (TDI)

Figure 6.5 Chemical structures of the organic dye-based FRET pairs listed in Table 6.1.

a sum over all the pairwise interactions between the individual atomic transition charges. The electronic coupling promoting the energy transfer can then be written as:

$$V_{\mathrm{DA}}^{\mathrm{Coulomb}} = \frac{1}{4\pi\varepsilon_0} \sum_{p}^{D} \sum_{q}^{D} \frac{\varrho_D(p)\varrho_A(q)}{r_{pq}} \quad (6.9)$$

The summation runs over all atomic positions p on the donor and q on the acceptor, r_{pq} is the interatomic center-to-center distance between p and q and $\varrho_D(p)$ and $\varrho_A(q)$ represent the atomic transition densities on, respectively, site p and site q for the lowest optical excitation on the donor and the acceptor. More information on this improved Förster model can be found in references [22, 23], where this model is applied to the study of interchain and intrachain energy transfer in rigid rod-shaped polyindenofluorene polymers endcapped with perylene derivatives. Conjugated polymeric materials are typical examples where the classical Förster model does not hold, because of the important through-bond electronic interactions between donor and acceptor.

6.2.2
Measuring FRET Efficiencies Through Excited-State Lifetimes

As illustrated in Equation 6.7 the efficiency of resonance energy transfer can be calculated by measuring the ratio of the donor excited state lifetime in the presence and absence of an acceptor. However, by far not all fluorophores used for FRET experiments have a single-exponential decay of the singlet excited state. In the case of multi-exponential decay, an average decay time can be used as an approximation (see Equation 6.10):

$$\tau_{\mathrm{average}} = \sum_{i} \alpha_i \cdot \tau_i \quad (6.10)$$

where

τ_i represents the characteristic time of the ith decay component
α_i is its corresponding weight factor.

In the presence of acceptor molecules the donor decay rate can also be multi-exponential in bulk experiments when several donor–acceptor distances are possible, or even in single-pair experiments when fast fluctuations in the interchromophoric distances occur. In these cases Equation 6.10 can also be applied to calculate an average $\tau_{D,\mathrm{FRET}}$.

Furthermore, it is worth noting that Equation 6.7 is based on the assumption that there is no influence on the donor decay other than the FRET influence. However, this condition is not fulfilled in all instances. For example, in biomolecules, allosteric interactions between donor and acceptor can induce changes in the excited state decay behavior or other photophysical parameters that do not result from energy transfer.

6.2.3
Spin Rules for FRET

For the acceptor molecule of the FRET pair, only transitions in which no change of electron spin is involved are allowed (singlet-to-singlet or triplet-to-triplet). However, the donor molecule can make a spin-forbidden transition, for instance from a singlet to a triplet state or vice versa, provided that this donor has a sufficiently high quantum yield for intersystem crossing [24]. In general there are three ways in which Förster transfer can occur, these being the normal singlet–singlet transfer $[D(S_1)+A(S_0) \rightarrow D(S_0)+A(S_1)]$, singlet–singlet annihilation $[D(S_1)+A(S_1) \rightarrow D(S_0)+A(S_2)]$, and singlet–triplet annihilation $[D(S_1)+A(T_1) \rightarrow D(S_0)+A(T_2)]$. As indicated in Figure 6.6, upon singlet–singlet or singlet–triplet annihilation the excitation is transferred to an acceptor that already resides in an excited singlet or triplet state. These processes are called annihilation processes as there is a net decrease in the number of excited states upon energy transfer.

Singlet–triplet transfer, being the transfer from a singlet excited state of the donor to yield a triplet excited acceptor $[D(S_1)+A(S_0) \rightarrow D(S_0)+A(T_1)]$ is spin-forbidden for Förster transfer, but can occur via the Dexter exchange mechanism. Also triplet–triplet transfer, from an excited donor in the triplet state yielding a triplet excited acceptor $[D(T_1)+A(S_0) \rightarrow D(S_0)+A(T_1)]$, can only occur through the Dexter type of energy transfer, as the acceptor makes a spin-forbidden transition. Triplet–singlet transfer $[D(T_1)+A(S_0) \rightarrow D(S_0)+A(S_1)]$, on the other hand, can occur by the Förster mechanism. Although the donor molecule makes a spin-forbidden transition, the low rate for this process will be compensated by the large lifetime of the donor triplet state.

6.2.4
Homo-FRET and FRET-Induced Depolarization

Fluorescence resonance energy transfer does not necessarily require two spectrally distinct fluorophores as a donor–acceptor pair. In particular, dyes with a rather small Stokes shift often have a significant overlap of their absorbance and emission spectra. So in principle, such a fluorophore can transfer its excitation energy to a neighboring molecule of the same species, which is denoted as homo-FRET or homotransfer [25]. Because energy transfer between identical chromophores is reversible, this mechanisms is also termed energy hopping, while the irreversible transfer from a donor to a red-shifted acceptor molecule is often referred to as energy trapping. In fact, the Förster theory was originally based on observations of homotransfer [1]. In this instance, no spectral distinction can be made between donor and acceptor, which makes it difficult to measure the FRET efficiency directly through the ratio of the acceptor and donor fluorescence intensities. Luckily other techniques are available for quantification of the degree of FRET, for instance, by measuring the degree of depolarization.

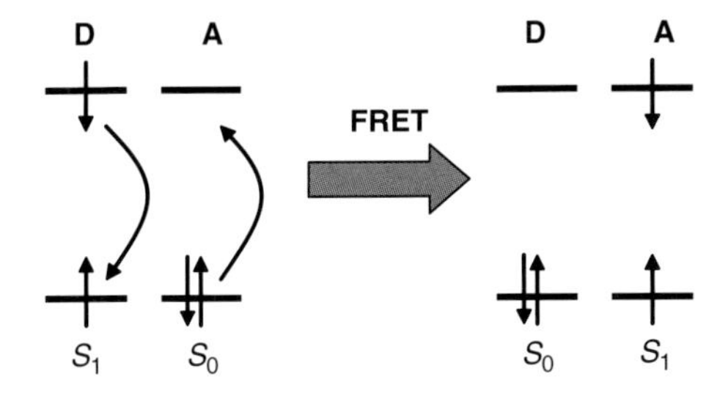

Singlet – singlet transfer

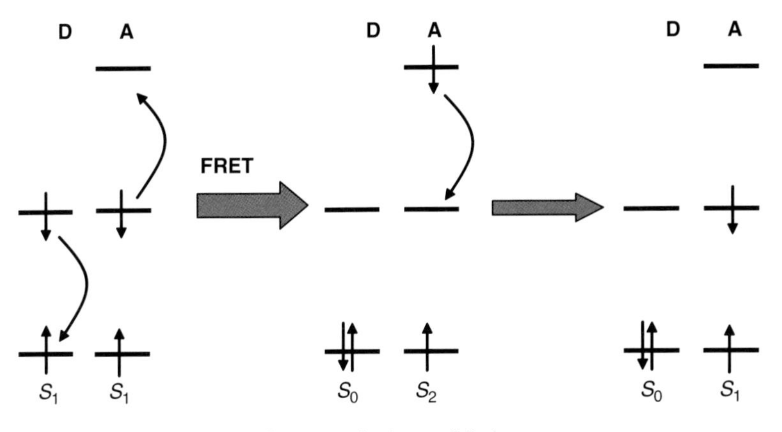

Singlet – singlet annihilation

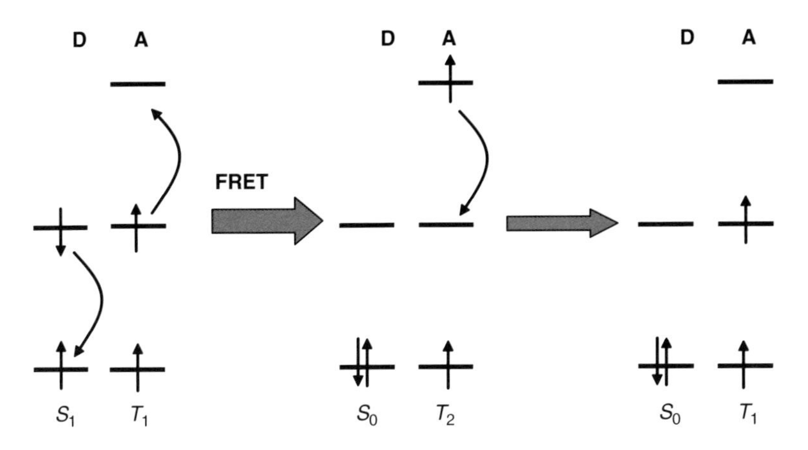

Singlet – triplet annihilation

Figure 6.6 Examples of different types of FRET transitions.

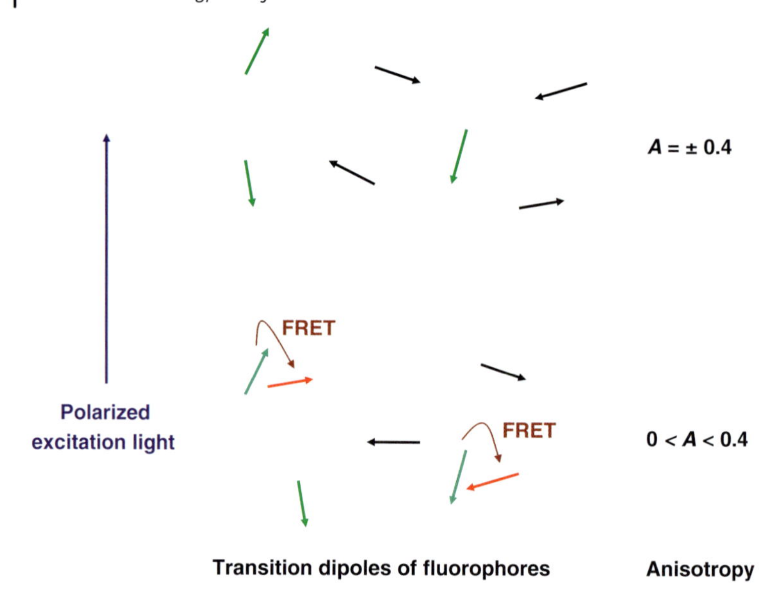

Figure 6.7 Depolarization caused by FRET. Upon excitation by plane polarized light (blue arrow), the broad angular allowance for FRET causes a net decrease in emission polarization (measured as a decrease in polarization anisotropy *A*). The black, green, and red arrows represent the orientation of the transition dipoles of the dye molecules. The dyes with a dipole parallel to the polarized light (green arrows) have the highest probability for excitation. Through FRET, the excitation energy can be transferred to a neighboring molecule with a transition dipole more perpendicular to the polarized excitation light (red arrows).

Indeed, FRET causes a net decrease in the steady-state anisotropy of the emission. When a donor molecule is excited by plane polarized light, its emission will also be polarized corresponding to the orientation of the emission dipole, provided that the rotational diffusion of the molecule is slower than the singlet excited state lifetime. However, when the donor fluorophore transfers its excitation energy to an acceptor, the acceptor emission will be relatively depolarized because of the broad range of allowed angles for the FRET transition (Figure 6.7). Usually the depolarization is measured by a decrease in the fluorescence anisotropy, A,

$$A = \frac{I_{\parallel} - I_{\perp}}{I_{\parallel} + 2 \cdot I_{\perp}} \tag{6.11}$$

where

$I_{\parallel}$ and $I_{\perp}$ represent the intensities of the light polarized parallel and perpendicular to the incident polarization, respectively.

For randomly oriented transition dipoles in the absence of energy transfer, A equals 0.4. When homo-FRET (or hetero-FRET) occurs, a significant depolarization is observed, yielding an anisotropy closer to 0. A general formula for the FRET efficiency based on polarization anisotropy can be given by:

$$E = 1 - \frac{A_{FRET}}{A_0} \tag{6.12}$$

Here A_{FRET} and A_0 are the anisotropy of the fluorescence in the presence and absence of energy transfer, respectively. Also, time-resolved anisotropy measurements are possible for correlation of the decrease in anisotropy to the FRET-rate. As FRET-quenching is, in most instances, an extremely fast process, characteristic depolarization times are found to be in the subnanosecond range. The response time of 30 ps for most TCSPT (time-correlated single photon timing) set-ups is therefore often a limiting factor in time-resolved depolarization measurements.

The advantage of depolarization measurements over intensity measurements for quantification of FRET efficiencies is the independence of depolarization measurements from dynamic processes other than FRET that may cause fluctuations in emission intensity. Homo-FRET and depolarization measurements have been successfully applied for the study of structural arrangements of biomolecules (for instance lipid rafts) in the membranes of (living) cells [26–28].

6.3 Experimental Approach for Single-Pair FRET-Experiments

When aiming at elucidating FRET properties of a donor–acceptor pair, the most powerful approach is observing the process at the single-molecule level. By doing so, one is no longer hampered by heterogeneities, for instance, in the distance and orientation of the various FRET pairs, causing the analysis of the FRET signals to be fairly complicated. In fact, as the excited volume contains two dyes, a donor and an acceptor, the term single-pair FRET experiments is more correct than single-molecule FRET experiments. Single-pair FRET experiments allow for the monitoring of time-resolved dynamics of the chromophoric system and as there is only one donor–acceptor distance, it simplifies the analysis of the effect of FRET on, for instance, the donor excited state lifetime. A general scheme for such experiments is highlighted in Figure 6.8.

6.3.1 Single-Laser Excitation

A confocal microscope set-up with pulsed laser excitation and two detection channels is most convenient, as FRET efficiencies can be monitored by comparing the intensities in the two channels *and* by monitoring the lifetime of the donor excited state. In short, the chromophoric system is excited by a pulsed laser with a wavelength matching the absorption spectrum of the donor, but *not* of the acceptor. The emission light is sent via a dichroic mirror to the detection part of the set-up where a second dichroic mirror allows the donor emission to be separated from the more red-shifted acceptor emission. In this way, the intensity originating from both dyes can be

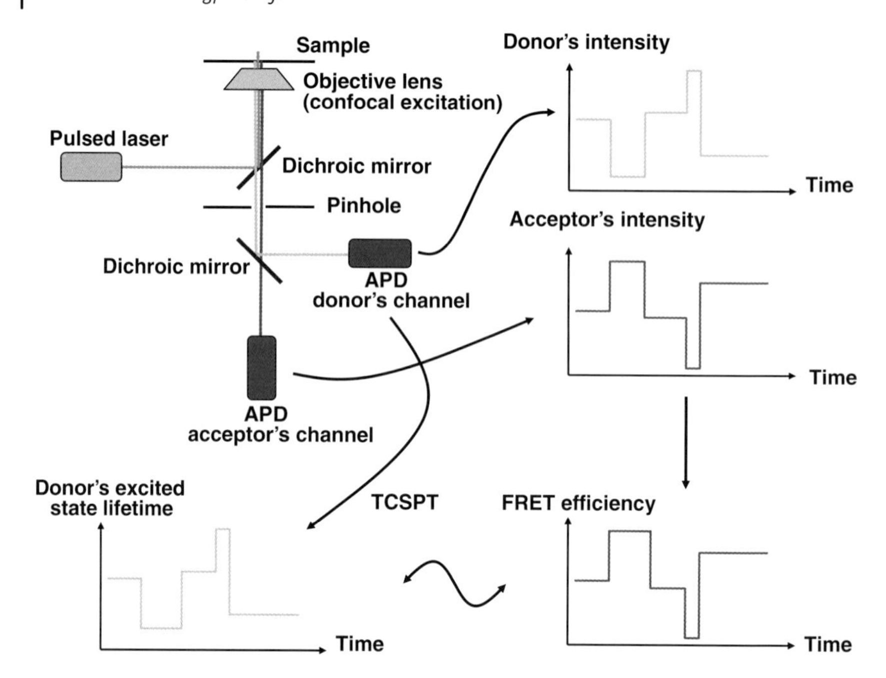

Figure 6.8 Scheme for single-pair FRET experiments using a confocal laser microscope and two detection channels. A dichroic mirror distinguishes between fluorescence from the donor and from the acceptor. In this way their respective intensities can be monitored separately (see simplified time traces), and the FRET efficiency can be calculated. Analysis of the time-resolved lifetime of the donor excited state must give a similar (but inverse) result.

separately monitored by using two avalanche photodiode (APD) detectors. Although in principle the FRET efficiency can be calculated in a straightforward way from the decrease in donor intensity upon FRET, as indicated in Equation 6.7, in real experiments strong intensity fluctuations may occur that do not originate from the energy transfer. Therefore, it is preferable to calculate E from the ratio of the acceptor fluorescence over the sum of the donor and acceptor intensities.

When measuring the acceptor fluorescence, one should keep in mind that, in most instances, the observed intensity in this channel does not originate exclusively from FRET-induced acceptor emission. Firstly, the long emission tail of the donor molecule often causes a small but considerable emission in the acceptor channel. Secondly, the acceptor absorption often shows short-wavelength tailing resulting in a (small) absorption cross-section at the donor excitation wavelength. Consequently, direct excitation of the acceptor may occur to a small extent, also leading to an increased non-FRET based emission level in the acceptor detection channel (Figure 6.9). Depending on the extent of these background phenomena, corrections need to be made to provide a reasonable estimation of the acceptor intensity originating from FRET. If it is possible to measure the emission spectrum over the total wavelength range of the donor and acceptor emissions, the precise FRET

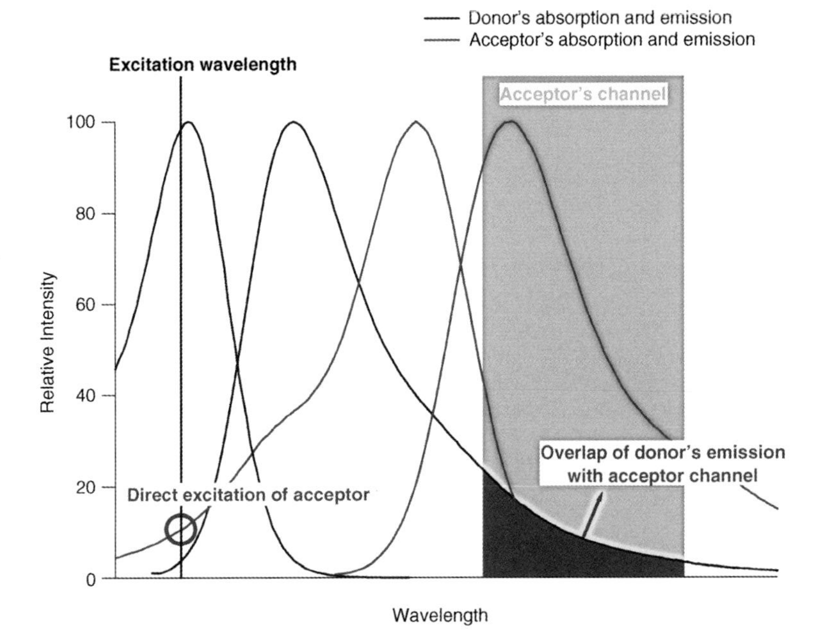

Figure 6.9 Origins of non-FRET induced emission in the acceptor channel. Short-wavelength tailing of the acceptor absorption results in direct excitation of the acceptor molecule. Also, long-wavelength tailing of the donor emission can overlap with the wavelength range of the acceptor channel.

efficiency can be calculated by curve fitting combined with complex numerical computer methods [29], based on the reasonable assumption that the shape of the emission spectra of the donor and acceptor do not change upon energy transfer. Alternatively, one can correct the acceptor intensity according to Equation 6.13.

$$I_{\mathrm{A,FRET}}^{\mathrm{corr}} = I_{\mathrm{A,FRET}} - \beta I_{D,\mathrm{FRET}} - \alpha I_{A}^{\mathrm{dir}} \tag{6.13}$$

Here the second term on the right-hand side of the equal sign corrects for the contribution from donor emission leaking into the acceptor channel (β is the crosstalk correction factor, defined by the ratio of donor signal detected in the acceptor channel to the donor signal in the donor channel). The last term corrects for direct acceptor excitation at the donor excitation wavelength (α is the ratio of acceptor intensity upon donor excitation, in the absence of donor to the acceptor intensity, when excited at the acceptor excitation wavelength).

In another approach the excited state lifetime of the donor molecule, which is also related to the FRET efficiency, can be calculated by time-correlated single photon timing (TCSPT). To measure depolarization effects, a similar set-up can be used. In this instance, however, the second dichroic mirror should be replaced by a polarized beam splitter to distinguish between the parallel and perpendicular polarized light.

Correlating the transfer efficiency with a precise interchromophoric distance is not always straightforward because of the uncertainty on the orientational factor κ^2. Indeed, for a single FRET pair, the exact orientation of donor and acceptor is often not known, and the typical estimation of 2/3 implies a certain error on the calculated distances. By combining Equations 6.7 and 6.4 it can be easily derived that

$$r = \sqrt[6]{\frac{\kappa^2}{2/3}} \cdot r_{\kappa^2=2/3} \tag{6.14}$$

The inverse sixth power of κ^2 indicates that the error on the orientation factor fortunately results only in minor changes to the calculated distance. Moreover, upper and lower limits can be obtained for κ^2 by estimating the donor and acceptor transition dipole orientation using polarization anisotropy measurements [30]. It can thus be concluded that FRET allows for accurate distance measurements in the nm regime relative to the calculated Förster distance, R_0.

6.3.2
Alternating-Laser Excitation (ALEX)

One of the problems with the single-laser approach is that it cannot discriminate between those situations where the distance between the two fluorophores happens to be larger than the few nanometers required for FRET, or where the donor fluorophore is present alone, as a result of incomplete labeling. When one is studying, for example, the interaction between different groups or parts of a molecule, then the following problem may occur. If one observes only donor and no acceptor emission, then one might wrongly conclude that there is no acceptor present, when in fact the particular conformation of the system might be such that the distance between the dyes is too large for significant energy transfer. Conversely, one might assume that this distance is too large, when in fact there is no acceptor present at all. Because the experiment does not discriminate between these scenarios there is a significant risk of misinterpretation of the data.

One might feel that a possible way to address this would be to use two lasers simultaneously, where one laser is tuned to the absorption band of the donor and the other to the absorption band of the acceptor. Thus we can obviously detect and separate the single fluorophores from the FRET pairs. This approach fails, however, when one wants to separate the FRET emission from the emission caused by the direct excitation of the emitter.

In fact, a solution to these problems can be found somewhere in between these cases. Similar to the previous idea, it makes use of two overlapping lasers of a different wavelength. However, both of the lasers are pulsed and synchronized such that the laser pulses are introduced intermittently, an excitation scheme known as alternating laser excitation (ALEX) [31] or pulsed-interleaved excitation (PIE) [32] (Figure 6.10). Moreover, by keeping track of which laser pulse induces which fluorescence photon, we can reliably distinguish between direct emission and energy transfer, and hence between scenarios of single emitters, single acceptors, and FRET

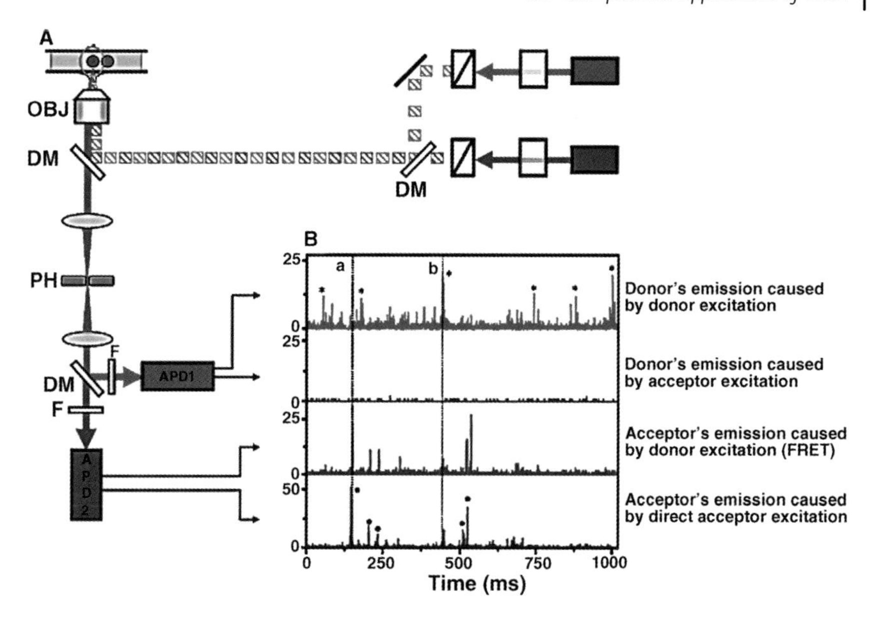

Figure 6.10 General scheme for alternating laser excitation microscopy. The sample is excited alternatingly by two lasers. This allows emission photons excited by the two lasers in each of the detection channels to be distinguished. DM, dichroic mirror; OBJ, objective lens; PH, pinhole; F, focusing lens; APD, avalanche photodiode detector. (Copyright Kapanidis *et al.* (2005) *Acc. Chem. Res.*, **38**, 523 [31].)

pairs with a "long" interdistance. Similar to other experiments, the FRET efficiency can be calculated by the (corrected) ratio of intensities of donor to acceptor as indicated in Equation 6.7. However in the case of alternating laser excitation, only the emission as a result of excitation of the donor channel should be considered. We can also define a distance-independent ratio *S*, which reports on the donor–acceptor relative stochiometry:

$$S = \frac{I_{D-\text{exc}}}{I_{D-\text{exc}} + I_{A-\text{exc}}} \tag{6.15}$$

where

$I_{D-\text{exc}}$ and $I_{A-\text{exc}}$ refer to the total emission as a result of excitation in the donor and acceptor absorption areas, respectively.

6.4 Examples and Applications of FRET

To prove the relevance of FRET in (multi–)chromophoric systems, some examples of FRET applications are outlined throughout this section. However, this list is by no means exhaustive and is only intended to give a short and clear overview of the possibilities of implementing FRET as a quasi non-invasive *in situ* tool for characterizing various types of processes. More examples of applications of FRET in biochemical research can be found in references [15, 33].

6.4.1
FRET Processes in Bulk Experiments

6.4.1.1 FRET-Based Molecular Biosensors

Sensing the presence of specific compounds or enzyme activity in living cells attracts a great deal of interest from biochemical researchers as this allows for an *in situ* real-time characterization of cellular properties. For this purpose, several FRET-based indicators have been developed over the last 15 years. A general scheme for such FRET-based molecular sensors is depicted in Figure 6.11: a FRET pair is linked through a peptide or protein with a specific binding domain for the target molecule. The interaction between this target and the binding domain triggers a conformational change in the peptide linker, resulting in a displacement of the two dyes with respect to each other. As this will affect the FRET efficiency, binding or unbinding of specific molecules can be probed *in situ*. By using fluorescent proteins such as (e)CFP and (e)YFP as the FRET pair, these biosensors are genetically encodable, thereby improving to a large extent their applicability in living cells [34].

This scheme is, for instance, successfully applied in an intracellular calcium indicator, termed a cameleon [35, 36]. The linker consists, in this case, of a calmodulin moiety and an M13-sequence. Binding of Ca^{2+} increases the affinity between calmodulin and M13, giving rise to the necessary conformational change. Similarly, a maltose sensor was constructed for studying maltose uptake in living yeast cells [37]. This process is of considerable importance in several food processes, such as the brewing of beer.

The group working with Ting used this sensor approach to investigate the ability of cells to perform post-translational modifications on histone peptides [38, 39]. Modification of histones is a key control point in gene transcription. Phosphorylation

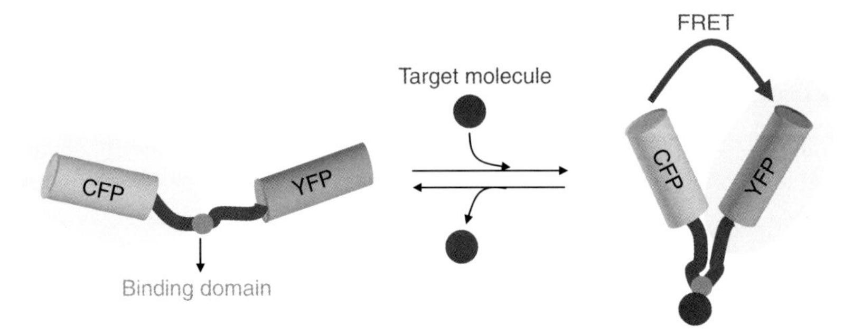

Figure 6.11 General scheme for FRET-based molecular sensors. Upon interaction of a target molecule (blue) with the binding domain of a specific bridging peptide, the donor (CFP) and acceptor (YFP) dyes move closer towards each other resulting in a high FRET efficiency.

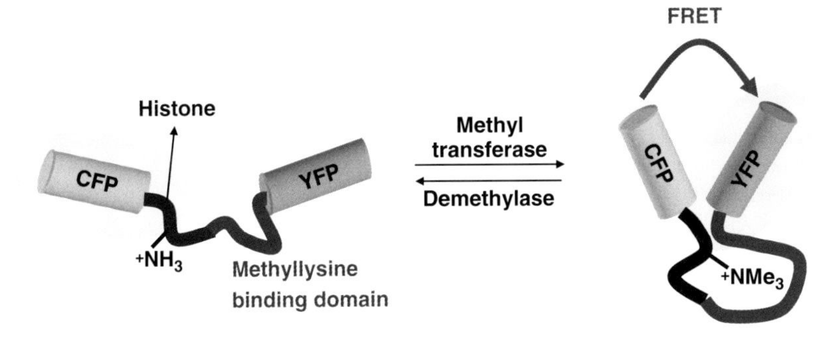

Figure 6.12 FRET-based molecular sensor for probing methyltransferase activity. Upon enzymatic methylation of the lysine residues in the histone moiety, the interaction with the methyllysine binding domain causes a conformational change resulting in enhanced FRET efficiency.

of serine-hydroxyl groups or methylation of lysine groups in the histones are two examples of such post-translational modifications. By incorporating the histone peptide and a phosphoserine or methyllysine binding domain in the linker between CFP and YFP, an efficient biosensor for phosphorylation and methylation, respectively, is produced. Indeed, the post-translational modification of the histone moiety induces an increased interaction with the methyllysine or phosphoserine binding domain, causing the FRET pair to move closer towards each other (Figure 6.12).

As formidable as this scheme may seem, one must take care when applying fluorescent proteins based on the GFP (green fluorescent protein) structure as reporter dyes. In bulk these dyes have a pronounced pH-dependency, which narrows their applicability to a well defined pH-range. For instance, the commonly used Förster acceptor (e)YFP with a pK_a of 7.0 only shows its bright fluorescence in the deprotonated form [40]. Moreover, GFP and its mutants have very complex photophysical properties, including excited state proton-transfer involving bright and dark states [41–44]. This complicates the interpretation of both bulky and single-molecule measurements of these dyes, and of the above mentioned cameleon sensors based on these dyes [45].

An extensive characterization of an ECFP–(e)YFP-based cameleon calcium indicator was performed by Habuchi *et al.* [40]. These workers could identify several subpopulations of the calcium indicator each with different FRET properties. (e)YFP was found to be an efficient acceptor molecule for Förster transfer only in its deprotonated form. Therefore, at a pH of 7.4 a first subpopulation comprising 28% of the total population with a FRET efficiency of zero was identified as the protonated (dark) state of the chromophore. For the calcium bound cameleons, three other subpopulations were found based on the folding of the calmodulin-M13 linker. In its completely stretched form (34% of the population) the transfer efficiency also equals zero. FRET can occur only via a compact form of the linker ($E = 0.62–0.67$; 23% of the population) or via a completely folded form with a transfer efficiency of 0.95 comprising 15% of the population. When no calcium is bound to the linker, only

Table 6.2 Contribution of the subpopulations of the ECFP-EYFP cameleon.

Transfer efficiency	Protonated form $E = 0$	Extended form $E = 0$	Compact form $E = 0.62{-}0.67$	Folded form $E = 0.95$
Ca^{2+}-bound sample	28%	34%	23%	15%
Ca^{2+}-free sample	28%	47%	25%	

the extended form (47%) and the compact form (25%) could be identified in the sample (Table 6.2). Their results proved the complexity of these systems, making them, at the same time, extremely interesting research objects in their own right. It also pinpoints the need for improving and developing new sensors composed of different GFP FRET pairs.

Note that the above mentioned examples all consist of reversible processes, in which the FRET probe can switch between the bound (high FRET efficiency) and the unbound (low FRET efficiency) state. However, one can also envision a scheme in which the linker is irreversibly cleaved by the action of the target molecule, causing the FRET pair to move away from each other. Efficient and accurate virus monitoring systems have recently been developed that can rapidly detect very low numbers of infectious enteroviruses in living cells by using a genetically expressed FRET-based indicator analogous to the one depicted in Figure 6.11 [46, 47]. In this case, the linker consists of a specific peptide that is exclusively cleaved by the viral 2A protease upon viral infection (Figure 6.13). The resulting loss of interaction between the donor and acceptor dye of the FRET pair leads to an increase in the donor fluorescence and to the loss of the acceptor fluorescence (Figure 6.14).

6.4.1.2 **Energy Hopping and Trapping in Chromophore-Substituted Polyphenylene Dendrimers**

Dendrimers are highly branched structures with a huge potential for use as building blocks in photonic devices [48]. Moreover, they are studied extensively because they

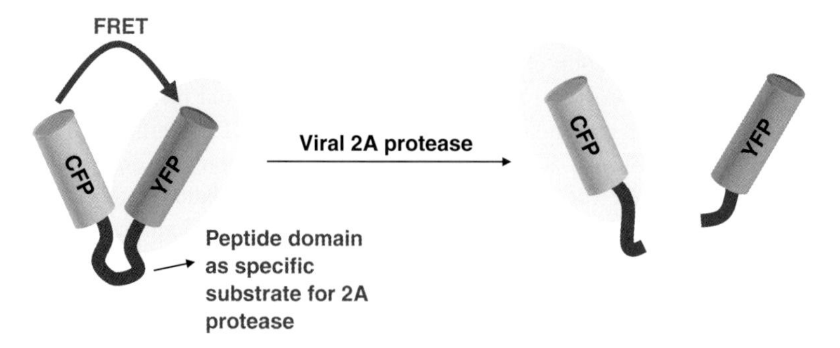

Figure 6.13 FFRET-based molecular sensor for probing enteroviral infection. Upon enteroviral infection, the viral 2A protease cleaves exclusively the linker of the genetically expressed FRET-based sensor, resulting in a dramatic decrease of FRET efficiency.

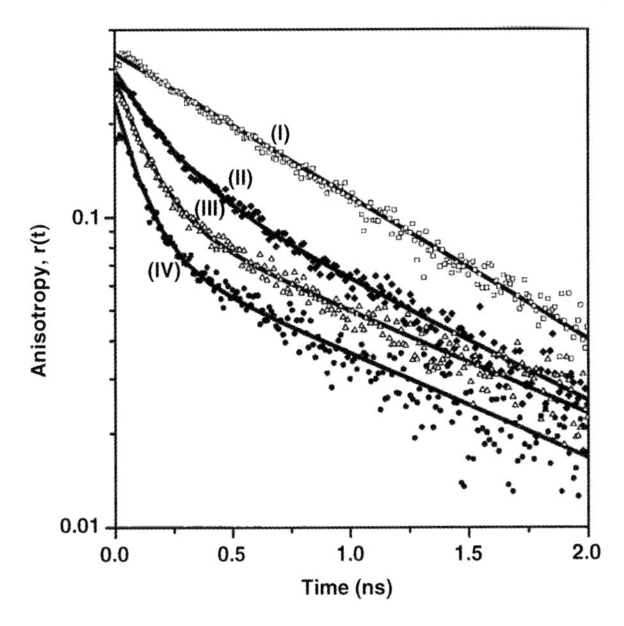

Figure 6.14 Anisotropy decay curves for G1R1 (I), G1R2 (II), G1R3 (III), and G1R4 (IV). The more chromophores are attached to the dendrimer, the more pronounced the fast decay component is. (Copyright Maus *et al.* (2001) *J. Phys. Chem. A*, **105**, 3961 [50].)

represent mimics of natural light antenna systems when decorated with chromophores, as will be elaborated on further in this chapter. Nowadays synthetic procedures are well established for creating a wide variety of symmetric or even asymmetric chromophore-derivatized dendrimeric structures [49].

Recently some detailed studies concerning intramolecular energy hopping and energy trapping in polyphenylene dendrimers with multiple chromophores were reported [50, 51]. The process of energy hopping was, for instance, studied by Maus *et al.* by using a first generation polyphenylene dendrimer with from 1 to 4 peryleneimide chromophores attached at the *meta*-positions of the outer phenyl group (G1Rx with $1 \leq x \leq 4$; Figure 6.15) [50]. The rigid tetrahedral central core ensures a spherical outer surface and thus a tetrahedral packing of the chromophores at equal distances. The energy hopping kinetics were analyzed by fitting the anisotropy decay curve according to the kinetic model given in the following equation:

$$r(t) = \frac{r_0}{i}\left[e^{-t/\theta_{\text{rot}}} + (i-1)e^{-\left(i \cdot k_{\text{hopping}} + \theta_{\text{rot}}^{-1}\right)t}\right] \tag{6.16}$$

where

i stands for the amount of peryleneimide dyes bound to the dendrimer (1–4)
θ_{rot} represents the rotational relaxation time
r_0 is the initial anisotropy.

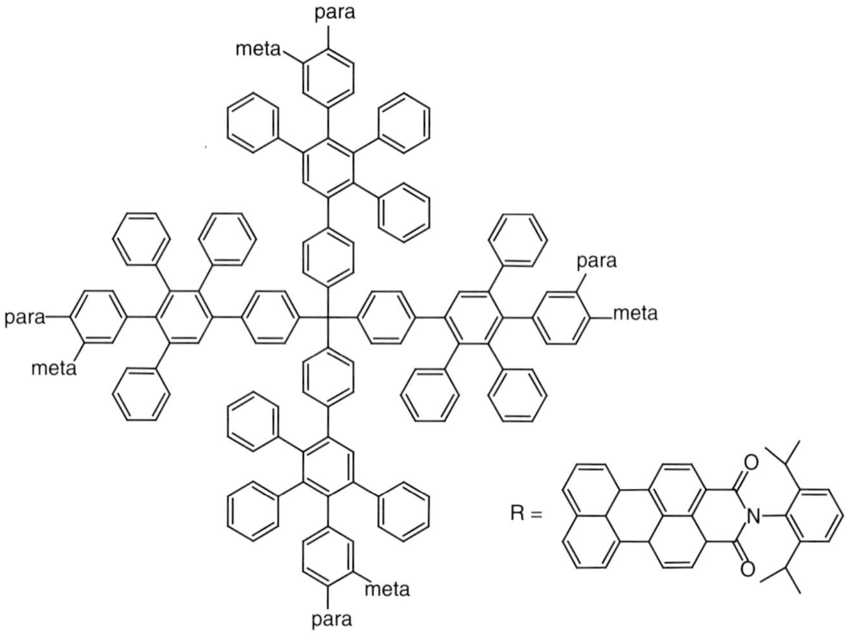

Figure 6.15 Structure of the first order polyphenylene dendrimer. The positions where the chromophores (R) are attached in the G1Rx series and in the G1R3Ter-p system are indicated by *meta* and *para*, respectively.

It is easily seen from the formula that at long times, in the absence of rotational diffusion the leveling-off value is r_0/i: the more chromophores are bound to the dendrimer, the more hopping occurs yielding a stronger anisotropy decrease. For the G1R1, no energy hopping is possible, and the anisotropy decays according to a single exponential with a time constant of approximately 1 ns, corresponding to the rotational relaxation. For the dendrimers containing more than one chromophore this rotational component has a slightly higher time constant because of the bigger molecular dimensions. Moreover, these molecules have a second fast component in the anisotropy decay with a time constant ranging from 200 ps for G1R2 to 110 ps for G1R4, which is within the typical FRET time range (Figure 6.14 and Table 6.3).

Table 6.3 Fitting parameters of the anisotropy decay curves for G1R1–G1R4.

Compound	r_0	θ_{rot} (ns)	$k_{hopping}$ (ns^{-1})	$(i-1)/i$[a]
aG1R1	0.38	0.95		
G1R2	0.31	1.1	2.05	48% (50%)
G1R3	0.28	1.2	2.29	63% (67%)
G1R4	0.24	1.3	2.08	66% (75%)

a) The values between the parentheses in the column of $(i-1)/i$ correspond to the theoretical values.

Most of the data were in excellent agreement with the theoretical kinetic model, although for the G1R4 a small deviation from the theoretical value was seen in the contribution of the FRET effect in the anisotropy decay. According to Equation 6.16 this contribution is given by $(i-1)/i$ and thus theoretically equals 50, 67 and 75% for G1R2, G1R3, and G1R4, respectively. The respective experimental results were 48, 63 and 66%. These workers attributed the smaller contribution for G1R4 to the formation of a preformed dimer (excimer) that acts as an energy trap, thereby preventing further energy hopping. This hypothesis was further supported by the observation of a second decay component in the fluorescence lifetime of the multi-chromophoric dendrimers. Fast energy trapping by an excimer can also explain the lower starting anisotropies (r_0) for dendrimers containing more peryleneimide chromophores: the anisotropy decay time of this very fast process is below the experimental time resolution of 10 ps.

In a similar study, energy trapping was investigated together with energy hopping in a polyphenylene dendrimer with three peryleneimide moieties as energy donors and one terryleneimide as energy acceptor [51]. To minimize the possibility of excimer formation, as was probably the case in the above mentioned G1Rx systems, the chromophores were attached to the *para-* instead of the *meta*-positions of the outer phenyl ring (Figure 6.15). This structure is subsequently abbreviated as G1R3Ter-p. Owing to the large spectral overlap between the terryleneimide absorption and the peryleneimide emission, the terryleneimide moiety can be expected to be a very efficient energy sink for the energy harvested by the peryleneimides. Through global analysis performed on the time-resolved emission spectra of the dendrimeric system, these workers found evidence for two FRET-pathways from the peryleneimides to the terryleneimide. In Table 6.4 the global fit results for the fluorescence decay at emission wavelengths of 600 and 750 nm, corresponding to, respectively, the donor and the acceptor emission, are summarized. The parameters τ_1 and τ_2 correspond to the decay of the perylene dyes as a result of energy transfer and are thus visible as decay components at 600 nm (emission wavelength of peryleneimide) and as rise components at 750 nm (emission wavelength of terryleneimide). τ_3 represents the decay of the terryleneimide acceptor and thus its contribution in the 600 nm channel is zero. The observation of two decay components for the donor molecule indeed supports the existence of two transfer pathways. As the contribution of the slow and the fast component are, respectively, 2/3 and 1/3 in the donor channel,

Table 6.4 Fitting parameters by global analysis of the fluorescence decay at 600 and 750 nm for G1R3Ter-p. At 600 nm only fluorescence from peryleneimide is detected, while the decay at 750 nm only contains fluorescence from the terryleneimide chromophore.

			λ_{fluo} = 600 nm			λ_{fluo} = 750 nm		
τ_1 (ns)	τ_2 (ns)	τ_3 (ns)	α_1 (%)	α_2 (%)	α_3 (%)	α_1 (%)	α_2 (%)	α_3 (%)
0.052	0.175	2.51	62.1	37.9	0	−30.9	−15.7	53.4

and $-2/3$ and $-1/3$ in the rise part of the acceptor channel, it is suggested that one of the three peryleneimides takes a much better through-space orientation or a smaller interchromophoric distance towards the terryleneimide acceptor compared with the other two peryleneimide chromophores.

These results obtained from bulk FRET experiments revealed the extent of the information that can be gained for synthetic light-harvesting systems. In the next section an example is given on how single-pair FRET studies can be applied to examine similar light-harvesting chromophore associations in even more detail.

6.4.2
Single-Molecule Observation of FRET

6.4.2.1 Light-Harvesting Systems: Phycobilisomes and Allophycocyanins

Photosynthesis is by far the most well known light-induced chemical reaction in nature. Essentially, solar energy is used by plants and several microorganisms as a driving force for biochemical reactions. In most organisms this reaction is initiated by the excitation of a special pair chlorophyll molecule with a maximal light absorbance in the far-red to near-infrared region. However, the relatively small absorption band of this chlorophyll can only collect a minor fraction of the incident sunlight. Therefore nature developed a complex light-harvesting system (partially) based on FRET to direct higher energy photons towards the active special pair chlorophyll. For instance, normal chlorophylls and carotenes that absorb at lower wavelengths (visible part of the solar spectrum) can serve as light antennae and transfer their excitation energy to the reaction center, thereby greatly enhancing the efficiency of the photosynthetic system. A similar mechanism is found in cyanobacteria and red algae that live deep in the ocean. As blue and red light have a very limited penetration depth in sea water, due to absorption by the water itself and by microorganisms in the top layer (this also causes the greenish color of the ocean), the yellow and green light needs to be efficiently converted into red light by the marine photosynthetic organisms. In this case, the job is done by the phycobilisomes containing phycoerythrins and phycocyanins organized in rods and a core of allophycocyanins. The rods surrounding the core serve as antennae that catch yellow and green light and direct this light by FRET processes towards the reaction center (Figure 6.16).

The allophycocyanin (APC) dyes themselves have been the subject of extensive research because this protein is actually a trimer of which each monomer contains two dyes, $\alpha84$ and $\beta84$ (Figure 6.17). Therefore this protein is a multichromophoric system with interesting properties of its own. Both dyes have the same chemical structure but the $\beta84$ chromophore has a strong steric and electronic interaction with a nearby tryosine residue. This interaction extends the conjugation of the chromophore and causes a small shift to the red of its absorption and emission spectrum compared with $\alpha84$. Within one monomer the distance between $\alpha84$ and $\beta84$ is roughly 5.1 nm. In the trimer, however, there is a closer distance of about 2.1 nm between the $\alpha84$ and $\beta84$ chromophore of the neighboring monomers (Figure 6.17) [52].

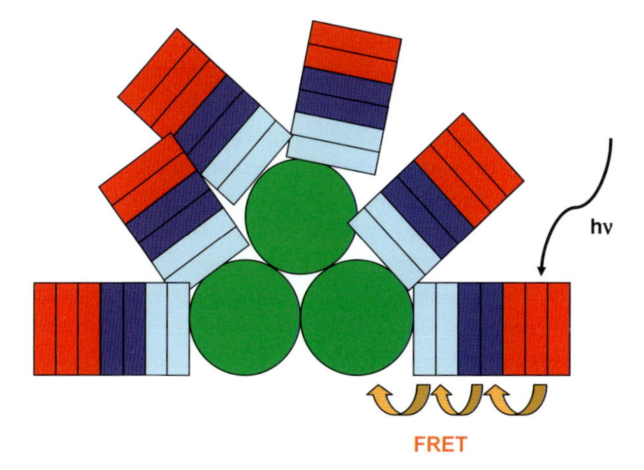

Figure 6.16 Schematic representation of phycobilisomes as a light harvesting system. The red and the blue parts represent rods containing phycoerythrins and phycocyanins, respectively. They tunnel yellow and green light through FRET towards the center, containing allophycocyanins (green circles).

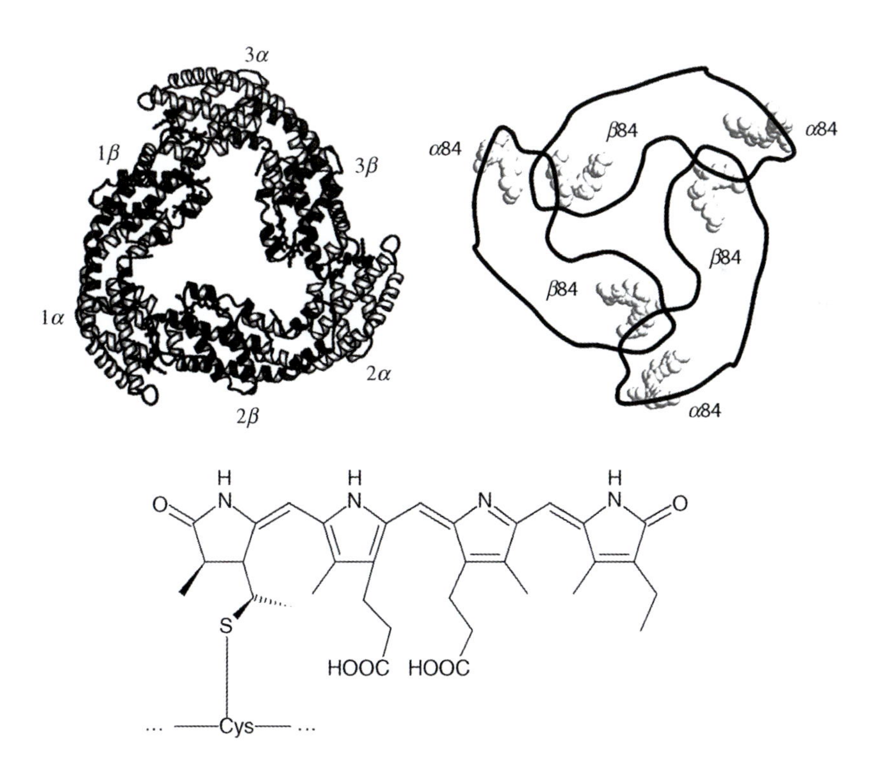

Figure 6.17 The structure of allophycocyanins (APC). APC is a trimer protein, in which the monomers contain two chromophores, $\alpha84$ and $\beta84$. In the trimer the $\alpha84$ dyes interact closely with the nearby $\beta84$ chromophore of the neighboring monomer. The structure of these chromophores is drawn in the lower part of the figure.

There is still a lot of uncertainty about the nature of the interaction between these close lying chromophores. Some workers claim to have proof of strong excitonic coupling between the dyes [53, 54]. This would imply that the excitation energy is spread out over both chromophores and is therefore completely delocalized. The system then behaves as a single absorber (and emitter). On the other hand, Loos *et al.* recently reported single-molecule fluorescence data that suggests only weak coupling between the absorbers, this being a Förster type of interaction [55]. α84 then acts as an energy donor that transfers its excitation energy to the more red-shifted β84. A transfer efficiency of almost 99.5% was calculated, based on a Förster distance of 5 nm, as reported by Brejc and coworkers [52]. The existence of up to six intensity levels in the fluorescence intensity trajectories of single APC trimers indeed suggests the presence of six distinct absorbers rather than three, as would be the case in the molecular exciton model (Figure 6.18).

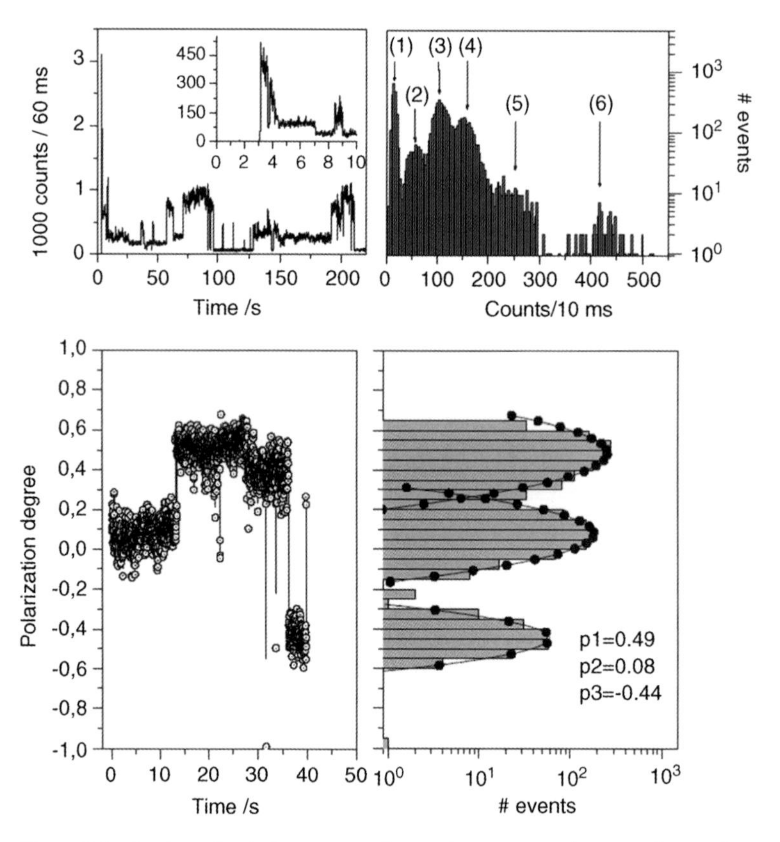

Figure 6.18 Single-molecule fluorescence data on single APC trimers. Upper graphs: the fluorescence time transient of a single APC trimer shows six distinct intensity levels, indicating the existence of six absorbers. Lower graphs: the time transient of the emission polarization of a single APC trimer shows discrete jumps between three polarization values. The well-defined polarization values are an indication of efficient energy hopping to one of the three acceptor dyes that serves as an emissive trap. (Copyright Loos *et al.* (2004) *Biophys. J.*, **87**, 2598 [55].)

Aside from heterotransfer between $\alpha84$ and $\beta84$, also homotransfer or energy hopping can be seen in this multichromophoric system. The individual $\beta84$ molecules are located at a center-to-center distance of about 3.1 nm, which is smaller than the Förster radius for homotransfer of 4.5 nm as calculated by Loos *et al.* Experimental proof for this energy hopping was found by measuring time transients of the emission polarization p defined by

$$p = 2\cos^2(\theta) - 1 \tag{6.17}$$

where

θ stands for the angle between the transition dipole moment of the dye and the direction of the electric field vector of the plane polarized light.

In Figure 6.18 discrete jumps between three states of this polarization p can be identified. This means that for a certain time (in the seconds range) emission always originates from the same chromophore that acts as an emissive trap for the others. From time to time the emissive trap is shifted to one of the other acceptor molecules. The existence of mainly three emitters also proves that energy is very efficiently transferred from the donors ($\alpha84$) to the acceptors ($\beta84$). Energy hopping between these acceptors subsequently tunnels the energy towards one of the $\beta84$ molecules that serves as an emissive trap.

The allophycocyanine system has also been studied at the single-molecule level by using an anti-Brownian electrokinetic (ABEL) trap, thereby minimizing the effect of surface immobilization of individual allophycocyanine molecules [56].

Inspired by the beauty of nature, many attempts have been made to mimic light-harvesting properties in synthetic multichromophoric systems [57–59]. For instance in a T2P8 dendrimer the energy from the excited perylenemonoimide (PMI) moieties at the outer rim is transferred with very high efficiency to the central terrylenediimide (TDI) (Figure 6.19).

Extensive single-molecule studies on this system have revealed detailed information on the different types of energy transfer mechanisms within this light-harvesting system [60]. Figure 6.20 shows typical emission trajectories of two single T2P8 molecules upon pulsed excitation with emission detection in two channels: a green one detecting the donor fluorescence and a red one detecting only the acceptor fluorescence. The upper graph is a time transient representative of the majority of molecules: initially no emission from the donor molecules is seen because of very efficient energy transfer to the central TDI molecule. Fast energy hopping among the different donor molecules ensures that even when an unfavorably oriented donor molecule is present, the excitation will still end up in the acceptor molecule. The stepwise decrease of the acceptor fluorescence is the result of successive bleaching of the different donor molecules. Only after bleaching of the acceptor can the donor emission (green) be observed. The inset shows the fractional emission of the acceptor, which equals 1 for the non-bleached acceptor, or 0 after bleaching of the acceptor.

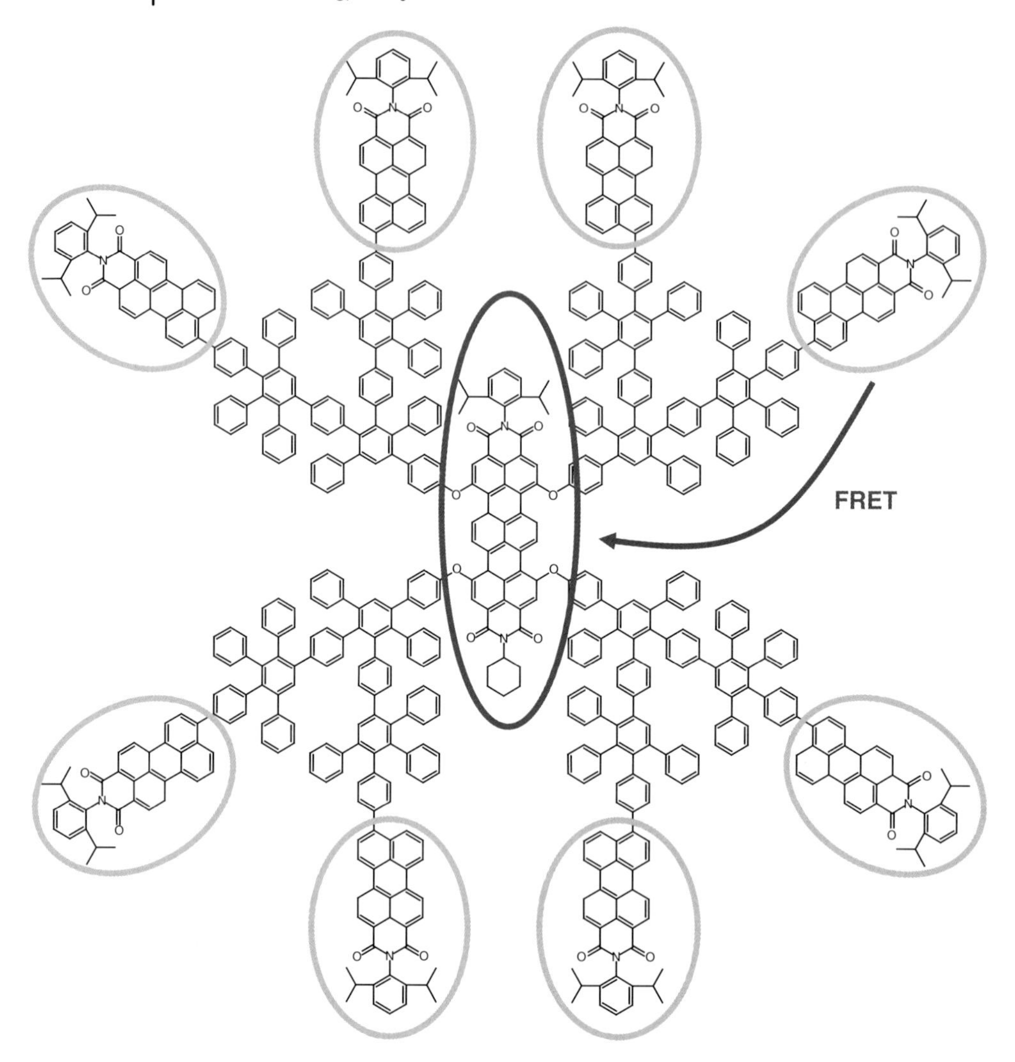

Figure 6.19 Structure of T2P8 as a synthetic light-harvesting system. The eight perylenemonoimides at the outer rim transfer their excitation energy very efficiently to the central terrylenediimide by Förster energy transfer.

However in 10–15% of the molecules studied, simultaneous emission from donor and acceptor is seen after bleaching of some of the donor molecules, yielding intermediate fractional emission values for the acceptor. This observation was previously attributed to the so-called exciton blockade phenomenon [61]. This means that when multiple donor molecules are excited by the same laser pulse, only one of them can transfer its energy to the acceptor as the excited acceptor cannot serve as the "acceptor" for an additional excitation. Because of the very similar fluorescence lifetimes of PMI and TDI, they will both emit their fluorescent photons at very similar

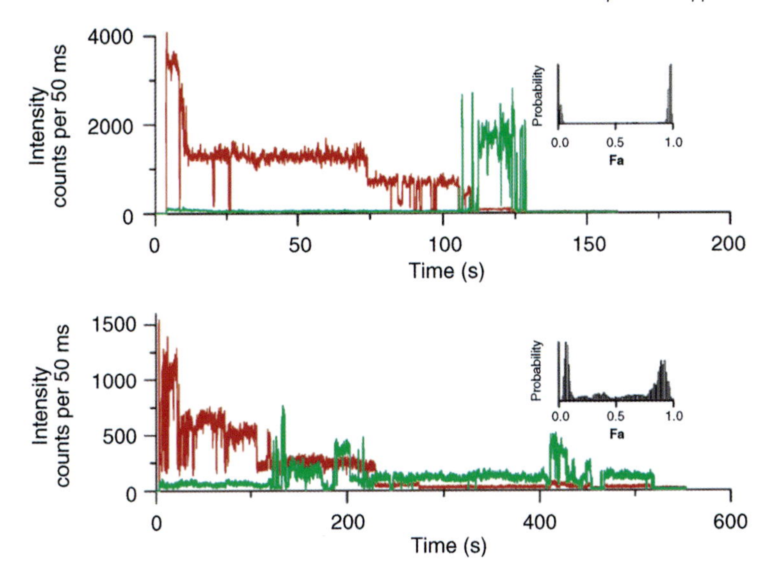

Figure 6.20 Typical fluorescence transients of T2P8 molecules. The red and green curves represent, respectively, the acceptor (TDI) and the donor (PMI) emission. The inset shows a histogram of the fractional intensities of the acceptors. The upper graph corresponds to the majority of molecules, in which donor emission is only seen after bleaching of the acceptor. The lower graph represents a minor group (10–15%) that shows simultaneous emission from donor and acceptor. The stepwise intensity decrease in the red channel is due to bleaching of the donor molecules. (Copyright Melnikov *et al.* (2007) *J. Phys. Chem. B*, **111**, 708 [60].)

times, yielding highly correlated emission photons. However, recent single-molecule investigations proved that this explanation is not very plausible for several reasons. Firstly, the simultaneous existence of excited PMI and excited TDI would result in a very efficient singlet–singlet annihilation, which would out-compete the fluorescence of PMI. Secondly, simultaneous emission of donor and acceptor under continuous-wave (CW) excitation was observed to the same extent as under pulsed excitation, although the probability of exciting two donor molecules simultaneously is extremely low under CW excitation. Thirdly, simultaneous excitation of different PMI molecules becomes less probable as more PMI molecules are bleached, which is inconsistent with the observation that simultaneous emission of PMI and TDI only occurs after bleaching part of the donor molecules. Finally, cross-correlation between the arrival times of the photons emitted by the donor and the acceptor molecules yields three equally intense peaks at time lags zero, $-t_{\text{pulse}}$, and $+t_{\text{pulse}}$ (t_{pulse} is the time between two successive excitation pulses (Figure 6.21). This cross-correlation profile corresponds to two non-correlated emitters, whereas in the case of the exciton blockade two correlated emitters are present, which would result in a much higher peak at zero time lags.

Through defocused wide-field imaging [62] (Figure 6.22), revealing the orientation of the emitter transition dipole, it was found that in the case of simultaneous

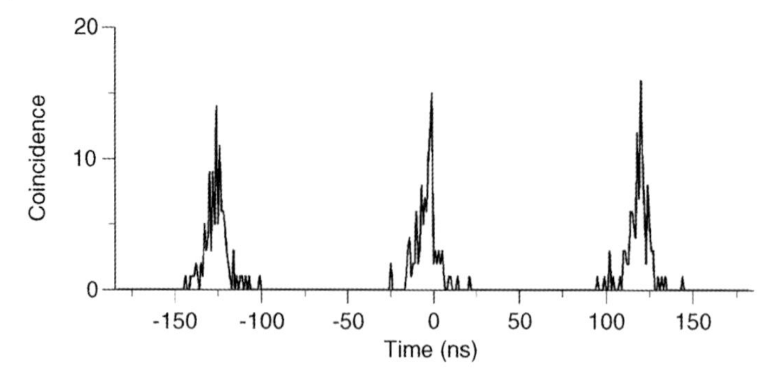

Figure 6.21 Cross-correlation curve between the donor and acceptor emission. Three equally intense peaks at time lags 0, −122 and + 122 ns indicates that both chromophores have an uncorrelated emission behavior. A time lag of 122 ns corresponds with the inter-pulse time of the excitation. (Copyright Melnikov *et al.* (2007) *J. Phys. Chem. B*, **111**, 708 [60].)

green and red emission, the two corresponding dipole moments were oriented perpendicular with respect to each other, yielding a FRET orientation factor κ of 0. Based on this observation, the hypothesis was suggested that after bleaching part of the donor molecules, some unfavorable oriented PMI chromophores will become "isolated" and thus will not undergo FRET nor singlet–singlet annihilation with the acceptor. The only relaxation pathway for such an isolated donor is the emission of a fluorescence photon. The fact that only a small fraction of molecules displays this behavior can be explained by the existence of different conformational isomers.

As stated previously, while energy transfer is usually considered between an excited donor and an acceptor in the ground state, other scenarios (e.g., excited donor transfer to excited acceptor or excited acceptor in the triplet state) are also possible. In single-molecule experiments on multichromophoric systems these other energy transfer pathways can be easily visualized and identified. Experimentally this was demonstrated for the tetra-PMI-substituded dendrimer discussed previously [50, 51]. If the multichromophoric system undergoes optical excitation under high photon flux, two excited states (S_1) can be present in a single molecule at the same time. If the S_1–S_0 transition of one chromophore is in resonance with a transition of S_1 of the other chromophore to a higher excited singlet state, that is, an S_1–S_n transition, energy transfer between the excited singlet states can occur. The process results in only one excited state remaining in the multi-chromophoric system and is often referred to as singlet–singlet annihilation. As depicted in Figure 6.23, there is a large overlap between the S_1–S_n absorption spectrum for PMI (the singlet absorption spectrum was measured by femtosecond transient absorption measurements). The presence of this process has been proven at the ensemble level, by means of femtosecond fluorescence upconversion and time-resolved polychromatic femtosecond transient absorption measurements [65]. At the single-molecule level, singlet–singlet annihilation ensures that multichromophoric dendrimer systems

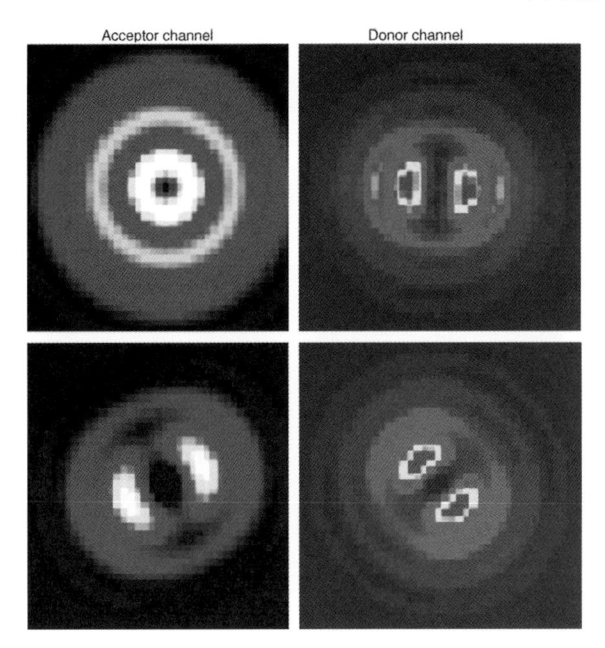

Figure 6.22 Defocused wide-field images of two T2P8 molecules showing simultaneous donor (right images) and acceptor (left images) emission. The first molecule (upper two images) has an out-of-plane orientation of the acceptor transition dipole, while that of the donor is perfectly in-plane. The second molecule has in-plane but perpendicular dipole moments for donor and acceptor. In both cases κ equals zero. The images shown are matching simulated patterns of the detected images. (Copyright Melnikov *et al.* (2007) *J. Phys. Chem. B*, **111**, 708 [60].)

act as excellent single-photon sources. Indeed, increasing the number of chromophores at the nanometer scale results in a higher absorption cross-section. In the case where pulsed laser light is used (higher excitation powers), just as in solution, multiple excitations can be generated per pulse. With efficient singlet–singlet annihilation, which depends on the spectral characteristics of the chromophores used and the distance between them, the process represents a feedback mechanism that ensures the emission of only one photon per pulse, even though multiple excitations occurred. This concept has been used to develop deterministic single-photon sources that are sought after for application in quantum encryption and so on [63, 66–72].

The time traces of the fluorescence intensity, the fluorescence decay time, and the interphoton arrival time distribution, measured from a G1R4 molecule embedded in zeonex film, are shown in Figure 6.24. In addition, the time trace of the ratio of the central peak (N_C) to the average of the lateral peaks (N_L), that is, N_C/N_L, is given in Figure 6.24b. Details on how N_C/N_L and interphoton times are determined experimentally can be found in reference [67]. Several intensity levels in the fluorescence intensity time trace (Figure 6.24b) indicate successive

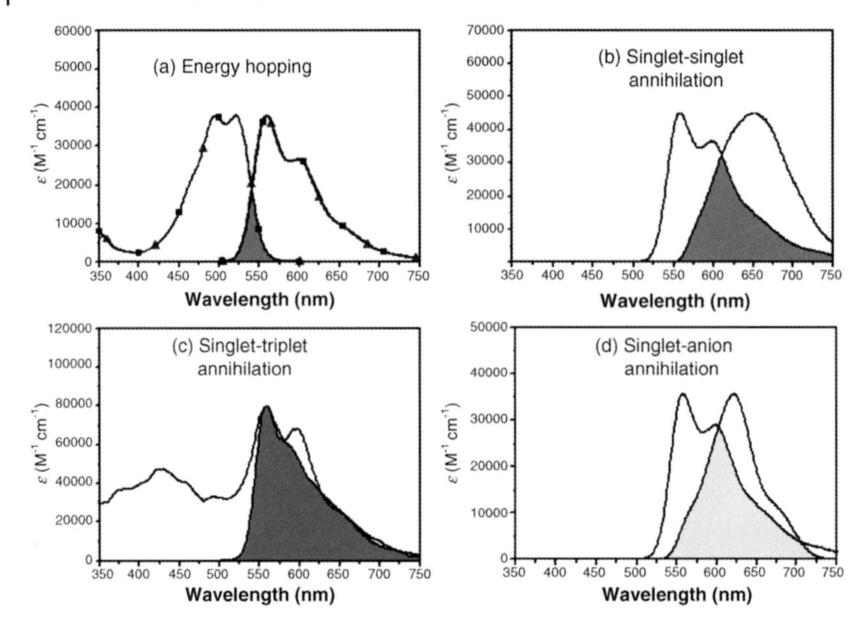

Figure 6.23 Four Förster type energy transfer processes that can occur in PMI-substituted polyphenylene dendrimers. (a) The spectral overlap of the PMI S_0 absorption and S_1 emission spectrum responsible for energy hopping among PMI chromophores. (b) The spectral overlap of the PMI S_1 absorption and S_1 emission spectrum responsible for singlet–singlet annihilation among PMI chromophores. (c) The spectral overlap of the PMI T_1 absorption and S_1 emission spectrum responsible for singlet–triplet annihilation among PMI chromophores. (d) The spectral overlap of the PMI radical anion absorption and S_1 emission spectrum responsible for singlet–radical anion annihilation among PMI chromophores [63, 64]. (Copyright Weil et al. (2010) *Angew. Chem. Int. Ed.*, doi 10.1002/anie 2009 02532.)

photobleaching of individual chromophores. As can be seen in Figure 6.24a, the fluorescence decay time shows a constant value of around 4 ns, which is characteristic of peryleneimide. Thus, the decay time trace indicates the absence of stronger interchromophoric interactions (excimer type interactions) in G1R4. Note that for all intensity levels, intensity drops to the background level can occasionally be observed. The origin of these on/off jumps, even when multiple chromophores are active, will be explained in the next paragraph. Figure 6.24c shows the interphoton arrival time distribution taken from the first intensity level of the time trace of G1R4 in Figure 6.24b, and Figure 6.24d shows an attenuated excitation laser beam (Ti:sapphire laser) that had approximately the same count rate. The center peak of the distribution at the time, $t = 0$ ns, corresponds to photon pairs induced by the same excitation laser pulse. In all other cases, interphoton arrival times are distributed around a multiple of the laser repetition period; that is, the peak appears every 122 ns. The peak width is determined by the fluorescence lifetime, and it can be seen that the peak width of G1R4 is broader than that of the

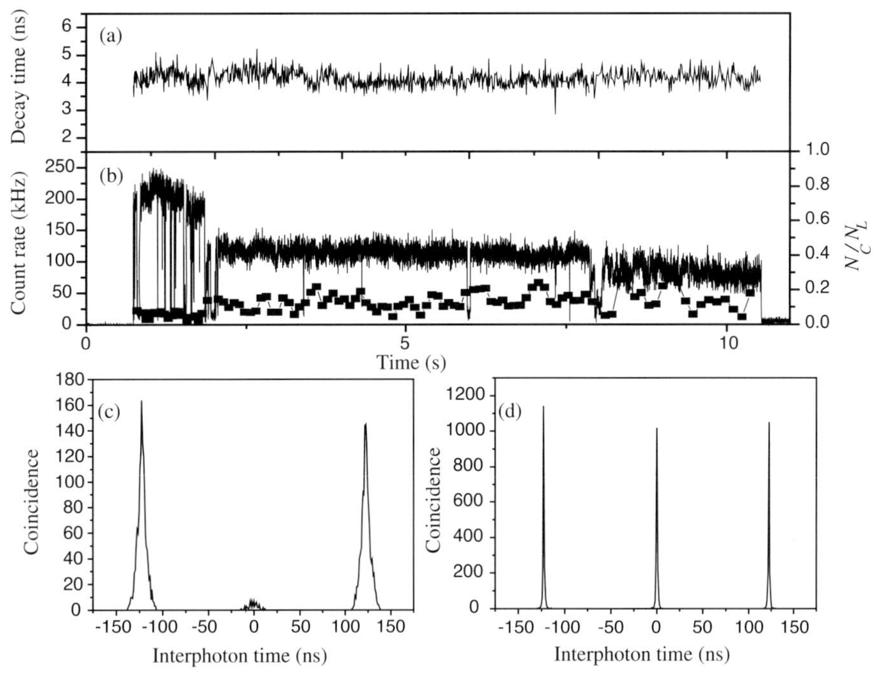

Figure 6.24 Typical time traces of fluorescence intensity (b, solid line), fluorescence decay time (a), and the ratio of the central peak to the average of the lateral peaks (N_C/N_L; b, squares) measured from a single G1R4 embedded in zeonex. (c) Interphoton arrival time distribution obtained from the first fluorescence intensity level of the time trace in panel (b). (d) Interphoton arrival time distribution measured from an attenuated Ti:sapphire laser. (Copyright Masuo *et al.* (2004) *J. Phys. Chem. B*, **108**, 16686 [67].)

pulsed laser beam. It has been shown, previously, that the N_C/N_L ratio can be used to estimate the number of independently emitting chromophores [73]. Neglecting background, N_C/N_L values of 0.0 and 0.5 are expected for one and two emitting chromophores, respectively. For the distribution shown in Figure 6.24c, N_C and N_L are 66 and 1376.5, respectively, giving an N_C/N_L value of 0.05. This indicates that G1R4 behaves as a single-photon emitter as the signal-to-background ratio means that the coincidences in the center peak can be attributed to background–background and background–signal photon pairs [72]. On the other hand, as can be seen in Figure 6.24d, the central peak of the interphoton arrival time distribution taken from the attenuated Ti:sapphire laser is identical in intensity to the lateral peaks. This distribution indicates typical Poissonian light behavior. A step-by-step analysis of the N_C/N_L ratio calculated every 20 000 photons is shown in Figure 6.24b. The consistently low value of the ratio proves that G1R4 behaves as a single-photon emitter throughout the trace.

Direct excitation of one of the chromophores in a multichromophoric assembly can also lead to intersystem crossing (ISC) of that chromophore to the triplet state.

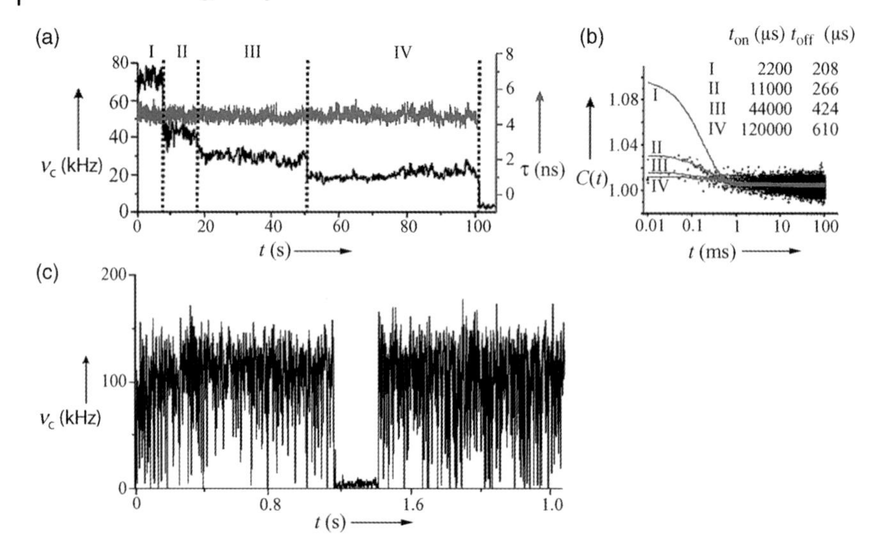

Figure 6.25 (a) Fluorescence intensity [count rate c (kHz) versus time t (s)] and lifetime [fluorescence lifetime, τ (ns) versus time t (s)] trajectory of a single molecule of the G1R4 dye. (b) Second-order intensity correlation of the various intensity levels clearly demonstrating different off times (triplet related) for the various intensity levels. (c) Expanded view of the first intensity level of a single molecule of G1R4, demonstrating short off times (around 200 μs, triplet related) and one long off event (400 ms, related to the formation of an anion). (Copyright Copyright Weil et al. (2010) *Angew. Chem. Int. Ed.*, doi 10.1002/anie 2009 02532.)

Because of the relatively long lifetime of the triplet state (for polymer-immobilized single molecules up to several hundreds of microseconds), a second chromophore can be excited while a first chromophore is still in the triplet state. As a result, two excited states are present, an S_1 and a T_1 state. If the triplet state exhibits transitions into higher excited triplet states, T_n, that are in resonance with the S_1 to S_0 transition, energy transfer from the excited singlet state to the energetically lower lying triplet state can occur [63]. This process is often called singlet–triplet annihilation. Such a process is not easy to see in bulk experiments but manifests itself directly in single-molecule fluorescence trajectories as collective on/off steps [63]. Figure 6.23 shows the spectral overlap of this process for the structure represented in Figure 6.15, and in Figure 6.25 an example is presented for collective on/off jumps. Note that careful analysis of the on/off jumps corresponding to different intensity levels in multichromophoric systems can result in quantitative information on the intersystem crossing and even on intersystem crossing of higher excited states [68]. Besides the above mentioned on/off blinking process, there is an additional even rarer event leading to longer fluorescence intermittence than the triplet off times. The cause of this is the formation of a radical anion on one of the PMIs. This radical anion can quench the fluorescence of the other PMIs due to energy transfer. Figure 6.23d shows the spectral overlap of this process and an example of a long collective off event is presented in Figure 6.25c.

6.4.2.2 **Hairpin Ribozyme Dynamics and Activity**

The pronounced distance dependency of FRET at r close to R_0 (Equation 6.4 and Figure 6.4) allows for probing distances in the nm-regime between fluorophores with high accuracy. This has recently been exploited in studying the conformational dynamics of several biomolecules. By selectively labeling the target with a FRET pair in combination with single-molecule detection techniques, dynamics can be probed with high time resolution. Two examples of this scheme will be highlighted: this section deals with the recent observation of single ribozyme activity through FRET, while the next section gives a short overview on studies on protein folding and unfolding.

Hairpin ribozyme is a two-strand RNA molecule with catalytic activity towards the cleavage of other single-strand RNA molecules. As depicted in Figure 6.26 the hairpin conformation of the ribozyme is drastically altered in the course of the catalytic cycle. Substrate binding induces the ribozyme to adapt its undocked or open conformation, which is in equilibrium with the closed or docked conformation. This docked state is stabilized by tertiary interactions between loops A and B as shown in Figure 6.26. Cleavage occurs in the docked state, after which undocking is necessary for the products to unbind and diffuse away.

Zhuang *et al.* used FRET from a Cy3 molecule attached at the 3′-terminal of this ribozyme to a Cy5 molecule attached at the 5′-terminal to probe these conformational changes [74]. The distance between the FRET pair is then approximately 8 nm for the undocked state, 4 nm for the docked state, and 6 nm for the free state (without substrate). The Cy3–Cy5 pair with its characteristic Förster distance R_0 of about 5.4 nm is particularly well suited for probing distances in this range. However, one has to be cautious in interpreting results from a Cy3–Cy5 FRET pair, as it has recently been demonstrated that these dyes show complex photophysical properties, such as reversible photoswitching [75, 76]. By monitoring the ratio of intensities of both dyes in a single molecule set-up with two-color detection, the FRET efficiency – and thus

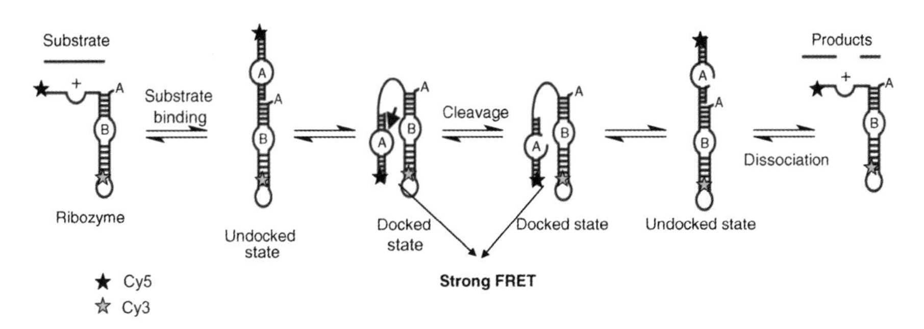

Figure 6.26 Schematic representation of the catalytic cycle of hairpin ribozyme. During the catalytic cycle a docked conformation is formed, which is stabilized by tertiary interactions between loops A and B. The conformational dynamics are monitored by labeling the two ends of the ribozyme with a Cy3 (grey star)–Cy5 (black star) FRET pair. The docked conformation can be identified by its high FRET efficiency, as the dyes are in close proximity in this state.

from this the conformational dynamics – can be calculated. In this way the docked state can easily be identified, because in this state the two dyes are in very close proximity giving rise to very high FRET efficiencies. Analysis of the dwell times between subsequent docked and undocked states revealed a heterogeneity that suggests the existence of four different docked states, of which one was predominantly present. Bulk measurements so far could only prove the existence of this one dominant docked state.

6.4.2.3 **Protein (Un)folding and Dynamics**

Recently a lot of attention has been paid to the study of folding and unfolding processes in proteins and DNA macromolecules by means of single-pair FRET fluorescence microscopy [77, 78]. The underlying concept is based on the fact that upon unfolding of a protein (often caused by denaturation) the distance between the residues increases drastically. This is, for instance, clearly elucidated for the ribonuclease H, labeled with a donor and acceptor dye [79]. A cartoon model of this enzyme with the localization of the FRET pair is given in Figure 6.27. For this study the denaturation process was induced by addition of guanidinium chloride. By measuring the interchromophoric distance for many molecules individually at different concentrations of the denaturing agent, these workers could extract information on the distribution of folded and unfolded states. As expected, they found a structurally well defined folded state, evidenced by the narrow distribution of E. However, the distribution of the transfer efficiency of the unfolded state is much broader, indicating a lack of structure for this denatured conformation. Moreover, the mean distance between donor and acceptor increases gradually upon increasing the guanidinium chloride concentration. Apparently in this type of denaturation, the protein backbone gradually expands more as the concentration of denaturing agent is increased.

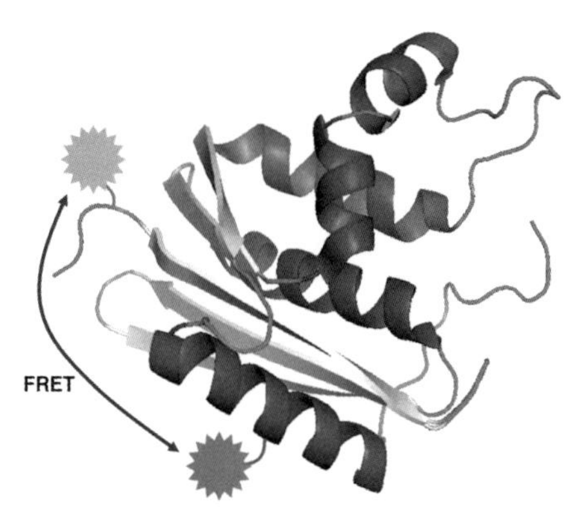

Figure 6.27 Structure of RNase H with a FRET pair for studying unfolding processes. By labeling two parts of the protein with an appropriate FRET pair, the distance between them can be tracked upon the denaturation process.

Even more interesting than unfolding of a protein is the reverse process: the folding of a protein from a highly random and disordered state into a well defined three-dimensional structure. It is postulated that specific pathways for the folding process must exist, but there is still a lot of debate about the nature of these pathways. Recently, the existence of a so-called collapsed state in which the protein is in a compact conformation with only small secondary structure elements was suggested. This state would facilitate the formation of the final three-dimensional structure [80]. Unfortunately, observing such transient states is extremely difficult because of their short lifetimes and their low equilibrium concentrations. However, thanks to the intrinsic property of single-molecule observations to distinguish between equilibrium subpopulations, ensemble averaging as happens in bulk experiments, can be avoided. Thus in principle this collapsed state should be observable.

Recent single-enzyme studies on the cold shock protein CspTm support the hypothesis of the collapsed state as an intermediate in protein folding [81–83]. In this case, distances between the amino acid residues where mapped by labeling specific sites in the enzyme with a FRET pair as indicated in Figure 6.28. FRET efficiencies in freely diffusing enzymes were then monitored as a function of guanidinium chloride concentration. To eliminate errors in the determination of the FRET efficiencies these values were calculated by combining intensity data with lifetime data. Similarly to the research on ribonuclease H, three subpopulations where identified based on the histogram of FRET efficiencies: a state with high FRET efficiency corresponding to the folded state, a lower FRET efficiency state, this being the collapsed state, and a population with (almost) no FRET originating from enzymes without acceptor dye or with a "damaged" acceptor dye. By using several enzyme mutants with dyes attached in different positions, these workers could investigate if the collapse was the result of a compaction of a small part of the protein rather than a global isotropic compaction. As indicated in Figure 6.29 the increase in FRET efficiency upon decreasing the guanidinium chloride concentration is similar for all the enzyme variants, which supports the isotropic collapse hypothesis.

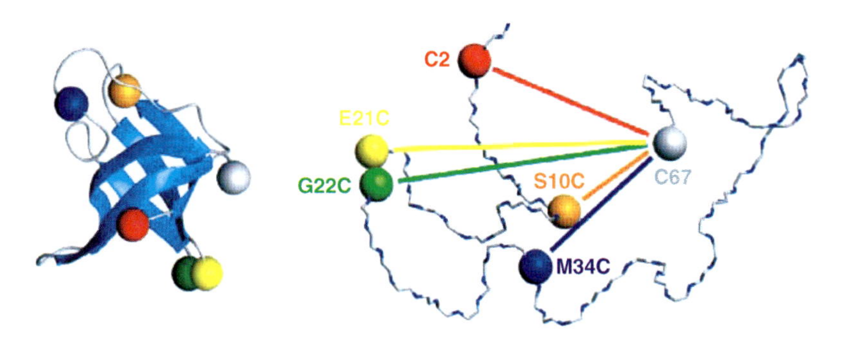

Figure 6.28 Labeling of the CspTm protein with FRET pairs. For all the enzyme variants one dye is attached to a cystein at position 67 (gray) while the other FRET dye is attached to one of the colored positions. Left: representation of the folded state; Right: representation of the denatured state. (Copyright Hoffmann *et al.* (2007) *Proc. Natl. Acad. Sci. USA*, **104**, 105 [82].)

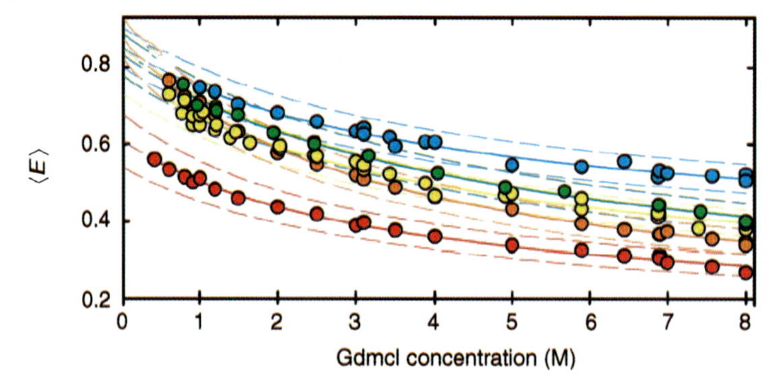

Figure 6.29 Efficiency of energy transfer as a function of denaturant concentration for the different enzyme variants. The colors correspond to those used in Figure 6.28 (Copyright Hoffmann *et al.* (2007) *Proc. Natl. Acad. Sci. USA*, **104**, 105 [82].)

However, they could prove the existence of small secondary structure elements in the collapsed state by combining the single-molecule measurements with bulk circular dichroism experiments. Here the collapsed state was kinetically populated by rapid dilution (in the microsecond time scale) of an enzyme solution with a high concentration of the denaturing agent. Because of the dilutions, all the enzymes will undergo the folding transition and this can be followed by measuring time-resolved circular dichroism spectra. Considerable amounts of β-sheet structures were found, indicating that the collapse is not completely isotropic in the very small distance range (<0.5 nm).

Aside from collapsing, these workers could also identify fast chain dynamics by using single-photon statistics. For this purpose, autocorrelation was performed on the donor intensity with corrections for triplet-state components, as described in the supplementary information of reference [83] (Figure 6.30). At a time lag τ of close to

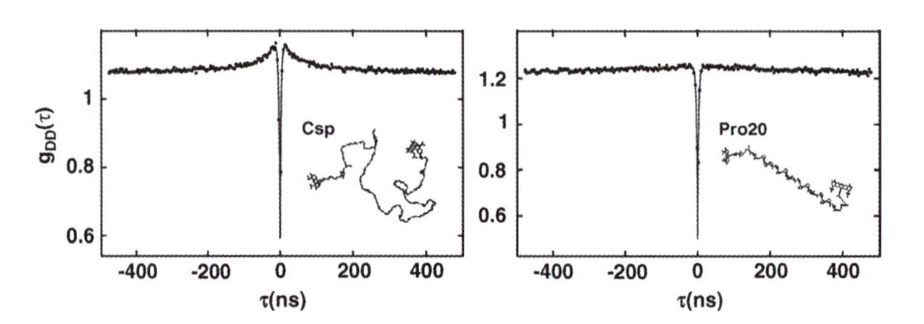

Figure 6.30 Donor intensity autocorrelation for unfolded CspTm (left) and for a stiff polyproline peptide (right) labeled with a FRET pair at the two terminals. In the case of CspTm, a clear bunching effect is seen for short time lags (<50 ns) due to chain dynamics on this time scale. The autocorrelation graph is calculated from the corresponding interphoton time distribution. Corrections are made for pile-up and triplet-state components as described in the supplementary information of reference [83]. (Copyright Nettels *et al.*, (2007) *Proc. Natl. Acad. Sci. USA*, **104**, 2655 [83].)

zero, a strong antibunching (negative amplitude) is seen that is a typical feature of single fluorophores, which can only emit one photon at a time. More interestingly, a weak bunching component (positive amplitude) is seen with a characteristic time constant of about 50 ns. This positive correlation is ascribed to chain dynamics in the unfolded state. Consider a specific inter-dye distance of the FRET pair at time $t = 0$. In very short periods of time, there will be hardly any difference in this inter-dye distance, and thus there will be a high correlation in donor intensity, as the degree of FRET will not have changed very much. However, after longer times (>50 ns) the correlation in distance will be lost due to the chain dynamics. As a comparison, the autocorrelation for a rigid polyproline peptide is shown in the right graph of Figure 6.30, lacking the bunching component.

This study serves as a basis for further extensive research on the complicated process of protein folding and related phenomena, such as enzyme activation. High-accuracy single-molecule techniques such as single-pair FRET experiments will be a major tool for elucidating these types of dynamics.

Besides the folding and unfolding mechanisms, also small reversible conformational dynamics can be efficiently monitored by the FRET approach. Single-molecule enzymology proved to bean important influence for such dynamics on the activity of these biocatalysts (Chapter 9). The dynamics of *Staphylococcus nuclease* (SNase) are for instance studied by specifically conjugating the enzyme with tetramethylrhodamine (TMR) as the FRET donor and Cy5 as the FRET acceptor [14, 84]. Fluctuations in FRET efficiency were found that could not be explained by normal fluorophore reorientations or spectral shifts, and thus must originate from conformational fluctuations of the protein backbone. Distinct patterns of dynamics were found in the millisecond time range.

As the time resolution of such experiments is mainly limited by the amount of signal photons that can be collected over a certain time, a 1 ms resolution is at the moment the best that can be achieved. However, lots of the protein motions known to date occur on a much faster time scale [85], so one of the key challenges for future advances will rely on the improvement of the signal-to-noise ratio in single-molecule experiments. There is no doubt that this will reveal new insights into protein dynamics.

References

1 Förster, T. (1948) Zwischenmolekulare energiewanderung und fluoreszenz. *Ann. Phys.*, **437** (1), 55–57.

2 Dexter, D.L. (1953) A theory of sensitized luminescence in solids. *J. Chem. Phys.*, **21**, 836–850.

3 Cario, G. (1922) Uber Entstehung wahrer Lichtabsorption un scheinbare Koppelung von Quantensprüngen. *Z. Physik.*, **10**, 185–199.

4 Cario, G. and Franck, J. (1922) Uber Zerlegugen von Wasserstoffmolekülen durch angeregte Quecksilberatome. *Z. Physik.*, **11**, 161–166.

5 Franck, J. (1922) Einige aus der theorie von klein und rosseland zu ziehende folgerungen über fluoreszenz, photochemische prozesse und die elektronenemission glühender Körper. *Z. Physik.*, **9**, 259–266.

6 Clegg, R.M. (2006) The history of FRET, In: *Reviews in Fluorescence* (eds. C.D. Geddes and J.R. Lakowicz), Springer, New York, pp. 1–45.

7 Förster, T. (1946) Energiewanderung und fluoreszenz. *Naturwissenhschaften*, **6**, 166–175.

8 Förster, T. (1947) Fluoreszenzversuche an farbstoffmischungen. *Angew. Chem. A*, **59**, 181–187.

9 Stryer, L. and Haugland, R.P. (1967) Energy transfer: a spectroscopic ruler. *Proc. Natl. Acad. Sci. USA*, **58** (2), 719–726.

10 Lakowicz, J.R. (1999) *Principles of Fluorescence Spectroscopy*, 2nd edn, Kluwer Academic/Plenum Publisher, Dordrecht, New York, pp. 368–378.

11 Clegg, R.M. (1996) Fluorescence resonance energy transfer, In: *Fluorescence Imaging Spectroscopy and Microscopy* (eds. X.F. Wang and B. Herman), John Wiley & Sons Inc., New York, pp. 179–752.

12 Valeur, B. (2001) *Molecular Fluorescence. Principles and Applications*, Wiley-VCH Verlag GmbH, Wienheim, pp. 113–124.

13 Steinberg, I.Z. (1997) Long-range nonradiative transfer of electronic excitation energy in proteins and polypeptides. *Annu. Rev. Biochem.*, **272**, 13270–13274.

14 Ha, T.J., Ting, A.Y., Liang, J., Caldwell, W.B., Deniz, A.A., Chemla, D.S., Schultz, P.G., and Weiss, S. (1999) Single-molecule fluorescence spectroscopy of enzyme conformational dynamics and cleavage mechanism. *Proc. Natl. Acad. Sci. USA*, **96** (3), 893–898.

15 Wu, P. and Brand, L. (1994) Resonance energy transfer: Methods and applications. *Anal. Biochem.*, **218**, 1–13.

16 Berlman, I.B. (1973) *Energy Transfer Parameters of Aromatic Compounds*, Academic Press, New York.

17 Yuan, F., Griffin, L., Phelps, L., Buschmann, V., Weston, K., and Greenbaum, N.L. (2007) Use of a novel Förster resonance energy transfer method to identify locations of site-bound metal ions in the u2-u6 snrna complex. *Nucleic Acids Res.*, **35** (9), 2833–2845.

18 R(0)-Values (FRET). ATTO-TEC GmbH. http://www.atto-tec.com/index.php?id=65&L=1 (Latest access: 20 Nov. 2010).

19 Kong, H.J., Polte, T.R., Alsberg, E., and Mooney, D.J. (2005) Fret measurements of cell-traction forces and nano-scale clustering of adhesion ligands varied by substrate stiffness. *Proc. Natl. Acad. Sci. USA*, **102** (12), 4300–4305.

20 Hubner, C.G., Ksenofontov, V., Nolde, F., Mullen, K., and Basche, T. (2004) Three-dimensional orientational colocalization of individual donor-acceptor pairs. *J. Chem. Phys.*, **120** (23), 10867–10870.

21 Patterson, G.H., Piston, D.W., and Barisas, B.G. (2000) Forster distances between green fluorescent protein pairs. *Anal. Biochem.*, **284** (2), 438–440.

22 Hennebicq, E., Pourtois, G., Scholes, G.D., Herz, L.M., Russell, D.M., Silva, C., Setayesh, S., Grimsdale, A.C., Mullen, K., Bredas, J.L., and Beljonne, D. (2005) Exciton migration in rigid-rod conjugated polymers: An improved Forster model. *J. Am. Chem. Soc.*, **127** (13), 4744–4762.

23 Beljonne, D., Pourtois, G., Silva, C., Hennebicq, E., Herz, L.M., Friend, R.H., Scholes, G.D., Setayesh, S., Mullen, K., and Bredas, J.L. (2002) Interchain vs. intrachain energy transfer in acceptor-capped conjugated polymers. *Proc. Natl. Acad. Sci. USA*, **99** (17), 10982–10987.

24 Turro, N.J. (1991) *Modern Molecular Photochemistry*, University Science Books, Sausalito, CA, pp. 296–361.

25 Rao, M. and Mayor, S. (2005) Use of Forster's resonance energy transfer microscopy to study lipid rafts. *BBA-Mol. Cell Res.*, **1746** (3), 221–233.

26 Kenworthy, A.K., Petranova, N., and Edidin, M. (2000) High-resolution FRET microscopy of cholera toxin B-subunit and GPI-anchored proteins in cell plasma membranes. *Mol. Biol. Cell*, **11** (5), 1645–1655.

27 Silvius, J.R. (2003) Fluorescence energy transfer reveals microdomain formation at physiological temperatures in lipid mixtures modeling the outer leaflet of the plasma membrane. *Biophys. J.*, **85** (2), 1034–1045.

28 Lidke, D.S., Heintzmann, R., Arndt-Jovin, D., Barisas, B.G., and Jovin, T.M. (2003) Detection of membrane protein association using homotransfer RET (emFRET) imaging. *Biophys. J.*, **84** (2), 486A–486A.

29 Clegg, R.M., Murchie, A.I.H., Zechel, A., Carlberg, C., Diekmann, S., and Lilley, D.M.J. (1992) Fluorescence resonance energy-transfer analysis of the structure of 4-way DNA junction. *Biochemistry*, **31** (20), 4846–4856.

30 Dale, R.E. and Eisinger, J. (1974) Intramolecular distances determined by energy transfer. dependence on orientational freedom of donor and acceptor. *Biopolymers*, **13**, 1573–1605.

31 Kapanidis, A.N., Laurence, T.A., Lee, N.K., Margeat, E., Kong, X.X., and Weiss, S. (2005) Alternating-laser excitation of single molecules. *Acc. Chem. Res.*, **38** (7), 523–533.

32 Muller, B.K., Zaychikov, E., Brauchle, C., and Lamb, D.C. (2005) Pulsed interleaved excitation. *Biophys. J.*, **89** (5), 3508–3522.

33 Bräuchle, C., Lamb, D.C., and Michaelis, J. (eds.) (2010) *Single Particle Tracking and Single Molecule Energy Transfer*, Wiley-VCH Verlag GmbH, Weinheim.

34 Zhang, J., Campbell, R.E., Ting, A.Y., and Tsien, R.Y. (2002) Creating new fluorescent probes for cell biology. *Nat. Rev. Mol. Cell Biol.*, **3** (12), 906–918.

35 Romoser, V.A., Hinkle, P.M., and Persechini, A. (1974) Detection in living cells of ca^{2+}-dependent changes in the fluorescence emission of an indicator composed of two green fluorescent protein variants linked by a calmodulin-binding sequence. *J. Biol. Chem.*, **13**, 1573–1605.

36 Miyawaki, A., Llopis, J., Heim, R., McCaffery, J.M., Adams, J.A., Ikura, M., and Tsien, R.Y. (1997) Fluorescent indicators for Ca^{2+} based on green fluorescent proteins and calmodulin. *Nature*, **388** (6645), 882–887.

37 Fehr, M., Frommer, W.B., and Lalonde, S. (2002) Visualization of maltose uptake in living yeast cells by fluorescent nanosensors. *Proc. Natl. Acad. Sci. USA*, **99** (15), 9846–9851.

38 Lin, C.W. and Ting, A.Y. (2004) A genetically encoded fluorescent reporter of histone phosphorylation in living cells. *Angew. Chem. Int. Ed.*, **43** (22), 2940–2943.

39 Lin, C.W., Jao, C.Y., and Ting, A.Y. (2004) Genetically encoded fluorescent reporters of histone methylation in living cells. *J. Am. Chem. Soc.*, **126** (19), 5982–5983.

40 Habuchi, S., Cotlet, M., Hofkens, J., Dirix, G., Michiels, J., Vanderleyden, J., Subramaniam, V., and De Schryver, F.C. (2002) Resonance energy transfer in a calcium concentration-dependent cameleon protein. *Biophys. J.*, **83** (6), 3499–3506.

41 Lossau, H., Kummer, A., Heinecke, R., PollingerDammer, F., Kompa, C., Bieser, G., Jonsson, T., Silva, C.M., Yang, M.M., Youvan, D.C., and MichelBeyerle, M.E. (1996) Time-resolved spectroscopy of wild-type and mutant Green Fluorescent Proteins reveals excited state deprotonation consistent with fluorophore-protein interactions. *Chem. Phys.*, **213** (1–3), 1–16.

42 Creemers, T.M.H., Lock, A.J., Subramaniam, V., Jovin, T.M., and Volker, S. (1999) Three photoconvertible forms of green fluorescent protein identified by spectral hole-burning. *Nat. Struct. Biol.*, **6** (6), 557–560.

43 Winkler, K., Lindner, J.R., Subramaniam, V., Jovin, T.M., and Vohringer, P. (2002) Ultrafast dynamics in the excited state of green fluorescent protein (wt) studied by frequency-resolved femtosecond pump-probe spectroscopy. *Phys. Chem. Chem. Phys.*, **4** (6), 1072–1081.

44 Habuchi, S., Ando, R., Dedecker, P., Verheijen, W., Mizuno, H., Miyawaki, A., and Hofkens, J. (2005) Reversible single-molecule photoswitching in the GFP-like fluorescent protein dronpa. *Proc. Natl. Acad. Sci. USA*, **102** (27), 9511–9516.

45 Brasselet, S., Peterman, E.J.G., Miyawaki, A., and Moerner, W.E. (2000) Single-molecule fluorescence resonant energy transfer in calcium concentration dependent cameleon. *J. Phys. Chem. B*, **104** (15), 3676–3682.

46 Hwang, Y.C., Chen, W., and Yates, M.V. (2006) Use of fluorescence resonance energy transfer for rapid detection of enteroviral infection in vivo. *Appl. Environ. Microbiol.*, **72** (5), 3710–3715.

47 Hsu, Y.Y., Liu, Y.N., Wang, W.Y., Kao, F.J., and Kung, S.H. (2007) In vivo dynamics of enterovirus protease revealed by fluorescence resonance emission transfer (FRET) based on a novel FRET pair. *Biochem. Biophys. Res. Commun.*, **353** (4), 939–945.

48 Bosman, A.W., Janssen, H.M., and Meijer, E.W. (1999) About dendrimers: Structure, physical properties, and applications. *Chem. Rev.*, **99** (7), 1665–1688.

49 Weil, T., Wiesler, U.M., Herrmann, A., Bauer, R., Hofkens, J., De Schryver, F.C., and Mullen, K. (2001) Polyphenylene dendrimers with different fluorescent chromophores asymmetrically distributed at the periphery. *J. Am. Chem. Soc.*, **123** (33), 8101–8108.

50 Maus, M., Mitra, S., Lor, M., Hofkens, J., Weil, T., Herrmann, A., Mullen, K., and De Schryver, F.C. (2001) Intramolecular energy hopping in polyphenylene dendrimers with an increasing number of peryleneimide chromophores. *J. Phys. Chem. A*, **105** (16), 3961–3966.

51 Maus, M., De, R., Lor, M., Weil, T., Mitra, S., Wiesler, U.M., Herrmann, A., Hofkens, J., Vosch, T., Mullen, K., and De Schryver, F.C. (2001) Intramolecular energy hopping and energy trapping in polyphenylene dendrimers with multiple peryleneimide donor chromophores and a terryleneimide acceptor trap chromophore. *J. Am. Chem. Soc.*, **123** (31), 7668–7676.

52 Brejc, K., Ficner, R., Huber, R., and Steinbacher, S. (1995) Isolation, crystallization, crystal-structure analysis and refinement of allophycocyanin from the cyanobacterium spirulina-platensis at 2.3 angstrom resolution. *J. Mol. Biol.*, **249** (2), 424–440.

53 Maccoll, R., Csatorday, K., Berns, D.S., and Traeger, E. (1980) Chromophore interactions in allophycocyanin. *Biochemistry*, **19** (12), 2817–2820.

54 Canaani, O.D. and Gantt, E. (1980) Circular-dichroism and polarized fluorescence characteristics of blue-green-algal allophycocyanins. *Biochemistry*, **19** (13), 2950–2956.

55 Loos, D., Cotlet, M., De Schryver, F., Habuchi, S., and Hofkens, J. (2004) Single-molecule spectroscopy selectively probes donor and acceptor chromophores in the phycobiliprotein allophycocyanin. *Biophys. J.*, **87** (4), 2598–2608.

56 Goldsmith, R.H. and Moerner, W.E. (2010) Watching conformational- and photodynamics of single fluorescent proteins in solution. *Nat. Chem.*, **2** (3), 179–186.

57 De Schryver, F.C., Vosch, T., Cotlet, M., Van der Auweraer, M., Mullen, K., and Hofkens, J. (2005) Energy dissipation in multichromophoric single dendrimers. *Acc. Chem. Res.*, **38** (7), 514–522.

58 Cotlet, M., Vosch, T., Habuchi, S., Weil, T., Mullen, K., Hofkens, J., and De Schryver, F. (2005) Probing intramolecular Forster resonance energy transfer in a naphthaleneimide-peryleneimide-terrylenediimide-based dendrimer by ensemble and single-molecule fluorescence spectroscopy. *J. Am. Chem. Soc.*, **127** (27), 9760–9768.

59 Song, H.E., Kirmaier, C., Schwartz, J.K., Hindin, E., Yu, L.H., Bocian, D.F., Lindsey, J.S., and Holten, D. (2006) Mechanisms, pathways, and dynamics of excited-state energy flow in self-assembled wheel-and-spoke light-harvesting architectures. *J. Phys. Chem. B*, **110** (39), 19121–19130.

60 Melnikov, S.M., Yeow, E.K.L., Uji-i, H., Cotlet, M., Mullen, K., De Schryver, F.C., Enderlein, J., and Hofkens, J. (2007) Origin of simultaneous donor-acceptor emission in single molecules of peryleneimide-terrylenediimide labeled polyphenylene dendrimers. *J. Phys. Chem. B*, **111** (4), 708–719.

61 Cotlet, M., Gronheid, R., Habuchi, S., Stefan, A., Barbafina, A., Mullen, K., Hofkens, J., and De Schryver, F.C. (2003) Intramolecular directional Forster resonance energy transfer at the single-molecule level in a dendritic system. *J. Am. Chem. Soc.*, **125** (44), 13609–13617.

62 Uji-i, H., Melnikov, S.M., Deres, A., Bergamini, G., De Schryver, F., Herrmann, A., Mullen, K., Enderlein, J., and Hofkens, J. (2006) Visualizing spatial and temporal heterogeneity of single molecule rotational diffusion in a glassy polymer by defocused wide-field imaging. *Polymer*, **47** (7), 2511–2518.

63 Vosch, T., Cotlet, M., Hofkens, J., Van der Biest, K., Lor, M., Weston, K., Tinnefeld, P., Sauer, M., Latterini, L., Mullen, K., and De Schryver, F.C. (2003) Probing forster type energy pathways in a first generation rigid dendrimer bearing two perylene imide chromophores. *J. Phys. Chem. A*, **107** (36), 6920–6931.

64 Hofkens, J., Cotlet, M., Vosch, T., Tinnefeld, P., Weston, K.D., Ego, C., Grimsdale, A., Mullen, K., Beljonne, D., Bredas, J.L., Jordens, S., Schweitzer, G., Sauer, M., and De Schryver, F. (2003) Revealing competitive forster-type resonance energy-transfer pathways in single bichromophoric molecules. *Proc. Natl. Acad. Sci. USA*, **100** (23), 13146–13151.

65 De Belder, G., Schweitzer, G., Jordens, S., Lor, M., Mitra, S., Hofkens, J., De Feyter, S., Van der Auweraer, M., Herrmann, A., Weil, T., Mullen, K., and De Schryver, F.C. (2001) Singlet-singlet annihilation in multichromophoric peryleneimide dendrimers, determined by fluorescence upconversion. *ChemPhysChem*, **2** (1), 49–55.

66 Tinnefeld, P., Weston, K.D., Vosch, T., Cotlet, M., Weil, T., Hofkens, J., Mullen, K., De Schryver, F.C., and Sauer, M. (2002) Antibunching in the emission of a single tetrachromophoric dendritic system. *J. Am. Chem. Soc.*, **124** (48), 14310–14311.

67 Masuo, S., Vosch, T., Cotlet, M., Tinnefeld, P., Habuchi, S., Bell, T.D.M., Oesterling, I., Beljonne, D., Champagne, B., Mullen, K., Sauer, M., Hofkens, J., and De Schryver, F.C. (2004) Multichromophoric dendrimers as single-photon sources: A single-molecule study. *J. Phys. Chem. B*, **108** (43), 16686–16696.

68 Tinnefeld, P., Hofkens, J., Herten, D.P., Masuo, S., Vosch, T., Cotlet, M., Habuchi, S., Mullen, K., De Schryver, F.C., and Sauer, M. (2004) Higher-excited-state photophysical pathways in multichromophoric systems revealed by single-molecule fluorescence spectroscopy. *ChemPhysChem*, **5** (11), 1786–1790.

69 Sliwa, M., Flors, C., Oesterling, I., Hotta, J., Mullen, K., De Schryver, F.C., and Hofkens, J. (2007) Single perylene diimide dendrimers as single-photon sources. *J. Phys.-Condens. Mat.*, **19** (44) 445004.

70 Flors, C., Oesterling, I., Schnitzler, T., Fron, E., Schweitzer, G., Sliwa, M., Herrmann, A., van der Auweraer, M., de Schryver, F.C., Mullen, K., and Hofkens, J. (2007) Energy and electron transfer in ethynylene bridged perylene diimide multichromophores. *J. Phys. Chem. C*, **111** (12), 4861–4870.

71 Yasukuni, R., Asahi, T., Sugiyama, T., Masuhara, H., Sliwa, M., Hofkens, J., De Schryver, F.C., Van der Auweraer, M., Herrmann, A., and Muller, K. (2008) Fabrication of fluorescent nanoparticles of dendronized perylenediimide by laser ablation in water. *Appl. Phys. A-Mater.*, **93** (1), 5–9.

72 Yasukuni, R., Sliwa, M., Hofkens, J., De Schryver, F.C., Herrmann, A., Mullen, K., and Asahi, T. (2009) Size-dependent optical properties of dendronized perylenediimide nanoparticle prepared by laser ablation in water. *Jpn. J. Appl. Phys.*, **48** (6) 065002.

73 Tinnefeld, P., Muller, C., and Sauer, M. (2001) Time-varying photon probability distribution of individual molecules at room temperature. *Chem. Phys. Lett.*, **345** (3–4), 252–258.

74 Zhuang, X.W., Kim, H., Pereira, M.J.B., Babcock, H.P., Walter, N.G., and Chu, S. (2002) Correlating structural dynamics and function in single ribozyme molecules. *Science*, **296** (5572), 1473–1476.

75 Heilemann, M., Margeat, E., Kasper, R., Sauer, M., and Tinnefeld, P. (2005) Carbocyanine dyes as efficient reversible single-molecule optical switch. *J. Am. Chem. Soc.*, **127** (11), 3801–3806.

76 Bates, M., Blosser, T.R., and Zhuang, X.W. (2005) Short-range spectroscopic ruler based on a single-molecule optical switch. *Phys. Rev. Lett.*, **94** (10) 108101.

77 Deniz, A.A., Dahan, M., Grunwell, J.R., Ha, T.J., Faulhaber, A.E., Chemla, D.S., Weiss, S., and Schultz, P.G. (1999) Single-pair fluorescence resonance energy transfer on freely diffusing molecules: Observation of Forster distance dependence and subpopulations. *Proc. Natl. Acad. Sci. USA*, **96** (7), 3670–3675.

78 Nienhaus, G.U. (2006) Exploring protein structure and dynamics under denaturing conditions by single-molecule FRET analysis. *Macromol. Biosci.*, **6** (11), 907–922.

79 Kuzmenkina, E.V., Heyes, C.D., and Nienhaus, G.U. (2006) Single-molecule FRET study of denaturant induced unfolding of RNase h. *J. Mol. Biol.*, **357** (1), 313–324.

80 Daggett, V. and Fersht, A. (2003) The present view of the mechanism of protein folding. *Nat. Rev. Mol. Cell Biol.*, **4** (6), 497–502.

81 Schuler, B., Lipman, E.A., and Eaton, W.A. (2002) Probing the free-energy surface for protein folding with single-molecule fluorescence spectroscopy. *Nature*, **419** (6908), 743–747.

82 Hoffmann, A., Kane, A., Nettels, D., Hertzog, D.E., Baumgartel, P., Lengefeld, J., Reichardt, G., Horsley, D.A., Seckler, R., Bakajin, O., and Schuler, B. (2007) Mapping protein collapse with single-molecule fluorescence and kinetic synchrotron radiation circular dichroism spectroscopy. *Proc. Natl. Acad. Sci. USA*, **104** (1), 105–110.

83 Nettels, D., Gopich, I.V., Hoffmann, A., and Schuler, B. (2007) Ultrafast dynamics of protein collapse from single-molecule photon statistics. *Proc. Natl. Acad. Sci. USA*, **104** (8), 2655–2660.

84 Ha, T.J., Ting, A.Y., Liang, J., Deniz, A.A., Chemla, D.S., Schultz, P.G., and Weiss, S. (1999) Temporal fluctuations of fluorescence resonance energy transfer between two dyes conjugated to a single protein. *Chem. Phys.*, **247** (1), 107–118.

85 Benkovic, S.J. and Hammes-Schiffer, S. (2003) A perspective on enzyme catalysis. *Science*, **301** (5637), 1196–1202.

7
Photoinduced Electron Transfer (PET) Reactions

7.1
Fluorescence Quenching by PET

Charge transfer or electron transfer processes are of utmost importance in a variety of biochemical processes, with one of the most prominent examples being photosynthesis. While ensemble or bulk level charge transfer processes are fairly well understood [1–4], the study and characterization of these processes at the individual or single-molecule level is still in its infancy [5]. Electron transfer between one and another molecule occurs if one molecule can accept or donate electrons to the other molecule. Often electron transfer can be observed to or from electronically excited molecules, such as fluorophores. If excitation occurs through absorption of photons in a fluorophore (usually an organic molecule with a delocalized π-electron system), its redox properties change, which might enable so-called photoinduced electron transfer (PET) whereby the fluorescence of the fluorophore is quenched. Both fluorescence resonance energy transfer (FRET) [6–9] and PET [1–5, 10–14] are two mechanisms that lead to variation of fluorescence emission by distance-dependent fluorescence quenching between a fluorophore and a quenching moiety. Whereas FRET from a donor (D) to an acceptor (A) chromophore scales with $1/[1 + (R/R_0)^6]$, where R_0 is typically between 2 and 8 nm, PET can be designed in a way such that contact formation (van der Waals contact) is required for efficient quenching, with a separation between the fluorophore and quencher that can also be seen as an electron transfer donor and acceptor on the sub-nanometer length scale.

To interpret fluorescence quenching caused by PET, the transfer mechanism and its distance dependence has to be well understood. In general, the electron transfer rate is proportional to the square of the electronic coupling between the donor and the acceptor, which in turn depends exponentially on the donor–acceptor distance [3]. Thus changes in the PET rate can also directly report on changes of the donor–acceptor distance caused, for example, by conformational dynamics of a biopolymer (nucleic acid, peptide or protein). In standard electron transfer theory, quenching of fluorophores in the first excited singlet state by electron donors or acceptors results in charge separation with rate constant k_{cs} and the formation of a radical ion pair $D^{+\bullet}A^{-\bullet}$, which returns to the ground state via charge recombination with rate

Handbook of Fluorescence Spectroscopy and Imaging. M. Sauer, J. Hofkens, and J. Enderlein

ISBN: 978-3-527-31669-4

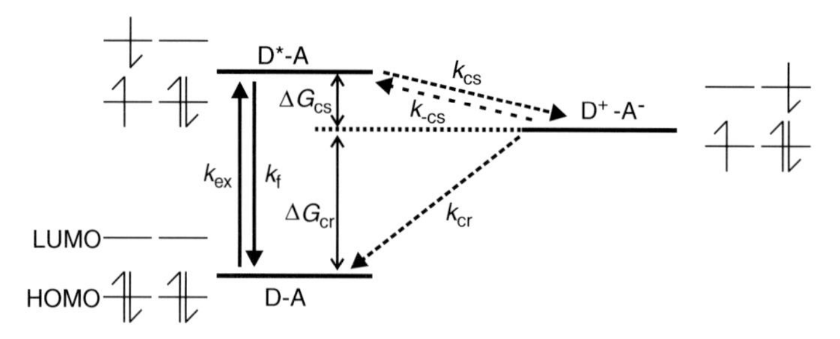

Figure 7.1 Kinetic and thermodynamic scheme for the intramolecular charge separation and charge recombination in donor–acceptor compounds. D represents the donor fluorophores, and A is the electron acceptor. Depending on the free energy of charge separation, ΔG_{cs}, and charge recombination, ΔG_{cr}, reversible electron transfer with rate constant k_{-cs} can occur to repopulate the locally excited state D*–A.

constant k_{cr} (Figure 7.1). The efficiency of charge separation and charge recombination is mainly controlled by the relationship between the free energy of the reactions, ΔG_{cs} and ΔG_{cr}, the reorganization energy λ, and the distance between donor and acceptor [1–5]. Typically the total reorganization energy is written as the sum of an inner contribution, λ_{in}, and an outer contribution, λ_{out}, attributed to nuclear reorganization of the redox partners and their environment (solvent), respectively.

Depending on the properties of the electron donor and acceptor, and the linker connecting both compounds, different charge transfer mechanisms of the donor–bridge–acceptor system have to be distinguished [1–5]. Charge separation can proceed by (i) coherent tunneling (super-exchange, with an exponential dependence on the separation distance), in which the electron or hole never resides on the bridge, or (ii) via thermally activated (or non-activated) reduction or oxidation of the bridge. When diffuse hopping between bridge sites becomes rate limiting, the distance dependence of PET is inversely proportional to the donor–acceptor distance. In general, tunneling and hopping pathways can operate in parallel.

A particular case of PET refers to the situation in which the locally excited state D*–A and the charge-separated state D^+–A^- (Figure 7.1) are relatively close in energy. Depending on the free energy for charge separation ΔG_{cs}, thermally activated reverse electron transfer can occur with rate k_{-cs} if the radiationless deactivation of the charge separated state D^+–A^- to the ground state D–A is inefficient [13, 15]. As a consequence, fluorescence emitted by the locally excited state D*–A is delayed but retains a high quantum yield. In some instances the stability of the charge separated state D^+–A^-, that is, ΔG_{cs}, and thus the efficiency of reverse charge separation can be sensitively tuned by the polarity of the solvent, for example, by destabilizing the charge separated state, changing to a solvent with lower polarity.

Neglecting through-bond interactions, PET quenching occurs transiently through molecular collisions, a process known as "dynamic quenching", or in molecular complexes that are stable for multiple excitation–emission cycles, a process called "static quenching". Complex formation between organic fluorophores and small aromatic compounds in aqueous solution are often driven by hydrophobic and

π-stacking interactions. In such complexes, the overall electron distribution changes, resulting in shifts in the absorption spectrum [14, 16]. The effect is well known from fluorophores that form homodimers at high concentration: depending on the interaction geometry, either fluorescent J-dimers with a red-shifted absorption spectrum, or non-fluorescent H-dimers, with blue-shifted absorption spectrum, can be found [17, 18]. In heterodimers, moreover, the respective redox potentials of the interacting compounds can enable PET and, consequently, efficient quenching of fluorescence occurs, such that a low quantum yield is combined with a red-shifted absorption spectrum [11, 16].

To estimate the efficiency of photoinduced charge separation, its change in free energy for charge separation, ΔG_{cs}, can be estimated using the Rehm–Weller equation (Equation 7.1) [19, 20],

$$\Delta G_{cs} = E_{ox} - E_{red} - E_{0,0} - C \tag{7.1}$$

where

E_{ox} and E_{red} are the first one-electron oxidation potential of the donor and the first one-eletron reduction potential of the acceptor, respectively, in the solvent under consideration
$E_{0,0}$ is the energy of the zero–zero transition to the lowest excited singlet state
C is the solvent dependent Coulomb interaction energy (which can be neglected in a moderately polar solvent).

Here the reaction rate for electron transfer follows an exponential D–A distance dependence, with a characteristic length scale of the order of a few Angströms [1, 5]. Evidence for PET reactions can be derived from experiments revealing the existence of transient charged-separated species, that is, radical ions. However, photoinduced electron transfer reactions between fluorophores and quenchers are often ultrafast, that is, in the femtosecond (fs) to picoseconds (ps) time regime, which complicates their unequivocal characterization. Zhong and Zewail reported electron transfer reaction times from tryptophan (D) to riboflavin (A^*) in the riboflavin binding protein in the order of $\sim$100 fs and charge recombination in the lower ps time scale [21]. Tryptophan and riboflavin are found to exist in stacked interaction geometries, similar to those revealed from simulations of the bimolecular fluorophore–tryptophan complex formation in water [22]. Experiments with denaturing agents showed that electron transfer rates decrease with changes in protein tertiary structure, demonstrating the necessity of well defined, stacked contact geometries between riboflavin and the aromatic amino acid residue for efficient electron transfer [21, 23].

Measured redox potentials allow estimation of PET efficiencies based on the Rehm–Weller formalism (Equation 7.1). Table 7.1 gives the one-electron redox potentials and zero–zero transitions energies for various standard fluorophores used in diverse fluorescence applications. As can be clearly seen by the exergonic free energy change for charge separation, ΔG_{cs}, for the reduction of the excited fluorophore by the electron donor guanosine, G, most rhodamine and oxazine derivatives are prone to photoinduced reduction [24–33]. With an oxidation potential of $E_{ox} = 1.24$ V versus SCE [10, 25] G exhibits the most pronounced electron donating properties among the naturally occurring nucleobases. The oxidation potential for tryptophan of $E_{ox} = 0.60$–1.00 V versus SCE is lower still. Therefore, tryptophan is an even more efficient electron donor and quenches the fluorescence of most organic fluorophores via PET [34–38]. The oxazine fluorophore MR121, for instance, has a reduction potential of $E_{red} = -0.42$ V versus SCE [39] measured in acetonitrile and a transition energy of $\sim$1.9 eV, such that PET from tryptophan or guanosine to MR121 is exergonic ($\Delta G_{cs} = -0.7$ and -0.2 eV, respectively) according to Equation 7.1. Furthermore, ascorbic acid exhibits a relatively low oxidation potential of $E_{ox} = 0.06$ V versus SCE [39, 40] and thus serves as a very efficient electron donor, that is, a quencher, even for dicarbocyanine dyes such as Cy5 with $\Delta G_{cs} = 0.98$ eV. On the other hand, oxidation of dicarbocyanine dyes is facilitated, compared with oxidation of rhodamine or oxazine derivatives (compare E_{ox} values in Table 7.1). Using a strong electron acceptor, however, such as methylviologen with a reduction potential of $E_{red} = -0.45$ V versus SCE [3] all fluorophores shown in Table 7.1 are efficiently quenched via PET with ΔG_{cs} varying between -0.46 eV for Cy5 and -0.10 eV for MR121.

Overall, the data in Table 7.1 are a good demonstration that structurally similar dyes show similar PET quenching behavior. Furthermore, it reveals that carbocyanine dyes are prone to oxidation whereas rhodamine and oxazine dyes are fairly easily reduced by potent electron donors. In other words, direct electronic interactions, that is, van der Waals contact, between rhodamine or oxazine and carbocyanine dyes, for example in FRET constructs with short donor/acceptor distance, have to be prevented to avoid undesirable fluorescence quenching or formation of non-fluorescent heterodimers [9]. The free energy change for charge separation between Cy5 (D) and Rhodamine 6G (A) is calculated as $\Delta G_{cs} = -0.29$ eV for excited rhodamine (A^*) and $\Delta G_{cs} = -0.01$ eV for excited carbocyanine. Thus, upon excitation of Rhodamine 6G in close proximity to Cy5, relatively efficient PET, that is, reduction of excited Rhodamine 6G by Cy5, can occur.

Finally, it has to be considered that PET reactions can also quench triplet states of fluorophores. Triplet quenching by electron transfer results in the formation of radical anions or cations depending on the charge of the ground state. Such ionized fluorophores can also be formed through other pathways, such as photoionization, and they represent additional potentially reactive intermediates in photobleaching pathways [39, 41–45]. Furthermore, triplet quenching by electron donors (e.g., thiols or ascorbic acid) and formation of thermally stable radical anions with lifetimes of up to several seconds in aqueous buffer represents an elegant method on which to base the development of super-resolution imaging methods such as *d*STORM (see Chapter 8) [46, 47].

7.2
Single-Molecule Fluorescence Spectroscopy to Study PET

In recent years, several innovations in fluorescence imaging and spectroscopy techniques have made possible, and in fact routine, the detection and study of

Table 7.1 First one-electron reduction potential E_{red} (V), first one-electron oxidation potentials E_{ox} (V), zero–zero transition energy $E_{0,0}$ (eV), and free energy of charge separation ΔG_{cs} (eV) for the reduction of excited fluorophores by the DNA nucleobase guanosine using $E_{ox} = 1.24$ V versus SCE [10, 25]. Redox potentials were measured by various methods and converted into V versus SCE. ATTO655 exhibits identical spectroscopic characteristics to MR121 and is therefore supposed to have a similar structure and redox properties [48]. Rhodamine 630 exhibits the basic rhodamine structure of fluorophores such as Alexa Fluor 546, ATTO565, Alexa Fluor 568, ATTO590, Alexa Fluor 594, and related rhodamine derivatives. The molecular structure of ATTO647N is taken from reference [49].

Fluorophore	Molecular structure	E_{red} (V/SCE)	E_{ox} (V/SCE)	$E_{0,0}$ (eV)	ΔG_{cs} (eV)	Ref.
5-FAM	O, OH, COOH, COOH	−0.71	—	2.46	−0.52	[31]
Bodipy FL	H_3C, H_3C, N, N^+, B^-, F, F, COOH	−1.07	—	2.43	−0.11	[37, 42]

(*Continued*)

Table 7.1 *(Continued)*

Fluorophore	Molecular structure	E_{red} (V/SCE)	E_{ox} (V/SCE)	$E_{0,0}$ (eV)	ΔG_{cs} (eV)	Ref.
R6G	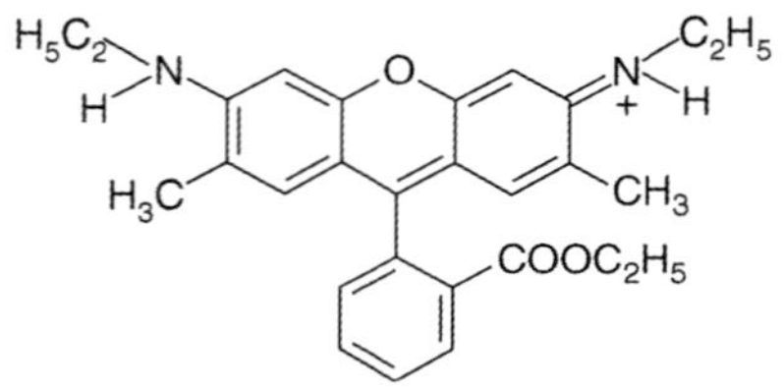	−0.95	1.39	2.28	−0.07	[9, 24]
Rhodamine 630	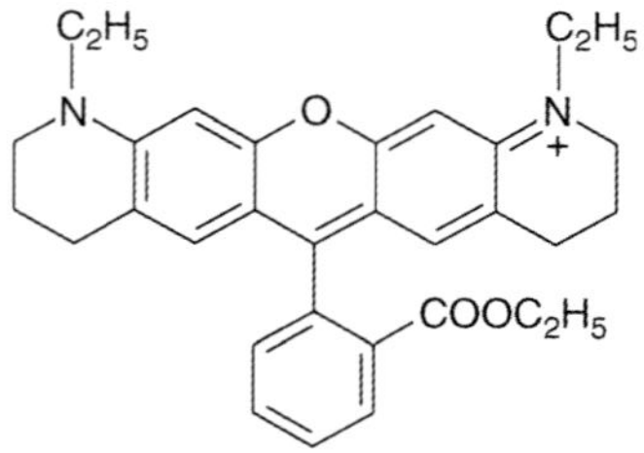	−0.82	1.03	2.16	0.01	[24]
Texas Red	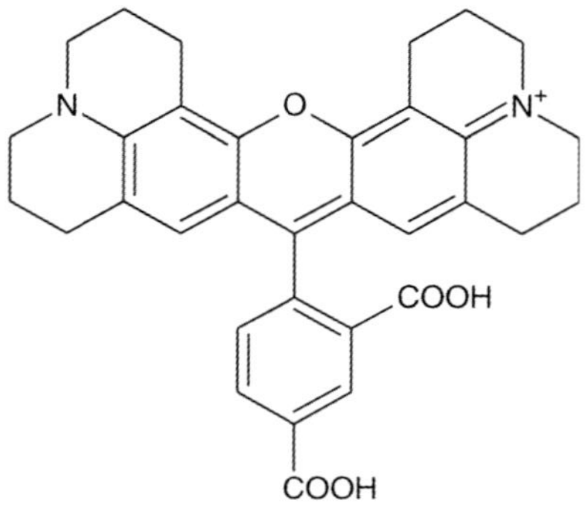	−1.12	—	2.08	0.28	[31]

JA133	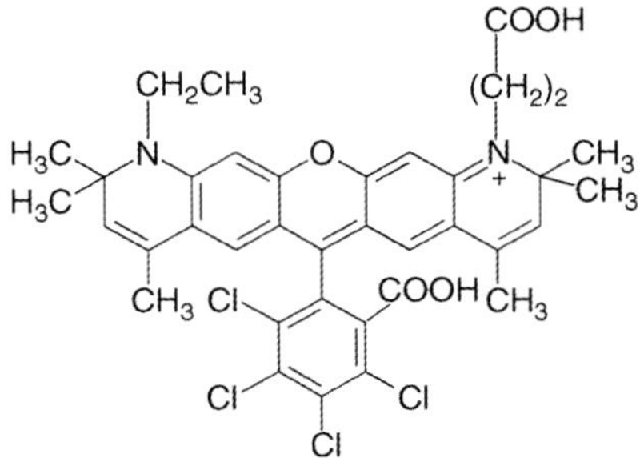 COOH, $(CH_2)_2$, CH_2CH_3, H_3C, H_3C, N, O, N^+, CH_3, CH_3, CH_3, Cl, COOH, CH_3, Cl, Cl, Cl	−0.71	1.20	1.96	0.00	[9, 24]
Cy5	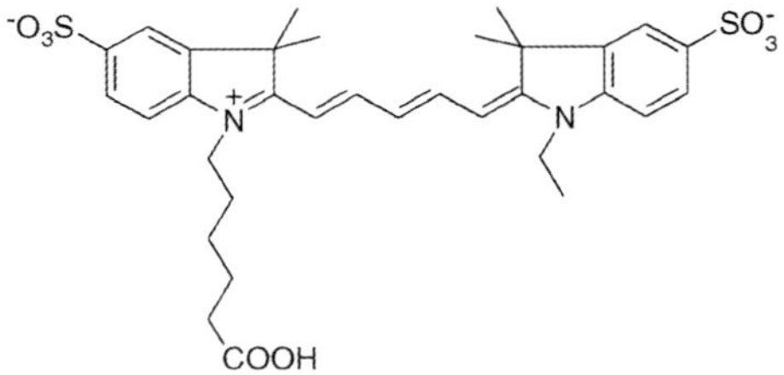 ^-O_3S, SO_3^-, N^+, N, COOH	−0.84	0.97	1.88	0.20	[9, 39]
ATTO647N	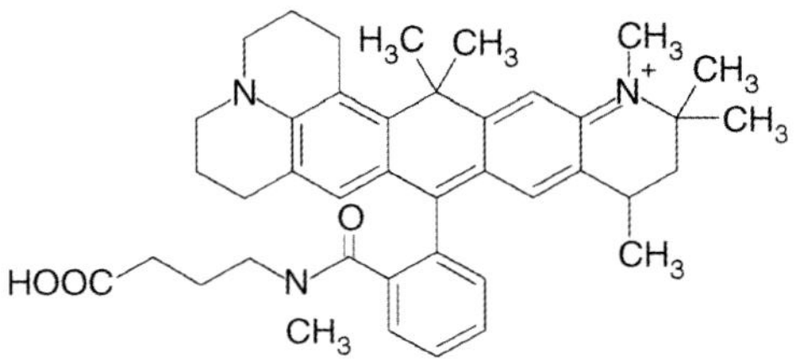 N, H_3C, CH_3, CH_3, N^+, CH_3, CH_3, CH_3, O, HOOC, N, CH_3	−0.64	1.11	1.90	−0.01	[39]

(*Continued*)

Table 7.1 *(Continued)*

Fluorophore	Molecular structure	E_{red} (V/SCE)	E_{ox} (V/SCE)	$E_{0,0}$ (eV)	ΔG_{cs} (eV)	Ref.
JA66	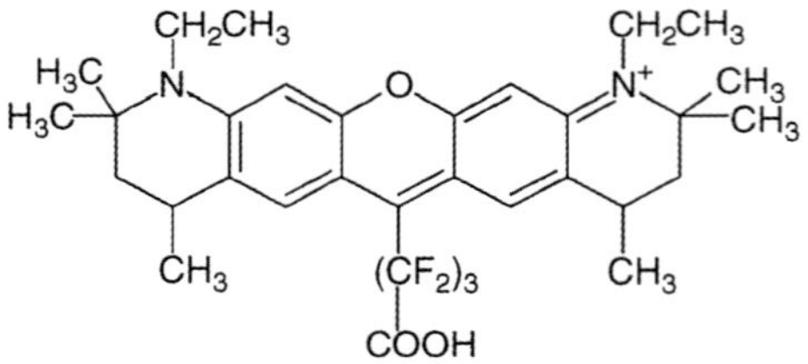	−0.56	1.18	1.92	−0.21	[24]
MR121	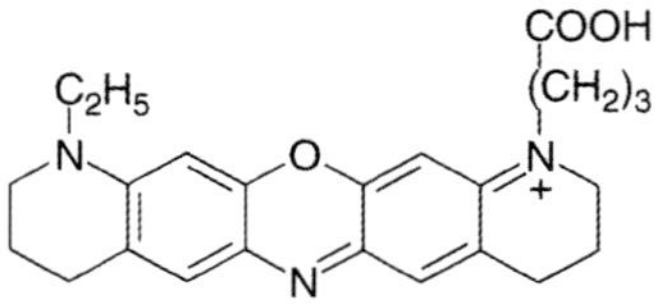	−0.42	1.31	1.86	−0.20	[14, 16, 22, 39]

individual fluorophores at room temperature [8, 50–53]. Single-molecule fluorescence studies allow one to observe fluctuations in fluorescence properties of individual molecules over time (dynamic disorder) in addition to variations in fluorescence properties of chemically identical molecules located in different environments within the same heterogeneous sample (static disorder). Some phenomena that would not necessarily be predicted based on our knowledge of ensemble measurements are readily observed by looking at single molecules, for example, blinking due to intersystem crossing into long-lived off-states. In addition, single-molecule experiments are ideally suited to unravelling subpopulations with slightly different reaction pathways, that is, phenomena that are hidden in ensemble experiments due to averaging. Finally, we should keep in mind that single-molecule observations allow the direct monitoring of conformational fluctuations, for example, protein folding, without synchronization under equilibrium conditions.

All single-molecule experiments imply that the molecule to be observed is labeled with a fluorophore whose fluorescence properties report specifically on the environment and intra- as well as inter-molecular changes. For example, distance changes might alter the electron transfer rate between a fluorophore and a quencher. Thus, fluorescence intensity and lifetime measurements represent complementary methods to measure electron transfer rates of individual molecules. However, for very efficient electron transfer ($k_{cs} \gg k_f$), the fluorophore will always be quenched upon excitation by charge separation and radiationless deactivation to the ground state via charge recombination. In order to be reflected in the fluorescence lifetime measured from a single fluorophore, k_{cs} has to be of a similar order of magnitude to k_f [12]. If, on the other hand, charge separation is very inefficient ($k_f \gg k_{cs}$) the fluorescence lifetime or intensity of a single fluorophore will barely be affected.

Often fluorescence quenching is diffusion limited, that is, each diffusional collision described by k_+ (Figure 7.2) between the fluorophore and quencher is effective at quenching the fluorescence in a so-called "dynamic" quenching process. Dynamic quenching can be observed through the linear dependence of the fluorescence lifetime on the quencher concentration, resulting in a reduced dynamic fluorescence quenching yield $QY_{dyn} = \tau/\tau_0$. Furthermore, the diffusive encounter of fluorophore/quencher pairs can depend on the hydrophobicity of their structure which induces, for example, the formation of hydrophobic complexes in aqueous solution [14, 22]. Owing to the short distance between fluorophore and quencher, k_{cs} is much larger than the intrinsic fluorescence emission rate k_f and the measured fluorescence intensity drops almost to zero. If the quenched state lasts much longer than the unquenched excited state, an effectively non-fluorescent species is formed and the quenching process can be considered as "static". In static quenching no change in the fluorescence lifetime of a mixed ensemble with quenched and fluorescent species is detected and quenching is reflected solely in a decreased fluorescence intensity. Whereas the dynamic quantum yield QY_{dyn} is determined from fluorescence lifetime measurements, a steady state fluorescence measurement in a spectrofluorometer provides the steady-state quantum yield, QY_{ss}.

In fact, it is static quenching in stacked complex interaction geometries that provides on–off switching of fluorescence with sufficient contrast to be analyzed in

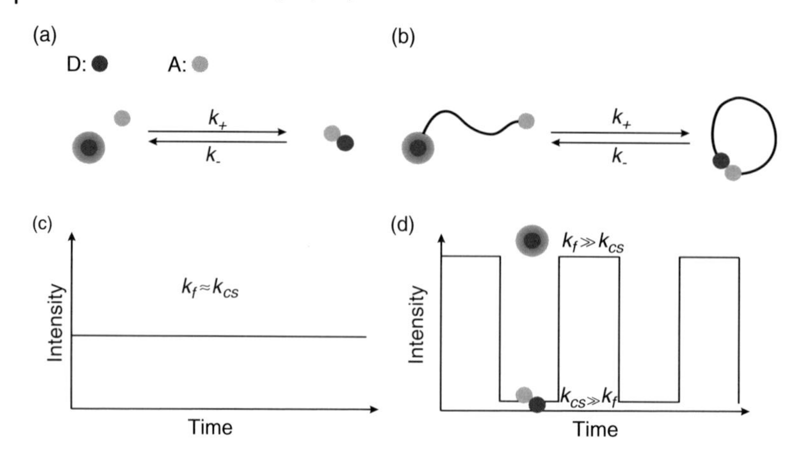

Figure 7.2 (a) Intermolecular and (b) intramolecular fluorescence quenching of a donor fluorophore by an electron acceptor can occur transiently through molecular collisions or complex formation. Both dynamic and static quenching is determined by the rate constants k_+ and k_-. (c) In the case of dynamic quenching in the range of $k_f \approx k_{cs}$, PET is reflected in reduced fluorescence intensity and shortened fluorescence lifetime. (d) Static quenching ($k_{cs} \gg k_f$) provides on–off switching of fluorescence between an unquenched and strongly quenched state. Thus, single molecule-fluorescence spectroscopy can be used advantageously to monitor the dynamics of individual molecules.

single-molecule fluorescence spectroscopy (Figure 7.2) [14, 16]. Aromatic ring systems are known to interact at short range through π-stacking, van der Waals, and in aqueous solution, through hydrophobic interactions, among others. The complex stability depends, besides individual non-covalent energetic contributions (e.g., dispersion and Coulomb interactions) and solvent compositions, on the interaction geometry between the compounds stabilized by hydrophobic forces. A precise quantitative description of such short-range interactions within macromolecules, however, is complicated by entropic effects from solvent and interacting molecules [54, 55].

Fluorescence intermittencies or blinking of individual fluorophores, that is, the reversible occupation of non-fluorescent off-states is generally accepted as proof for the observation of a single quantum system. Besides triplet-state dynamics [53, 56–58], molecular reorientation [59], spectral diffusion [60], and conformational changes [61, 62], intramolecular PET reactions can also be the source of such intensity fluctuations (Figure 7.3) [11–15, 27]. Furthermore, single-molecule studies on fluorophores immobilized on glass surfaces and in polymer matrices revealed that intermolecular charge separation between the excited state of the fluorophore and traps present in the inhomogeneous surrounding environment can cause blinking of individual fluorophores [63–67].

Several studies revealed long-lived dark states with durations ranging from milliseconds to hundreds of seconds. Interestingly, the resulting on- and off-time distributions often obey a power law function, which can be explained by a charge tunneling model where a charge is transferred between the fluorophore and localized

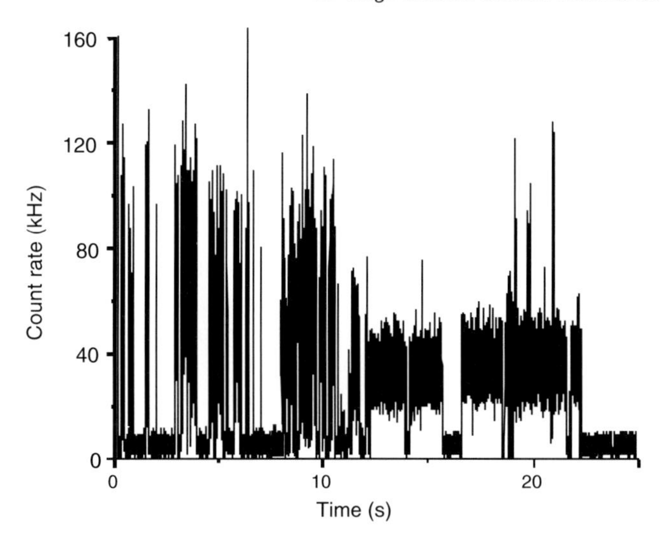

Figure 7.3 Fluorescence trajectory of an individual oxazine fluorophore immobilized on a dry cover glass. The fluorescence intensity shows frequent jumps between a bright on- and a non-fluorescent off-state. Besides spectral and fluorescence lifetime fluctuations, the fluorophore shows a remarkable change in blinking frequency after about 10 s [68].

states in the polymer matrix or surface material [69–71]. Subsequent charge recombination by back charge tunneling restores the fluorescence. The dependency of the on- and off-time distributions upon power law (i.e., the distribution can be fitted well to a straight line when plotted on a log–log scale) manifests itself in the exponential distribution of charge tunneling rates that arise from the exponential distributions of both the spatial locations and energies of the trap states present. These results led to the suggestion that power law behavior of off-state distributions might appear as a universal feature for single emitters undergoing fluorescence blinking [42]. However, other studies demonstrated that the blinking behavior changes, which are dependent on the environment and on- and off-time distributions derived from fluorophores immobilized under aqueous conditions in the presence of a homogeneously distributed oxidizing agent, can be fitted by an exponential function [68].

7.3 Single-Molecule Sensitive Fluorescence Sensors Based on PET

The development and investigation of molecular systems and methodologies that enable the monitoring of specific recognition and binding events is crucial for a better understanding of molecular signaling and information transfer in biological systems [72]. In the ideal case, the molecular system signals selective recognition or binding events via a change in a characteristic property, for example, fluorescence that can be conveniently transmitted into an electronic signal by the appropriate detector. Besides signaling, sensors have the potential for information processing if measurable characteristic properties can be switched between two distinguishable states through environmental stimuli. The use of fluorescence characteristics for sensing or switching is advantageous as single photons can easily be converted into electrical signals with a high efficiency using, for example, avalanche photodiodes as detectors.

Again, PET reactions can be used advantageously to transfer discrete and stoichiometric recognition events into an altered fluorescence signal of a fluorophore. Molecular systems in which the excited state of a fluorophore is controlled by the redox properties of a receptor module covalently attached to the fluorophore form an important class of chemosensory materials [72]. In these sensors a spacer holds the fluorophore and a receptor close to, but separate from, each other. The redox properties of the receptor module are altered upon guest complexation and decomplexation. As PET reactions are controlled by the relationship between the free energy of the reaction, the reorganization energy, and the distance between the donor and acceptor, careful selection of the optical, guest-binding, and redox properties of the components allow the optimization of the signaling parameters of the PET sensor. Hence, reversible guest-induced "off–on" and "on–off" fluorescence sensors and switches are both designable.

Usually, "chemosensors" are used in homogeneous solution to detect target concentrations as low as micromolar to nanomolar [72–74]. On the other hand, optical probing of individual binding events represents the ultimate degree of sensitivity for sensing and imaging. However, there have been only a few single-molecule fluorescence spectroscopy studies of potential chemo- or biosensors, that is, fluorophores that respond to binding events via changes in their spectroscopic properties. This is mainly due to the fact that most fluorophores used in sensors to date, for example, anthracene or coumarin derivatives, have to be excited in the UV and exhibit a relatively low photostability [75]. Optimally, new fluorescence sensors should be excitable in the visible range and exhibit a sufficient fluorescence quantum yield and photostability to ensure their detection at the single-molecule level.

In this context, perylene 3,4,9,10-tetracarboxyl bisimide fluorophores have emerged as useful alternative to probe local structure, reversible chemical reactions, and interfacial processes such as electron transfer (Figure 7.4) [76]. The perylene 3,4,9,10-tetracarboxyl bisimide fluorophore can be excited in the green wavelength range and exhibits a high photostability and fluorescence quantum yield (>99%). To realize a chemical sensor, a *p*-aminophenyl group conjugated to the perylene bisimide has to be attached, which can efficiently quench the fluorescence of the fluorophore in the unbound state via intramolecular PET. Benzylic amino moieties covalently linked to a fluorophore have been used widely to probe the presence of metal ions or protons [72]. The high-energy nonbonding electron pair of the nitrogen atom efficiently quenches the excited state of the fluorophore via PET (Figure 7.4). Thermally-induced back-electron transfer from the lowest unoccupied molecular orbital (LUMO) of the fluorophore follows as a self-repair mechanism to regenerate the potentially damaged PET sensor. Protonation of the nonbonding electron pair of

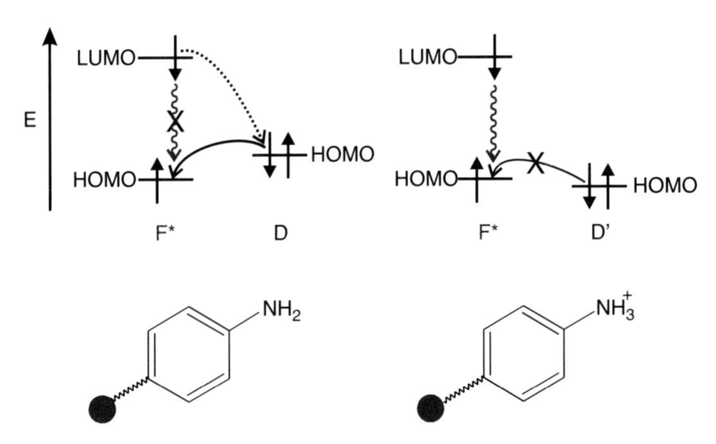

Figure 7.4 Molecular structure of a perylene 3,4,9,10-tetracarboxyl bisimide derivative and simplified frontier orbital energy diagrams illustrating the thermodynamics of intramolecular PET. Fluorescence is efficiently quenched by the benzyclic amino moiety (electron donor) covalently attached to the fluorophore. Thermally-induced charge recombination repopulates the singlet ground state of the fluorophore. Protonation of the nitrogen group or reaction with metal atoms lower the energy of the HOMO and prevent PET. Thus, fluorescence of the fluorophore is restored.

the amino group or titration with solutions of $ZnCl_2$ in THF (tetrahydrofuran), Pt $(SEt_2)_2Cl_2$ in $CHCl_3$, TiO_2 nanoparticles in THF, and aldehydes in $CHCl_3$ lower the energy of the electron pair. Thus, PET is no longer possible thermodynamically and fluorescence is restored [76]. Chemical PET sensors such as these can thus provide sensitive reports about their nanoenvironment, for example, the protonation state or pH value.

This somewhat classical method of transducing a metal-binding event into a fluorescent signal relies on metal binding induced alterations of the redox potential of a receptor molecule covalently attached to a fluorophore. Alternatively, a synthetic polypeptide template (zinc finger domain) and a covalently attached fluorescent reporter can be used as an efficient sensor for metal atoms [77]. Such sensors rely on site-specific labeling with appropriate fluorophores and conformationally induced alterations in energy transfer efficiency upon binding. Although fluorescence resonance energy transfer (FRET) has been successfully transposed into the single-molecule level only a few potential biosensors based on PET have been investigated at the single-molecule level.

On the other hand, new PET biosensors have been developed that use conformationally induced alterations in the PET efficiency upon binding for the specific detection of DNA or RNA sequences and antibodies, and for the study of biopolymer dynamics under equilibrium conditions at the single-molecule level [14, 16, 29, 32, 33, 38, 78–83]. The method takes advantage of specific properties of naturally occurring DNA nucleotides, and amino acids and will be explained in more detail in the next chapter.

7.4
PET Reporter System

In order to monitor conformational dynamics of biomolecules, such as the initial steps of protein folding or conformational transitions crucial for functional mechanisms, distance changes between certain amino acid or nucleic acid residues within the polymer chain have to be measured with high spatial and temporal resolution. From the spectroscopic point of view, any distance dependent energy transfer method, such as resonant or non-resonant energy transfer, proton transfer, or photoinduced electron transfer reactions, can be used to record conformational changes provided they are reflected in alterations in the interaction distance or geometry of extrinsic or intrinsic probe molecules, for example, fluorophores. From the technical point of view one is confronted with the problem of synchronization of conformational dynamics. Various techniques have been applied to induce the synchronous conformational change of an ensemble of molecules by ultra-rapid mixing [84, 85], temperature or pressure jumps [86], or photochemical triggering methods [87]. An elegant alternative to ensemble spectroscopic experiments is provided by single-molecule fluorescence spectroscopy. The observation of conformational fluctuations of an individual (bio)polymer is advantageous because the need for any synchronization procedure can be avoided.

On the other hand, a reporter system consisting of a fluorophore and a quencher has to be identified that can efficiently report on the structure formation and conformational dynamics with high specificity. For example, coupling of a fluorescein dye to a hydrophobic cavity of an intestinal fatty acid binding protein enabled the observation of characteristic fluorescence fluctuations by fluorescence correlation spectroscopy (FCS) [88]. As it is known that fluorescein is quenched fairly efficiently by several amino acids, in particular, by tryptophan residues [37, 89], it is likely that the intensity fluctuations observed are caused by fluctuations in the interaction geometry of the fluorescein moiety and a tryptophan residue located in close proximity. Hence, small conformational changes are directly reflected in the charge separation efficiency. That is, PET reactions enable the direct monitoring of sub-nanometer conformational fluctuations of a protein in its native state at the single-molecule level with microsecond time resolution. Furthermore, PET between a tyrosine residue and the flavin moiety could be directly observed in single-protein molecules [90]. Correlation of the fluorescence lifetime fluctuations measured from

Table 7.2 Relative fluorescence quantum yields, $\Phi_{f,\mathrm{rel}}$ of the oxazine fluorophore MR121 covalently attached to different peptides and oligonucleotides. Fluorescence is selectively quenched by tryptophan and guanine residues. Abbreviations for the amino acid residues are: A: alanine; D: aspartic acid; E: glutamic acid; F: phenylalanine; G: glycine; I: isoleucine; K: lysine; L: leucine; N: asparagine; P: proline; Q: glutamine; R: arginine; S: serine; T: threonine; W: tryptophan.

Sequence	$\Phi_{f,\mathrm{rel}}$
Peptide: MR121-SQETFSDLFKLLPEN	1.00
Peptide: MR121-SQETFSDL**W**KLLPEN	0.17
Peptide: MR121-SPDDIEQ**W**FTEDPGPDEAPR	0.28
Peptide: MR121-SPDDIEQFFTEDPGPDEAPR	1.00
Oligo: MR121-ACTAATTAATTAACC	1.00
Oligo: MR121-ACT**G**ATTAATTAACC	0.42

single flavin molecules enabled the monitoring of conformational dynamics with sub-millisecond time resolution.

As seen in Section 7.1. tryptophan (Trp) and guanine (G) are found to be the only compounds among all amino acids and nucleotides, respectively, that serve as potent electron donors and selectively quench most fluorophores via PET (Table 7.2). Thus, the incorporation of fluorophore/Trp or fluorophore/G pairs in polypeptides or nucleic acids and investigation of fluorescence fluctuations yields a universal single-molecule sensitive method to investigate fast conformational dynamics with nanosecond time resolution [14, 16, 29, 32, 33, 38, 78–83]. This requires site-specific modification of biopolymers with a fluorophore and a quencher, for example, through NHS-ester or maleimide reactive linkers. The naturally occurring compounds tryptophan and guanine represent intrinsic quenchers that can be incorporated by site-directed mutagenesis, or chemical peptide or oligonucleotide synthesis.

As previously mentioned, complex formation occurs between most organic fluorophores and tryptophan and guanine residues in aqueous solution. This complex formation is mostly due to intermolecular forces mediated by solvent properties, and to a first approximation is independent of the photophysical state. Alterations of the spectral characteristics are thus a likely indicator of, but not a prerequisite for, complex formation. A computational example of bimolecular complex formation between fluorophores (Rhodamine 6G and the oxazine derivative MR121) and Trp was given by Vaiana *et al.* [22]. Molecular dynamics (MD) simulations in explicit water were used to show formation of non-fluorescent hydrophobic complexes and estimated a binding energy of 20–30 kJ mol^{-1}.

To construct a suitable PET-reporter system, the fluorophore and quencher have to be attached to the (macro)molecule of interest in a well defined manner. Site-specific labeling necessarily requires a molecular linker between the aromatic ring system and the (macro)molecule. Most fluorophores are synthesized with a reactive group attached through a poly-carbon linker. This linker introduces a certain variation in the position and orientation of the fluorophore relative to the attachment point on the macromolecule. These degrees of freedom allow the fluorophore and quencher to align correctly to form a stacked complex as soon as the attachment points are closer than a certain threshold distance. With the stacked complex presenting a local energy minimum for the separation distance of the fluorophore and quencher, two distinct states, the complexed and non-complexed states with highly efficient and inefficient PET, respectively, can be distinguished. Any intermediate state in which the PET efficiency might have intermediate values is not heavily populated and therefore not detected in the experiment. PET efficiencies of 1 and 0, that is, less than the molecular contact or separation, can be clearly distinguished through the fluorescence emission of any fluorophore with a fluorescence emission rate lower than the largest PET rate. The PET interaction thus reports on the contact between the fluorophore–quencher pair with exquisite contrast in the observed fluorescence signal.

In an idealized system, the fluorophore and quencher interact as soon as the attachment points are closer than the reaction radius. The reaction radius generally influences the measured rate constants for the diffusional encounter of the attachment points. This is shown in various theoretical and computational treatments of first contact times in diffusion-limited reactions [91–94]. In order to characterize the effective reaction radius of the MR121/Trp fluorophore (F)/quencher pair, polyprolines F-Pro$_x$-Trp with $0 < x < 10$ were studied [80]. In aqueous solution, polyprolines form a rather rigid structure with a large PPII-helix content [95, 96], and the distance between the attachment points is fairly well defined. The dependence of static quenching as a function of the number of proline residues yields a steep transition from strong to very low quenching around $x = 4$ (albeit not quite zero due to prolyl-isomerization based subpopulations [80, 97]) reflecting an effective reaction radius of the order of ~1 nm.

In comparison with FRET systems, which can be seen as molecular ruler for probing distances continuously between 1 and 10 nm, the PET system provides a molecular measuring rod to distinguish between distances above and below ~1 nm. Whereas FRET determines the fluorescence readout continuously as a function of distance, PET yields a digital fluorescence readout with maximum contrast. Even though such a two-state signal provides less spatial information about the molecular system, transitions in between the two states can be analyzed in a straightforward manner and with high temporal resolution. Single-molecule fluorescence spectroscopy and FCS allow the analysis of transition rates between a few nanoseconds and minutes [11, 98, 99], provided that no other photophysical effects interfere. Such interferences can arise from fluorescence intermittency due to population of non-fluorescent triplet states (intersystem crossing). When monitoring fluorescence fluctuations it is important to use a well characterized fluorophore and to test the dependence on excitation power: fluctuations due to conformational dynamics as reported by PET-based quenching are independent of excitation power, whereas triplet states are heavier populated with increasing excitation rates. The oxazine dyes MR121 and ATTO655 exhibit a very low intersystem crossing rate [43, 53] and are thus particularly well suited for probing microsecond kinetics.

In a number of studies long-range electron transfer between electron D and A either through-space or through-bond have been observed [1, 5, 100]. Rate constants for such ET processes strongly depend on the connecting bonds and the environment around D and A [5]. However, it has been shown [78] that no such long-range ET significantly influences the dye–quencher reporter system discussed previously. This important conclusion is based on the following observations. (i) PET quenching between oxazine and rhodamine fluorophores and Trp spaced by polyprolines exhibit strong length dependence for the static quantum yield and the number of strongly quenched complexes (as measured by FCS) [78]. In other words, what changes with D–A distance is the ratio between quenched and non-quenched molecules or the efficiency of formation of the quenched complexes. At the same time only a weak length dependence for the dynamic quantum yield (estimated from fluorescence lifetime) and the brightness per molecule (measured by FCS) was found. In a distance-dependent long-range ET process, the quenching rate as reflected in the dynamic quantum yield would change continuously. (ii) A molecular construct with Rhodamine 6G directly labeled to the *N*-terminus of Trp is strongly fluorescent with a quantum yield close to that of the free dye. This can be rationalized by the fact that the local orientation of Rhodamine 6G is heavily restricted, such that the aromatic ring systems cannot enter into a stacked orientation. Any through-bond contribution that would have to be seen with maximum efficiency in such a small construct can thus be excluded.

7.5 Monitoring Conformational Dynamics and Protein Folding by PET

Intrachain contact formation in an unfolded polypeptide chain is thought to be the initial step in protein folding [101, 102]. In very early studies, the dynamics of unfolded polypeptide chains were investigated by Haas *et al.* using FRET from D- to A-fluorophores placed at the chain ends of an oligopeptide [103]. Later, intrachain diffusion in chemically unfolded heme proteins was estimated by probing ligation kinetics using photosensitization, or electron transfer kinetics [104, 105]. More recently, intrachain contact formation rates in unfolded polypeptides were investigated by triplet–triplet energy transfer (TTET) [102, 106, 107], triplet quenching [108, 109], and fluorescence quenching of long-lifetime fluorescence probes [110, 111]. In addition, contact-induced PET-quenching of rhodamine or oxazine dyes by Trp can be used successfully to monitor conformational dynamics of polypeptides [14, 22, 81]. Here, fluorescence fluctuations induced by contact-induced quenching can be analyzed by PET-FCS with nanosecond time resolution at the single-molecule level under equilibrium conditions (Figure 7.5).

The potential of the PET-FCS technique can be demonstrated studying end-to-end contact formation in unstructured, highly flexible polypeptides [102, 107]. These polypeptides consist of repetitive units of glycine–serine (GS) residues and were modified with the fluorophore MR121 and Trp at the amino- and carboxy-terminal ends, respectively. PET-based quenching interactions report on end-to-end contact formation of the polypeptide backbone (i.e., loop closure) mediated by intrachain diffusion (Figure 7.5c). From the resulting decay of the autocorrelation function in FCS experiments, the end-to-end contact formation rate constant of the polypeptide chain can be extracted [81]. The observed dependence of the contact rate on

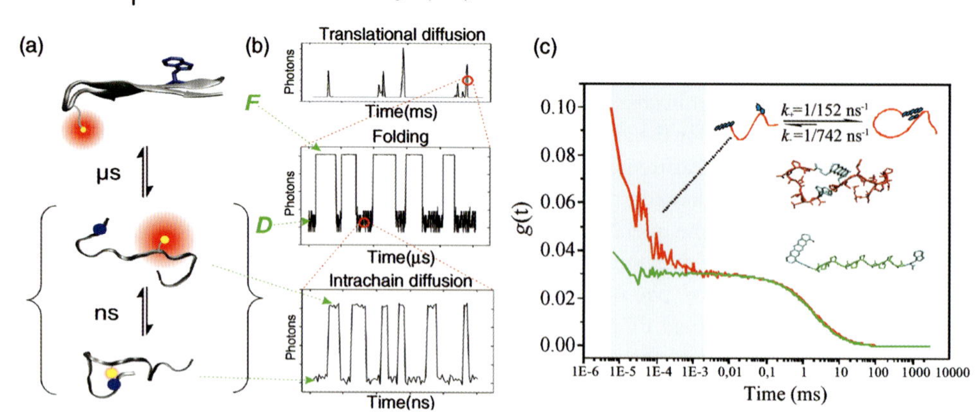

Figure 7.5 (a) Principle of studying fast -folding dynamics using PET-FCS. Reporter design: the extrinsic fluorophore (red) is site-specifically attached to the biopolymer (here a peptide forming a β-hairpin) such that fluorescence quenching contacts with Trp (blue) in the cross-strand hydrophobic cluster of the fold (*F*) cannot be formed. In contrast, in the denatured state (*D*) intramolecular fluorescence quenching contacts are formed mediated by intrachain diffusion. (b) Folding dynamics can be observed at the single-molecule level. Kinetics of β-hairpin folding and translational diffusion through the detection volume result in stochastic fluorescence fluctuations. Furthermore, intramolecular complex formation and dissociation of dye and Trp within the denatured state result in on–off fluorescence fluctuations on nanosecond time scales, reporting on the kinetics of intrachain diffusion. (c) Monitoring conformational dynamics in an unstructured (red) peptide by PET-FCS with nanosecond time resolution. The autocorrelation function is characterized by a decay in the millisecond time domain, reporting on translational diffusion, and a nanosecond to microsecond decay (gray area) reporting on internal conformational dynamics. Fluorescence of the dye attached to the peptide is quenched upon intramolecular contact formation with a single Trp residue in the peptide chain. Temporal fluorescence fluctuations directly reflect amino acid contact formation (k_+) and dissociation (k_-) dynamics (red curve). The green curve shows data recorded from a stiff, polyproline-based dye–peptide conjugate. In this peptide sequence, six proline residues space the dye conjugated to the amino terminal end from a single Trp residue at the carboxy terminus; intramolecular contact formation between the dye and Trp is prevented.

polypeptide length can be described using a Gaussian chain model [107]. Moreover, FCS measurements carried out under varying solvent viscosities demonstrated that formation of end-to-end contacts is purely diffusive. Therefore a poly-(GS)-peptide can serve as a model for an ideally unfolded protein chain that lacks non-random chain conformations caused by side-chain–side-chain or side-chain–backbone interactions. Recent experimental, computational, and theoretical work, however, suggest that backbone hydrogen-bond networks might be present [112–114].

The use of a reporter system that consists of a fluorophore and quencher that interact to form non-fluorescent complexes obviously has an effect on the energy landscape of the system under study. Investigation of the *bimolecular* interaction between oxazine or rhodamine fluorophores and Trp or guanosine revealed a binding constant of the order of 10–100 M^{-1} measured through changes in static quenching. The equivalent binding energy $\Delta G = \Delta H - T\Delta S = -RT\ln(K_S)$ is thus of the order of 6–13 kJ mol^{-1}.

However, an entirely different energy landscape arises for the reporter system when attached to a biopolymer. This is due to very different entropic contributions from solvent and linkers in the monomolecular system. Comparing MD simulations of GS-peptides with and without the reporter system reveals that the closed state is more populated due to stabilization by complex formation in the reporter system. However, the closed state is only stabilized by a few kJ mol^{-1}, up to an order of magnitude less than that observed in bimolecular interactions. Furthermore, with the stabilizing forces only being significant at short-range, the open configuration ensemble is not influenced, as shown by correct estimates of end-to-end contact rate constants for the polymers [81].

In conclusion, the use of a reporter system that is based on formation of stacked complexes inevitably influences the energy landscape. However, in many instances the influence is negligible in the context of the molecular process under study. In other cases, the altered molecular system directly illustrates the molecular processes, for example, as a validation link between experimental and computer simulation studies. Finally, a well designed molecular system with its own unique energy landscape can be used to study the relative effects under environmental or structural perturbations.

In contrast to unfolded poly-(GS) peptides, natural proteins consist of heterogeneous amino acid sequences that have been selected through evolution to spontaneously adopt characteristic three-dimensional structures tailored for function [101]. The unique fold of the protein is solely determined by its amino acid sequence. In the folded state the entropic contributions to the free energy of the originally unfolded chain in water is canceled by enthalpic contributions that arise from a large variety of intramolecular interactions within the polypeptide chain [55]. These include van der Waals, hydrogen bonding, and Coulombic interactions. Characterization of interaction networks and the possible residual structure in the unfolded state in addition to its dynamic nature is crucial for the understanding of folding mechanisms.

The introduction of the MR121/Trp reporter system at tailored positions in small proteins such as the 20-residue mini-protein Trp-cage, allows fast folding dynamics to be investigated by PET-FCS (Figure 7.6) [78]. The Trp-cage is one of the smallest proteins known to date [115]. A single Trp residue is buried in a small hydrophobic core, well shielded from solvent exposure. Site-specific modification of a lysine residue on the surface of the mini-protein yields a single-molecule sensitive reporter system for folding transitions (Figure 7.6). The stability of the protein remains essentially unperturbed by fluorescence modification [78, 115].

In the folded state MR121 is shielded from fluorescence quenching interactions by the Trp residue that is "caged" in the hydrophobic core, whereas in the denatured state Trp is solvent-exposed and accessible to MR121. As folding/unfolding transitions of the Trp-cage are in the microsecond time domain [116], they appear as monoexponential correlation decays in FCS experiments [78]. Moreover, dynamic properties of the denatured state of the Trp-cage can be extracted from nanosecond relaxations. In the denatured state, intrachain contact formation between MR121 and Trp is possible and the corresponding contact formation is mediated by the flexibility of the peptide backbone. In thermal denaturation experiments essentially

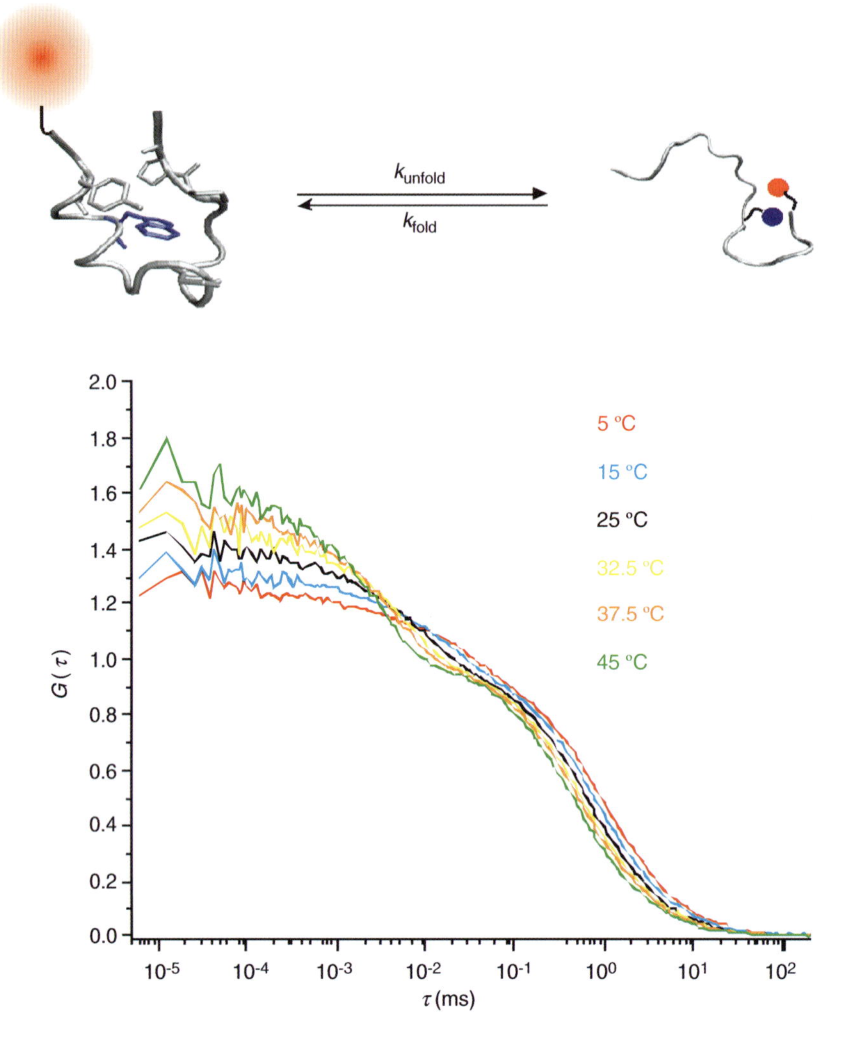

Figure 7.6 Temperature dependent PET-FCS data of fluorescently modified Trp-cage (pdb code 1L2Y). Selective fluorescence quenching of MR121 (red) in the denatured state of the Trp-cage by the single Trp (blue) residue present in the sequence results in "on"–"off" fluorescence switching correlated with folding–unfolding transitions and result in a microsecond decay of the autocorrelation function reporting on folding kinetics. With increasing temperature the "quenching amplitude" in the microsecond time range increases. Data analysis of PET-FCS curves reveals that folding occurs in a few microseconds. Measurements were performed in PBS, pH 7.4.

no such fluctuations are present in FCS data, even beyond the mid-point temperature of 35 °C. In contrast, upon chemical denaturation and introduction of a helix-breaking single-point mutation, nanosecond fluorescence fluctuations with considerable amplitude are evident in FCS data reporting on increased conformational flexibility of the denatured state ensemble. Therefore, PET-FCS reveals that under conditions that favor folding, the unfolded peptide chain first collapses to a conformationally confined denatured state ensemble from where the final folding transition proceeds.

The denatured state ensemble of Trp-cage is apparently considerably different from that observed for idealized model systems, such as poly-(GS) peptides. This observation addresses important and open questions in protein folding: how "random" is the conformational ensemble of the protein denatured states? Is collapse of the originally unfolded protein chain to a conformationally confined denatured state a general feature in protein folding, and is structure formation involved even during this early stage of folding? These examples demonstrate that PET-FCS combined with protein engineering represents a valuable tool to address fundamental questions of protein folding.

7.6 Biological and Diagnostic Applications

The extraordinary sensitivity of fluorescence-based methods is especially important for the application of "smart" fluorescent probes, used for the detection of minute amounts of target structures from biological samples. In the past PET sensors or probes have been developed that use conformationally induced alterations in PET efficiency upon binding, for the specific detection of DNA or RNA sequences, antibodies, and also proteases and nucleases at the single-molecule level [29, 32, 33, 37, 38, 83, 117]. Again, these PET probes take advantage of specific fluorescence quenching of fluorophores by selected quenchers such as the naturally occurring DNA nucleotides (G) and amino acids (Trp) to probe the presence of target molecules. With careful design of these conformationally flexible sensors, efficient single-molecule sensitive PET-probes can be produced. If quenching interactions between the fluorophore and G or Trp residue are deteriorated upon specific binding to the target, for example, due to binding of a complementary DNA sequence or antibody, or due to cleavage by an endonuclease or protease enzyme, fluorescence of the sensor is restored.

For example, selective contact-induced fluorescence quenching of rhodamine and oxazine derivatives has been successfully used to study end-to-end contact rates in unstructured poly-thymine with a length of 4 to 10 nucleotides [79]. End-to-end contact rates scale with polymer length according to a power law as is expected for unstructured polymers. Through extending the termini by complementary bases that form a closed stem due to base pairing, it has been demonstrated that as few as one or two complementary bases in the stem region reduce opening and closing rate constants by an order of magnitude [79]. Likewise, protein binding to DNA or RNA can be studied by PET-quenching interactions. Here, changes in conformational dynamics of RNA or DNA upon protein binding can be analyzed as the reporter for the binding process [82]. Such experiments yield structural information about various binding modes but also serve as diagnostic tools to detect protein–nucleotide

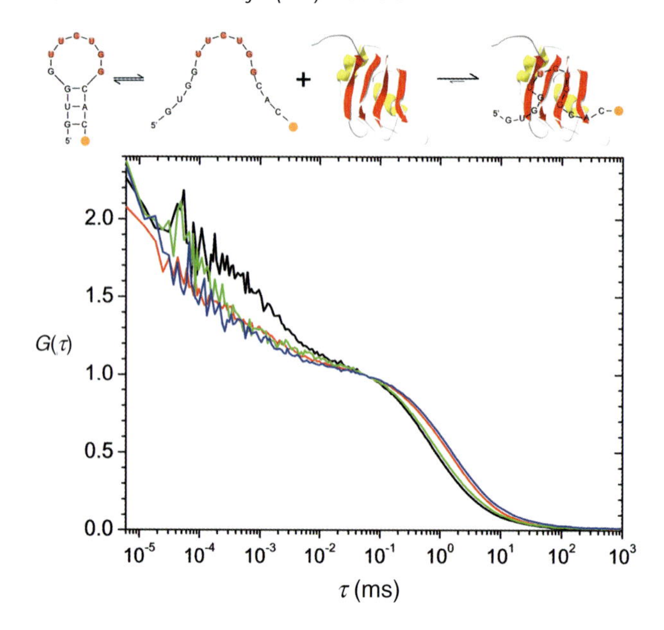

Figure 7.7 Normalized FCS curves measured from 10^{-9} M solutions of two MR121 labeled RNA sequences (native sequence, NS: MR121-5′-UUUG**UUCUGG**UUC-3′ and hairpin, HP: MR121-5′-GUGG**UUCUGG**CAC-3′) in the absence and presence of the RNA binding protein *At*GRP7 [82]. The RNA recognition sequence is given in bold letters; hairpin forming stem nucleotides are underlined. The fluorescence correlation function of the free RNA hairpin, HP, and native sequence, NS, is shown in black and green, respectively. The curve of the HP bound to the *At*GRP7 protein is shown in red, the NS bound to *At*GRP7 in blue. The proposed model of the conformation of a two-state hairpin folding that can be bound by the protein in the unfolded state is illustrated above. The protein binds preferentially to the fully stretched structure where the recognition sequence is not involved in hairpin formation. Binding of the RNA binding protein is reflected in an increase in diffusion time and a decrease in quenching amplitude (compare the black and red curves).

and also protein–protein interactions. For example (Figure 7.7), PET-FCS experiments demonstrate that the conformational flexibility of short single-stranded (ss) fluorophore-labeled oligonucleotides is reduced upon binding of the RNA binding protein *At*GRP7. In contrast to many other RNA recognition motif proteins, *At*GRP7 binds to ssRNA preferentially if the RNA is fully stretched and not embedded within a stable secondary structure. The results suggest that *At*GRP7 binding leads to a conformational rearrangement in the mRNA, arresting the flexible target sequence in an extended structure of reduced flexibility, which may have consequences for further post-transcriptional processing of the mRNA (Figure 7.7) [82].

With careful design of conformationally flexible PET probes a new class of DNA-hairpins – so-called "smart probes" – can be synthesized [29, 32, 33, 83]. Smart probes are single-stranded nucleic acids that adopt a stem–loop structure and are ideally suited for highly sensitive homogeneous and heterogeneous DNA binding or

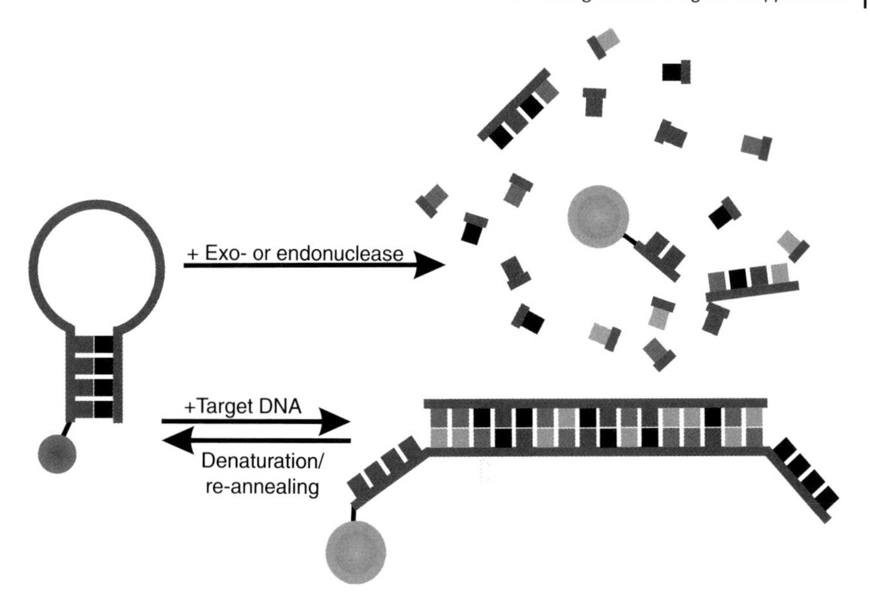

Figure 7.8 Scheme of the functional principle of smart probes. The fluorophore is attached to the 5′-end of the oligonucleotide and quenched by G residues in the complementary stem. Upon hybridization to the target sequence (complementary to the loop sequence) or exo- or endonucleolytic digestion fluorescence is restored.

cleavage assays (Figure 7.8). The loop consists of a probe sequence that is complementary to a portion of the target sequence, whereas the stem is formed by annealing of two complementary strands that are unrelated to the target sequence. In contrast to molecular beacons [118–120] where an additional quencher has to be attached to the oligonucleotide, in smart probes intrinsic guanosine residues terminating the 3′-end of the hairpin are responsible for strong fluorescence quenching of the dye labeled to the 5′-end. If quenching interactions between the fluorophore and the guanosine residue are deteriorated, for example, due to binding of a complementary DNA sequence, or due to cleavage by an endo- or exonuclease enzyme, fluorescence of the DNA-hairpin is restored (Figure 7.8). DNA-hairpins labeled with a single oxazine dye at the 5′-end increase fluorescence upon hybridization up to 20-fold [32], which provides the basis for a cost-effective, and highly sensitive DNA/RNA detection method. The design of efficient PET-based DNA-hairpins has been achieved by a careful study of various factors that influence the photoinduced intramolecular electron transfer efficiency. Among these are the selection of suitable fluorophores, the influence of the guanosine position in the complementary stem, the attachment of additional overhanging single-stranded nucleotides in the complementary stem, and the exchange of guanosine by more potent electron donors, such as 7-deazaguanosine [32]. In this way smart probes have also been successfully used for the sensitive and specific identification of mycobacterial strains [83].

In addition, smart probes can be used to efficiently detect the presence of single target DNA or RNA molecules, even in the sub-picomolar concentration range within a reasonable period of time [33]. This has been achieved by immobilization of smart probes on coated cover slips while maintaining their native conformation. Therefore, the method is ideally suited to the search for specific sequences using extremely low concentrations of target sequence. It is anticipated that optimization of reaction conditions (salt, temperature, etc.) and variations of the fluorophore in addition to the use of modified nucleotides can result in even higher sensitivities.

The extraordinary sensitivity of the PET-based quenching method can also be used for the detection of antibodies using specific peptide epitopes [38, 121]. Through the design and synthesis of peptide epitopes based on selective fluorescence quenching of an attached fluorophore by a single tryptophan residue, the detection of p53-autoantibodies was achieved with very high sensitivity in homogeneous solution. P53-autoantibodies are among the most encouraging universal tumor markers in cancer diagnosis. They are found in sera of a significant number of cancer patients who have different cancer types, and have the potential for early-stage diagnosis of certain tumors [122]. Common heterogeneous diagnostic tools for antibody detection, such as ELISA (enzyme-linked immunosorbent assay), are time consuming and suffer from a lack of specificity. Hence, the development of fast and simple homogeneous formats is favored in biomedical sciences. Owing to the unique quenching mechanism that requires contact formation between the fluorophore and the tryptophan residue, fluorescently labeled conformationally flexible peptides that serve as epitopes for the antibody can be used as efficient PET biosensors. Upon specific binding to a p53-autoantibody the peptide chain adapts to the shape of the antibody cleft. Consequently, contact formation between tryptophan and fluorophore is prevented (Figure 7.9). The resulting increase in fluorescence intensity can be used to signal binding events. In combination with the single-molecule sensitivity, the new assay allows for direct monitoring of p53-antibodies present in blood serum samples of cancer patients with picomolar detection sensitivity [38].

Alternatively, the selective quenching mechanism in tryptophan-containing peptides can be used advantageously for the detection of proteases. The interest in fast and sensitive assays for proteolytic enzymes, that is, enzymes that specifically cleave peptide bonds, has increased considerably in the last few years. In particular, there are two medically important facts that accelerated the development of proteolytic assays. Firstly, more and more diseases can be implicated using proteases. Because of the involvement of proteases in tumor progression and metastasis, for example, matrix metalloproteinases, urokinase plasminogen activator (uPA), and cathepsins, such as cathepsin B, and cathepsin D, proteolytic assays play a central role in cancer diagnosis and follow-up of malignant diseases [123–128]. Secondly, viral infections such as HIV can be diagnosed directly by detection and monitoring of their own, specific proteases. This underscores the need for new highly sensitive and fast assays for the specific detection of proteolytic enzymes.

As chemical modifications on peptide or nucleic acid substrates reduce the affinity of the resulting probe for the target molecule and subsequently the detection sensitivity of the assay, minimum chemical modification is favorable for the design of high-affinity molecular probes for proteases. Here again, contact-induced fluo-

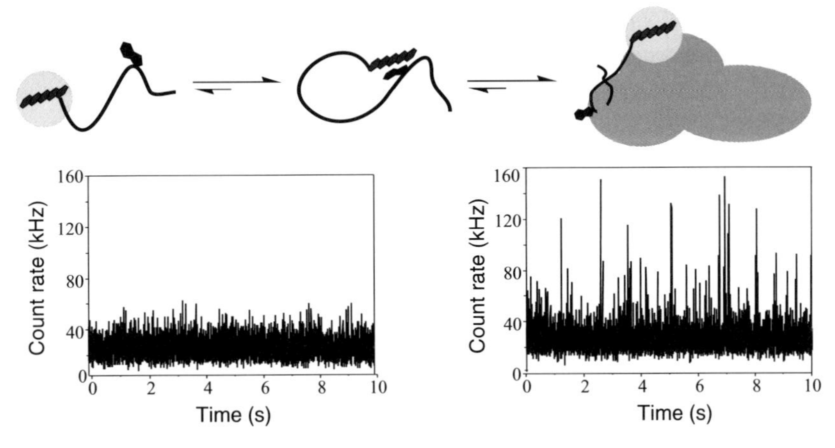

Figure 7.9 Principle of operation of peptide-based PET biosensor for the specific detection of p53-autoantibodies. Driven by the conformational flexibility, the peptide adopts conformations where the fluorophore is efficiently quenched by Trp upon contact formation. The quenched conformation coexists in equilibrium with an open, unquenched conformation. Upon specific binding to the antibody binding site, the peptide adopts a new conformation and charge transfer interactions between tryptophan and the dye are diminished. As the consequence, the fluorescence intensity increases (right-hand side). The fluorescence trajectories (1 ms integration time) observed from 1:10 diluted human serum samples of a healthy donor (left) and from a breast cancer patient containing 10^{-11} M peptides demonstrate the potential of such conformational flexibly PET biosensors [38].

rescence quenching via PET between fluorophores and the amino acid tryptophan or DNA base guanine emerge as methods for ultra-sensitive and specific detection of protease and nuclease enzymes in homogeneous solution (Figure 7.10) [117]. The PET-based approach to assay proteases and nucleases offers several advantages. Firstly, the synthesis of large quantities of quenched probes is uncomplicated and inexpensive compared with the use of doubly labeled probes. Secondly, modification of the peptide or nucleic acid substrate is reduced to a minimum, that is, a single reporter fluorophore, which potentially enables more accurate monitoring of enzyme activity. By monitoring the time-dependent fluorescence intensity, the presented technique permits real-time analysis of enzyme activity (see Chapter 9). Thirdly, the assay enables fast, specific, and highly sensitive detection of enzymes with a broad dynamic range (more than six orders of magnitude), and detection limits below the picomolar range (Figure 7.10) [117]. The capability of single-molecule experiments in the far-red spectral range to study biomolecular recognition and enzymatic activity in complex biological media, such as human blood serum, opens up the possibility of developing diagnostic tools for *in vivo* applications and near-patient testing in the near future.

Finally, it has to be pointed out that the smart probe method can also be combined with bead-based heterogeneous detection strategies. Although more sensitive methods, for example, single-molecule fluorescence spectroscopy (SMFS), are about to

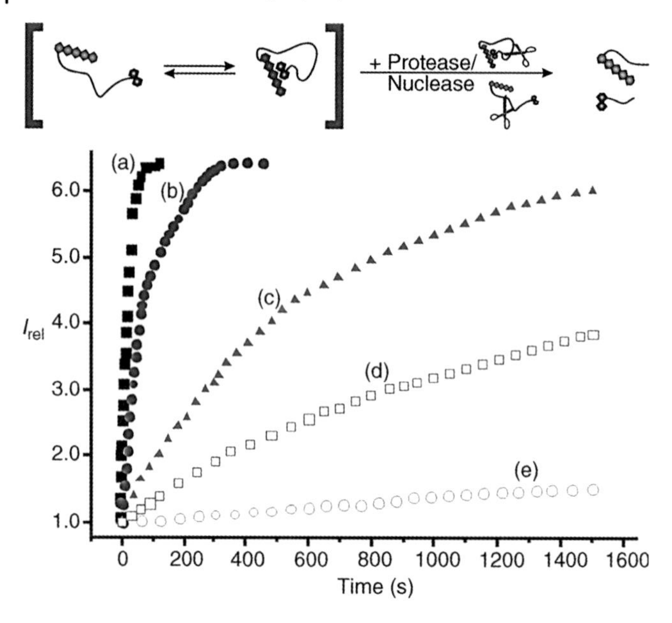

Figure 7.10 PET-quenched probes for the specific detection of proteases and nucleases in homogeneous assays. Owing to the conformational flexibility of the biopolymer (peptide or oligonucleotide) the fluorophore is efficiently quenched by, for example, a tryptophan residue located at the other end of the peptide via contact-induced PET quenching. In the presence of a protease that specifically recognizes the peptide sequence located between the fluorophore and tryptophan residue, contact formation and subsequent fluorescence quenching is prevented due to specific cleavage. Thus, the fluorescence intensity of the fluorophore increases. The lower graph shows the relative fluorescence intensities, I_{rel}, measured for the fluorescently labeled peptide MR121-Lys-Trp at a concentration of 10^{-7} M versus time, after addition of various amounts of carboxypeptidase A (CPA). (a) 10^{-6} M, (b) 10^{-9} M, (c) 10^{-11} M, (d) 10^{-13} M, and (e) 10^{-15} M CPA. Measurements were performed in pure water at 25 °C (excitation wavelength: 640 nm; emission wavelength: 690 nm). Data points were measured every 5–20 s switching between the various samples [117].

break the current barriers of sensitivity, alternative approaches, such as heterogeneous assays can be used to increase the sensitivity by use of less demanding techniques [129, 130]. Thus, the increase in fluorescence intensity of smart probes upon binding can be combined with the enrichment of the signal on the bead surface. To construct a smart probe bead assay for the specific detection of target DNA or RNA sequences capture, oligonucleotides have to be immobilized on micrometer-sized beads, for example, via biotin/streptavidin binding on paramagnetic beads, and appropriately designed smart probes have to be added to the sample. In the presence of specific target sequence a sandwich-type complex is formed and fluorescence is enriched on the beads. Thus, standard wide-field fluorescence imaging can be used to quantify the fluorescence brightness of the beads without the use of any washing step.

References

1 Marcus, R.A. and Sutin, N. (1985) *Biochim. Biophys. Acta*, **811**, 265–322.
2 Barbara, P.F., Meyer, T.J., and Ratner, M.A. (1996) *J. Phys. Chem.*, **100**, 13148–13168.
3 Kavarnos, G.J. (1993) *Fundamentals of Photoinduced Electron Transfer*, Wiley-VCH Verlag GmbH, Weinheim.
4 Jortner, J. and Bixon, M. (1999) *Adv. Chem. Phys.*, **106**, 1–734.
5 Adams, D.M., Brus, L., Chidsey, C.E.D., Creager, S., Creutz, C., Kagan, C.R., Kamat, P.V., Lieberman, M., Lindsay, S., Marcus, R.A., Metzger, R.M., Michel-Beyerle, M.E., Miller, J.R., Newton, M.D., Rolison, D.R., Sankey, O., Schanze, K.S., Yardley, J., and Zhu, X. (2003) *J. Phys. Chem. B*, **107**, 6668–6697.
6 Ha, T. (2001) *Methods*, **25**, 78–86.
7 Selvin, P.R. (2000) *Nat. Struct. Biol.*, **7**, 730–734.
8 Weiss, S. (1999) *Science*, **283**, 1676–1683.
9 Dietrich, A., Buschmann, V., Müller, C., and Sauer, M. (2002) *Rev. Mol. Biotechnol.*, **82**, 211–231.
10 Lewis, F.D., Letsinger, R.L., and Wasielewski, M.R. (2001) *Acc. Chem. Res.*, **34**, 159–170.
11 Neuweiler, H. and Sauer, M. (2004) *Curr. Pharm. Biotechnol.*, **5**, 285–298.
12 Holman, M.W. and Adams, D.M. (2004) *ChemPhysChem*, **5**, 1831–1836.
13 Cotlet, M., Masuo, S., Luo, G., Hofkens, J., van der Auweraer, M., Verhoeven, J., Müllen, K., Xie, X.S., and De Schryver, F. (2004) *Proc. Natl. Acad. Sci. USA*, **101**, 14343–14348.
14 Doose, S., Neuweiler, H., and Sauer, M. (2009) *ChemPhysChem*, **10**, 1389–1398.
15 Lor, M., Thielemans, J., Viaene, L., Cotlet, M., Hofkens, J., Weil, T., Hampel, C., Müllen, K., Verhoeven, J.W., Van der Auweraer, M., and De Schryver, F.C. (2002) *J. Am. Chem. Soc.*, **124**, 9918–9925.
16 Doose, S., Neuweiler, H., and Sauer, M. (2005) *ChemPhysChem*, **6**, 2277–2285.
17 Jelley, E.E. (1936) *Nature*, **138**, 1009–1010.
18 Scheibe, G. (1937) *Angew. Chem.*, **50**, 212–219.
19 Rehm, D. and Weller, A. (1970) *Israel J. Chem.*, **8**, 259–271.
20 Weller, A. (1982) *Z. Phys. Chem.*, **133**, 93–97.
21 Zhong, D. and Zwail, A.H. (2001) *Proc. Natl. Acad. Sci. USA*, **98**, 11867–11872.
22 Vaiana, A., Neuweiler, H., Schulz, A., Wolfrum, J., Sauer, M., and Smith, J.C. (2003) *J. Am. Chem. Soc.*, **125**, 14564–14572.
23 Zhong, D. and Zwail, A.H. (2001) *Proc. Natl. Acad. Sci. USA*, **98**, 11873–11878.
24 Sauer, M., Han, K.T., Müller, R., Nord, S., Schulz, A., Seeger, S., Wolfrum, J., Arden-Jacob, J., Deltau, G., Marx, N.J., Zander, C., and Drexhage, K.H. (1995) *J. Fluoresc.*, **5**, 247–261.
25 Seidel, C.A.M., Schulz, A., and Sauer, M. (1996) *J. Phys. Chem.*, **100**, 5541–5553.
26 Edman, L., Mets, U., and Rigler, R. (1996) *Proc. Natl. Acad. Sci. USA*, **93**, 6710–6715.
27 Sauer, M., Drexhage, K.H., Lieberwirth, U., Müller, R., Nord, S., and Zander, C. (1998) *Chem. Phys. Lett.*, **284**, 153–163.
28 Eggeling, C., Fries, J.R., Brand, L., Günther, R., and Seidel, C.A.M. (1998) *Proc. Natl. Acad. Sci. USA*, **95**, 1556–1561.
29 Knemeyer, J.P., Marmé, N., and Sauer, M. (2000) *Anal. Chem.*, **72**, 3717–3724.
30 Wallace, M.I., Ying, J.M., Balasubramanian, S., and Klenerman, D. (2000) *J. Phys. Chem. B*, **104**, 11551–11555.
31 Torimura, M., Kurata, S., Yamada, K., Yokomaku, T., Kamagata, Y., Kanagawa, T., and Kurane, R. (2001) *Anal. Sci.*, **17**, 155–160.
32 Heinlein, T., Knemeyer, J.P., Piestert, O., and Sauer, M. (2003) *J. Phys. Chem. B*, **107**, 7957–7964.
33 Piestert, O., Barsch, H., Buschmann, V., Heinlein, T., Knemeyer, J.P., Weston, K.D., and Sauer, M. (2003) *Nano Lett.*, **3**, 979–982.
34 Wagenknecht, H.A., Stemp, E.D., and Barton, J.K. (2000) *J. Am. Chem. Soc.*, **122**, 1–7.
35 DeFelippis, M.R., Murthy, C.P., Broitman, F., Weinraub, D., Faraggi, M.,

and Klapper, M.H. (1991) *J. Phys. Chem.*, **95**, 3416–3419.

36 Jovanovic, S.V., Harriman, A., and Simic, M.G. (1986) *J. Phys. Chem.*, **90**, 1935–1939.

37 Marmé, N., Knemeyer, J.P., Sauer, M., and Wolfrum, J. (2003) *Bioconjugate Chem.*, **14**, 1133–1139.

38 Neuweiler, H., Schulz, A., Vaiana, A.C., Smith, J.C., Kaul, S., Wolfrum, J., and Sauer, M. (2002) *Angew. Chem. Int. Ed.*, **41**, 4769–4773.

39 Vogelsang, J., Kasper, R., Steinhauer, C., Person, B., Heilemann, M., Sauer, M., and Tinnefeld, P. (2008) *Angew. Chem. Int. Ed.*, **47**, 5465–5469.

40 Lambert, C.R. and Kochevar, I.E. (1997) *Photochem. Photobiol.*, **66**, 15–25.

41 Widengren, J., Chmyrov, A., Eggeling, C., Lofdahl, P.A., and Seideln, C.A.M. (2007) *J. Phys. Chem. A*, **111**, 429–440.

42 Hoogenboom, J.P., van Dijk, E.M.H.P., Hernando, J., van Hulst, N.F., and Garcia-Parajo, M.F. (2005) *Phys. Rev. Lett.*, **95**, 097401.

43 Tinnefeld, P., Herten, D.P., Masuo, S., Vosch, T., Cotlet, M., Hofkens, J., Müllen, K., De Schryver, F.C., and Sauer, M. (2004) *ChemPhysChem*, **5**, 1786–1790.

44 Zondervan, R., Kulzer, F., Orlinskii, S.B., and Orrit, M. (2003) *J. Phys Chem. A*, **107**, 6770–6776.

45 Zondervan, R., Kulzer, F., Kol'chenko, M.A., and Orrit, M. (2004) *J. Phys Chem. A*, **108**, 1657–1665.

46 van de Linde, S., Endesfelder, U., Mukherjee, A., Schüttpelz, M., Wiebusch, G., Wolter, S., Heilemann, M., and Sauer, M. (2009) *Photochem. Photobiol. Sci.*, **8**, 465–469.

47 Heilemann, M., van de Linde, S., Mukherjee, A., and Sauer, M. (2009) *Angew. Chem. Int. Ed.*, **48**, 6903–6908.

48 Buschmann, V., Weston, K.D., and Sauer, M. (2003) *Bioconjugate Chem.*, **13**, 195–204.

49 Eggeling, C., Ringemann, C., Medda, R., Schwarzmann, G., Sandhoff, K., Polyakova, S., Belov, V.N., Hein, B., von Middendorff, C., Schönle, A., and Hell, S.W. (2009) *Nature*, **457**, 1159–1162.

50 Nie, S.M. and Zare, R.N. (1997) *Annu. Rev. Biophys. Biomol. Struct.*, **26**, 567–596.

51 Lu, H.P. and Xie, X.S. (1998) *Science*, **282**, 1877–1882.

52 Moerner, W.E. and Orrit, M. (1999) *Science*, **283**, 1670–1676.

53 Tinnefeld, P. and Sauer, M. (2005) *Angew. Chem. Int. Ed.*, **44**, 2642–2671.

54 Cooper, A. (1999) *Curr. Opin. Chem. Biol.*, **3**, 557–563.

55 Baldwin, R.L. (2007) *J. Mol. Biol.*, **371**, 283–301.

56 Yip, W.T., Hu, D.H., Yu, J., Vanden Bout, D.A., and Barbara, P.F. (1998) *J. Phys. Chem. A*, **102**, 7564–7575.

57 Tinnefeld, P., Buschmann, V., Weston, K.D., and Sauer, M. (2003) *J. Phys. Chem. A*, **107**, 323–327.

58 Köhn, F., Hofkens, J., Gronheid, R., van der Auweraer, M., and De Schryver, F.C. (2002) *J. Phys. Chem. A*, **106**, 4808–4814.

59 Ambrose, W.P., Goodwin, P.M., Martin, J.C., and Keller, R.A. (1994) *Phys. Rev. Lett.*, **72**, 160–163.

60 Lu, H.P. and Xie, X.S. (1997) *Nature*, **385**, 143–146.

61 Weston, K.D., Carson, P.J., Metiu, H., and Buratto, S.K. (1998) *J. Chem. Phys.*, **109**, 7474–7485.

62 Weston, K.D. and Buratto, S.K. (1998) *J. Phys. Chem. A*, **102**, 3635–3638.

63 Yeow, E.K.L., Melnikov, S.M., Bell, T.D.M., De Schryver, F.C., and Hofkens, J. (2006) *J. Phys. Chem. A*, **110**, 1726–1734.

64 Haase, M., Hübner, C.G., Reuther, E., Herrmann, A., Müllen, K., and Basché, T. (2004) *J. Phys. Chem. B*, **108**, 10445–10450.

65 Schuster, J., Cichos, F., and von Borczyskowski, C. (2005) *Appl. Phys. Lett.*, **87**, 051915.

66 Zondervan, R., Kulzer, F., Orlinski, S.B., and Orrit, M. (2003) *J. Phys. Chem. A*, **107**, 6770–6776.

67 Tinnefeld, P., Herten, D.P., and Sauer, M. (2001) *J. Phys. Chem. A*, **105**, 7989–8003.

68 Clifford, J.N., Bell, T.D.M., Tinnefeld, P., Heilemann, M., Melinkov, S.M., Hotta, J., Sliwa, M., Dedecker, P., Sauer, M., Hofkens, J., and Yeow, E.K.L. (2007) *J. Phys. Chem. B*, **111**, 6987–6991.

69 Tachiya, M. and Mozumder, A. (1975) *Chem. Phys. Lett.*, **34**, 77–79.
70 Kuno, M., Fromm, D.P., Johnson, S.T., Gallaher, A., and Nesbitt, D.J. (2003) *Phys. Rev. B*, **67**, 125304.
71 Verberk, R., van Oijen, A.M., and Orrit, M. (2002) *Phys. Rev. B*, **66**, 233202.
72 de Silva, A.P., Gunaratne, H.Q.N., Gunnlaugsson, T., Huxley, A.J.M., McCoy, C.P., Rademacher, J.T., and Rice, T.E. (1997) *Chem. Rev.*, **97**, 155–1566.
73 Burdette, S.C., Walkup, G.K., Spingler, B., Tsien, R.Y., and Lippard, S.J. (2001) *J. Am. Chem. Soc.*, **123**, 7831–7841.
74 Gawley, R.E., Pinet, S., Cardona, C.M., Datta, P.K., Ren, T., Guida, W.C., Nydick, J., and Leblanc, R.M. (2002) *J. Am. Chem. Soc.*, **124**, 13448–13453.
75 Eggeling, C., Brand, L., and Seidel, C.A.M. (1997) *Bioimaging*, **5**, 105–115.
76 Zang, L., Liu, R., Holman, M.W., Nguyen, K.T., and Adams, D.M. (2002) *J. Am. Chem. Soc.*, **124**, 10640–10641.
77 Walkup, G.K. and Imperiali, B. (1998) *J. Am. Chem. Soc.*, **118**, 3053–3054.
78 Neuweiler, H., Doose, S., and Sauer, M. (2005) *Proc. Natl. Acad. Sci. USA*, **102**, 16650–16655.
79 Kim, J., Doose, S., Neuweiler, H., and Sauer, M. (2006) *Nucl. Acids Res.*, **34**, 2516–2527.
80 Doose, S., Neuweiler, H., Barsch, H., and Sauer, M. (2007) *Proc. Natl. Acad. Sci. USA*, **104**, 17400–17405.
81 Neuweiler, H., Löllmann, M., Doose, S., and Sauer, M. (2007) *J. Mol. Biol.*, **365**, 856–869.
82 Schüttpelz, M., Schöning, J.C., Doose, S., Neuweiler, H., Peters, E., Staiger, D., and Sauer, M. (2008) *J. Am. Chem. Soc.*, **130**, 9507–9513.
83 Stöhr, K., Häfner, B., Nolte, O., Wolfrum, J., Sauer, M., and Herten, D.P. (2005) *Anal. Chem.*, **77**, 7195–7203.
84 Eatton, W.A., Munoz, V., Thompson, P.A., Henry, E.R., and Hofrichter, J. (1998) *Acc. Chem. Res.*, **31**, 745–753.
85 Bryngelson, J.D. and Wolynes, P.G. (1987) *Proc. Natl. Acad. Sci. USA*, **84**, 7524–7528.
86 Shakhnovich, E.I. (1997) *Curr. Opin, Struct. Biol.*, **7**, 29–40.
87 Matthews, C.R. (1993) *Annu. Rev. Biochem.*, **62**, 653–683.
88 Chattopadhyay, K., Saffarian, S., Elson, E.L., and Frieden, C. (2002) *Proc. Natl. Acad. Sci. USA*, **99**, 14171–14176.
89 Watt, R.M. and Voss, E.W. (1977) *Immunochemistry*, **14**, 533–541.
90 Yang, H., Luo, G., Karnchanaphanurach, P., Louie, T.M., Rech, I., Cova, S., Xun, L., and Xie, X.S. (2003) *Science*, **302**, 262–266.
91 Szabo, A., Schulten, K., and Schulten, Z. (1980) *J. Chem. Phys.*, **72**, 4350–4357.
92 Pastor, R.W., Zwanzig, R., and Szabo, A. (1996) *J. Chem. Phys.*, **105**, 3878–3882.
93 Wilemski, G. and Fixman, M. (1974) *J. Chem. Phys.*, **60**, 866–876.
94 Portman, J.J. (2003) *J. Chem. Phys.*, **118**, 2381–2391.
95 Helbecque, N. and Lochcheux-Lefebvre, M.H. (1982) *Int. J. Pept. Prot. Res.*, **19**, 94–101.
96 Jacob, J., Baker, B., Bryant, R.G., and Cafiso, D.S. (1999) *Biophys. J.*, **77**, 1086–1092.
97 Best, R.B., Merchant, K.A., Gopich, I.V., Schuler, B., Bax, A., and Eaton, W.A. (2007) *Proc. Natl. Acad. Sci. USA*, **104**, 18964–18969.
98 Widengren, J., Dapprich, J., and Rigler, R. (1997) *Chem. Phys.*, **216**, 417–426.
99 Hess, S.T., Huang, S., Heikal, A.A., and Webb, W.W. (2002) *Biochemistry*, **41**, 697–705.
100 Gray, H.B. and Winkler, J.R. (2005) *Proc. Natl. Acad. Sci. USA*, **102**, 3534–3539.
101 Dill, K.A., Ozkan, S.B., Shell, M.S., and Weikl, T.R. (2008) *Annu. Rev. Biophys. Bioeng.*, **37**, 289–316.
102 Bieri, O., Wirz, J., Hellrung, B., Schutkowski, M., Drewello, M., and Kiefhaber, T. (1999) *Proc. Natl. Acad. Sci. USA*, **96**, 9597–9601.
103 Haas, E., Katchalski-Katzir, E., and Steinberg, I.Z. (1978) *Biopolymers*, **17**, 11–31.
104 Hagen, S.J., Hofrichter, J., Szabo, A., and Eaton, W.A. (1996) *Proc. Natl. Acad. Sci. USA*, **93**, 11615–11617.
105 Chang, I.J., Lee, J.C., Winkler, J.R., and Gray, H.B. (2003) *Proc. Natl. Acad. Sci. USA*, **100**, 3838–3840.

106 Möglich, A., Krieger, F., and Kiefhaber, T. (2005) *J. Mol. Biol.*, **345**, 153–162.
107 Krieger, F., Fierz, B., Bieri, O., Drewello, M., and Kiefhaber, T. (2003) *J. Mol. Biol.*, **332**, 265–274.
108 Lapidus, L.J., Steinbach, P.J., Eaton, W.A., Szabo, A., and Hofrichter, J. (2002) *J. Phys. Chem. B*, **106**, 11628–11640.
109 Lapidus, L.J., Eaton, W.A., and Hofrichter, J. (2000) *Proc. Natl. Acad. Sci. USA*, **97**, 7220–7225.
110 Hudgins, R.R., Huang, F., Gramlich, G., and Nau, W.M. (2002) *J. Am. Chem. Soc.*, **124**, 556–564.
111 Huang, F. and Nau, W.M. (2003) *Angew. Chem., Int. Ed. Engl.*, **42**, 2269–2272.
112 Fierz, B., Satzger, H., Root, C., Gilch, P., Zinth, W., and Kiefhaber, T. (2007) *Proc. Natl. Acad. Sci. USA*, **104**, 2163–2168.
113 Tran, H.T., Mao, A., and Pappu, R.V. (2008) *J. Am. Chem. Soc.*, **130**, 7380–7392.
114 Rose, G.D., Fleming, P.J., Banavar, J.R., and Maritan, A. (2006) *Proc. Natl. Acad. Sci. USA*, **103**, 16623–16633.
115 Neidigh, J.W., Fesinmeyer, R.M., and Andersen, N.H. (2002) *Nat. Struct. Biol.*, **9**, 425–430.
116 Qiu, L., Pabit, S.A., Roitberg, A.E., and Hagen, S.J. (2002) *J. Am. Chem. Soc.*, **124**, 12952–12953.
117 Marmé, N., Knemeyer, J.P., Wolfrum, J., and Sauer, M. (2004) *Angew. Chem. Int. Ed.*, **43**, 3798–3801.
118 Tyagi, S. and Kramer, F.R. (1996) *Nat. Biotechnol.*, **14**, 303–308.
119 Kostrikis, L.G., Tyagi, S., Mhlanga, M.M., Ho, D.D., and Kramer, F.R. (1998) *Science*, **279**, 1228–1229.
120 Bonnet, G., Tyagi, S., Libchaber, A., and Kramer, F.R. (1999) *Proc. Natl. Acad. Sci. USA*, **96**, 6171–6176.
121 Scheffler, S., Sauer, M., and Neuweiler, H. (2005) *Z. Phys. Chem.*, **219**, 665–678.
122 Soussi, T. (1996) *Immunol. Today*, **17**, 354–356.
123 Simonetti, O., Lucarini, G., Brancorsini, D., Nita, P., Bernardini, M.L., Biagini, G., and Offidani, A. (2002) *Cancer*, **95**, 1963–1970.
124 Franchi, A., Santucci, M., Masini, E., Sardi, I., Paglierani, M., and Gallo, O. (2002) *Cancer*, **95**, 1902–1910.
125 Kugler, A. (1999) *Anticancer Res.*, **19**, 1589–1592.
126 Mochizuki, Y., Tsuda, S., Kanetake, H., and Kanda, S. (2002) *Oncogene*, **21**, 7027–7033.
127 Harbeck, N., Kates, R.E., Look, M.P., Meijer-Van Gelder, M.E., Klijn, J.G., Kruger, A., Kiechle, M., Janicke, F., Schmitt, M., and Foekens, J.A. (2002) *Cancer Res.*, **62**, 4617–4622.
128 Bajou, K., Lewalle, J.M., Martinez, C.R., Soria, C., Lu, H., Noel, A., and Foidart, J.M. (2002) *Int. J. Cancer.*, **100**, 501–506.
129 Horejsh, D., Martini, F., Poccia, F., Ippolito, G., Di Caro, A., and Capobianchi, M.R. (2005) *Nuc. Acids Res.*, **33**, e13.
130 Agrawal, A., Sathe, T., and Nie, S. (2007) *J. Agric. Food Chem.*, **55**, 3778–3782.

8
Super-Resolution Fluorescence Imaging

8.1
Diffraction Barrier of Optical Microscopy

Optical microscopy enables the non-invasive investigation of optically transparent material, such as cells, to be made under a variety of conditions in three dimensions (3D). Through the development of efficient fluorescent labels that can be covalently attached to antibodies and other biomarkers, and the associated possibility to specifically label cellular structures, the expansion of fluorescence microscopy in laboratory applications and research has been accelerated substantially [1–5]. Today, fluorescence microscopes are essential in biological and biomedical sciences for 3D non-invasive imaging of the interior of cells. Thus, diverse confocal and wide-field optical fluorescence microscopes are in operation at most major research institutes.

However, conventional fluorescence microscopy only enables two neighboring emitting objects to be spatially resolved when they are separated by approximately the wavelength of used light. The ability to spatially resolve a structure has a physical limit, which is caused by the wave nature of light. Because of diffraction by lenses, focusing of light always results in a blurred spot, the size of which determines the resolution achievable [6]. Already at the end of the nineteenth century Abbé had shown that the diffraction limit is proportional to the wavelength and inversely proportional to the angular distribution of the light observed [7]. Therefore, any lens-based microscope can not resolve objects that are closer together than half the wavelength of the light in the imaging plane, that is, for visible light in the region of ~200 nm. The resolution along the optical axis is even worse, that is, fluorescent objects can only be distinguished if their axial distance is larger than approximately 700 nm. Analogously, the resolution limit for optical microscopy can be described by the point-spread function (PSF), which describes the response of an imaging system to a fluorescent point object. Owing to diffraction of light passing through an aperture, the fluorescence signal of a single fluorophore produces an Airy disk in the image plane of a fluorescence microscope, a bright region in the center with a series of concentric rings called the Airy pattern. The dimensions of this image are much larger than the fluorophore itself, and are determined by the wavelength of the light and the size of the aperture (Figure 8.1).

Handbook of Fluorescence Spectroscopy and Imaging. M. Sauer, J. Hofkens, and J. Enderlein

ISBN: 978-3-527-31669-4

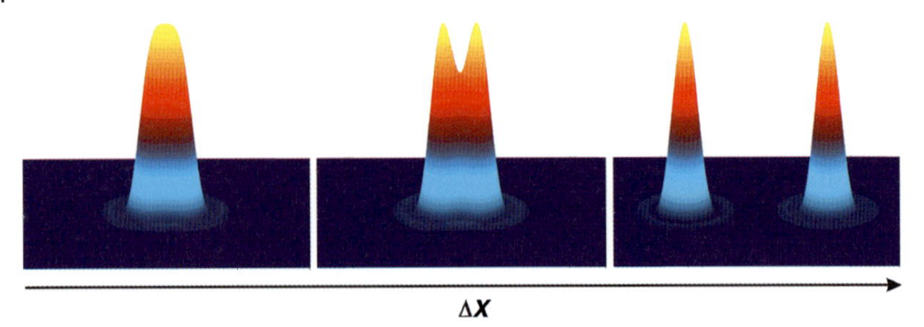

Figure 8.1 The diffraction barrier for optical microscopy. Owing to diffraction, focusing of light always results in a blurred spot. If two fluorophores are separated at distances shorter than the diffraction barrier, that is, $\Delta x < 200$ nm, the two fluorophores cannot be resolved (left-hand side). The image in the middle shows the limit of resolution, that is, the Rayleigh criterion when 80% of the Airy disks overlap. The image on the right-hand side shows the disks of two fluorophores spaced 2 μm from each other.

Owing to this "spreading" of the emission, the images of adjacent fluorophores overlap, and the ability to distinguish between them is lost. On the other hand, with the introduction of low-noise high-quantum yield charge-coupled device (CCD) cameras, it is possible to localize the center of mass of the PSF of a single fluorophore at sub-wavelength scales [8–10]. This enables colocalization or distance determination between individual emitters, that is, single fluorophores that are separated by a distance less than the diffraction limit, provided that their emission can be separated in some way. This can be achieved using any optically distinguishable characteristic [11] either by spectral [12–14] or time-resolved [15, 16] detection, and also by temporal means via subsequent photobleaching of the fluorophores [17, 18].

Unfortunately, the decryption of structural information with nanometer resolution is challenging, relying solely on colocalization methods. What is challenging scientists the world over is the development of techniques to improve the spatial resolution of far-field fluorescence microscopy, using alternative methods capable of resolving the structural details of sub-cellular structures. This might pave the way towards a refined understanding of how biomolecules are assembled within cells to form the fundamental molecular machines supporting living organisms [19–21]. As was suggested in 1928 by Synge, near-field techniques are qualified to enable high-resolution optical microscopy [22]. However, the use of a nanometer-sized mechanical tip to excite molecules or to detect their fluorescence signal has the disadvantage that it can only be used on surfaces and cannot image the interior of cells [23]. Although the spatial resolution cannot compete with that of electron microscopy, optical far-field microscopes allow us to study living samples, for example, cells or tissue, with minimal perturbation. However, compared with the biomolecular length scale, which is in the range of a few nanometres, one can immediately determine that structural details or the organization of biomolecular assemblies can not be adequately resolved by light microscopy. This fact motivated researchers worldwide to overcome the limit imposed by the diffraction barrier and to find ways to report on important details on biomolecular structures and interactions at the molecular level and inside the cell.

8.2 Multi-Photon and Structured Illumination Microscopy

The first concepts developed to improve axial resolution are based on the quadratic dependence of the fluorescence signal on excitation intensity. Two-photon microscopy as introduce by Denk *et al.* in 1990 [24] does not lead to a resolution enhancement *per se*: although the size of the excitation point-spread function (PSF) is reduced, the trade-off is that the wavelength has to be doubled to excite the fluorescent probes. However, and more importantly, two-photon microscopy reduces out-of-focus light considerably, and allows for true optical sectioning in the axial direction. At the same time, the imaging depth of two-photon microscopy is much larger because of the reduced scattering of the infrared light used to excite the sample. Therefore, two-photon microscopy is an ideal tool to study deep tissue and even living animals [25]. Two main concepts capable of effectively improving the axial resolution have been introduced independently, these are, 4Pi microscopy [26] and I^5M microscopy [27]. Both concepts use a set of opposing microscope lenses that sharpen the PSF along the optical axis through interference of the counter-propagating wave fronts. At the same time, the efficiency of collecting light is increased by the presence of two objectives. 4Pi microscopy is a spot-scanning method that achieved a four- to sevenfold increase in axial resolution using different experimental configurations [28]. On the other hand, I^5M microscopy is a wide-field method but it achieves a similar axial resolution of 100 nm [27, 29, 30]. Both 4Pi and I^5M microscopy have mainly been used to study the structure of sub-cellular organelles in 3D [27–32], but they have also been combined with other microscopic techniques that improve the lateral resolution. A concise comparison of both methods can be found in reference [33].

Structured illumination microscopy (SIM), on the other hand, is a concept that combines wide-field imaging and illumination of a sample with a known pattern of excitation light, and achieves a twofold resolution improvement [34]. Experimentally, a periodic illumination pattern of parallel stripes of excitation light is projected onto the sample with the help of a fine grating. A series of images is recorded where the light pattern is moved along the sample laterally and rotated into different angles. Structural features with spatial frequencies that are higher than the frequency of the illumination pattern are modulated by the latter, resulting in so-called Moiré fringes, and can be extracted mathematically (Figure 8.2). As such, a reconstructed image with increased spatial resolution can be obtained.

As a purely physical approach, SIM does not depend on any particular fluorophore properties, such as high photostability, or on any particular transitions between orthogonal states, and can therefore be applied generally. For example, multicolor SIM has been used to study the nuclear periphery of mammalian cells [35]. Furthermore, the acquisition time of SIM as a parallelized imaging method is substantially shorter than spot-scanning methods. As such, SIM is a well suited method to study dynamic processes in living cells [36].

The principle of SIM has been extended to three dimensions by two different experimental configurations. In the first approach, three coherent beams were used

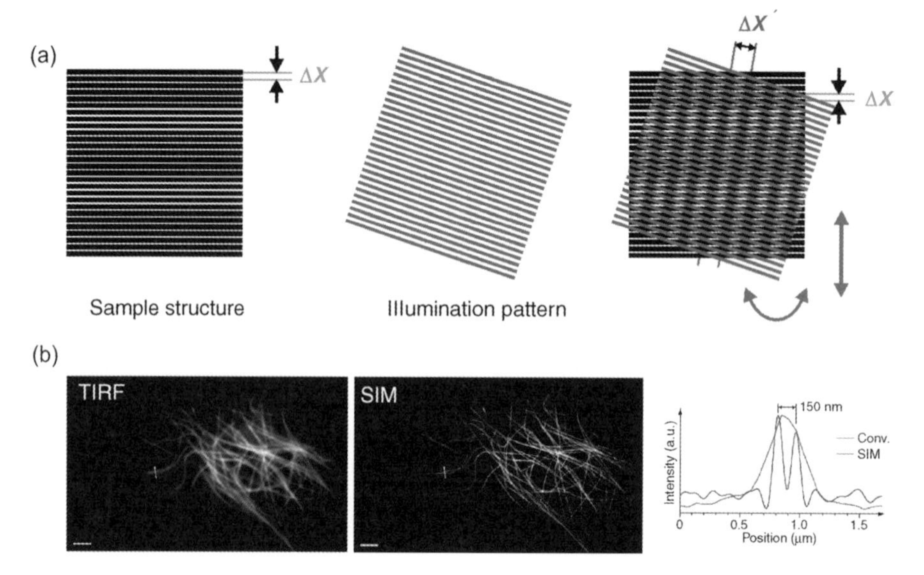

Figure 8.2 (a) Structured-illumination microscopy (SIM) illuminates an unknown structural feature of a sample with a known periodic pattern. The spatial frequencies of the original structure appear as a beat pattern with lower spatial frequencies and can be resolved. (b) Total internal reflection fluorescence (TIRF) (left) and reconstructed SIM image (middle) from the microtubular network of a mammalian cell (scale bar 1 μm). (Reprinted from Ref. [36]; reproduction with kind permission from Nature Publishing Group.)

to record an interference pattern varying both laterally and axially, yielding a twofold enhancement in resolution in all three dimensions after image reconstruction [three-dimensional SIM (3D-SIM)] [37]. Employing a nonlinear structural illumination scheme, saturated SIM (SSIM) experimentally demonstrated a lateral resolution of ~50 nm and is essentially unlimited [29]. The key feature that is exploited in SSIM is the nonlinear response of the fluorescence intensity with respect to the excitation intensity.

Alternatively, multi-photon absorption of semiconductor quantum dots and subsequent generation of multiple exciton states can be used for resolution enhancement [38, 39]. As opposed to two- or multi-photon microscopy, the generation of multi-excitonic states in quantum dots does not require the use of infrared light and high excitation intensities, such that a ~twofold resolution enhancement can be easily realized on any confocal microscope system equipped with a continuous-wave laser light source providing appropriate excitation wavelengths, with low excitation intensities (Figure 8.3). As a pure physical process, this so-called quantum dot triexciton imaging (QDTI) operates under any experimental conditions and in particular in living cells [39]. Although the resolution is less than that afforded by other techniques, QDTI is currently the only sub-diffraction–resolution imaging technique that can provide images of whole fixed and living cells with enhanced resolution in both axial and lateral directions, using standard fluorescence

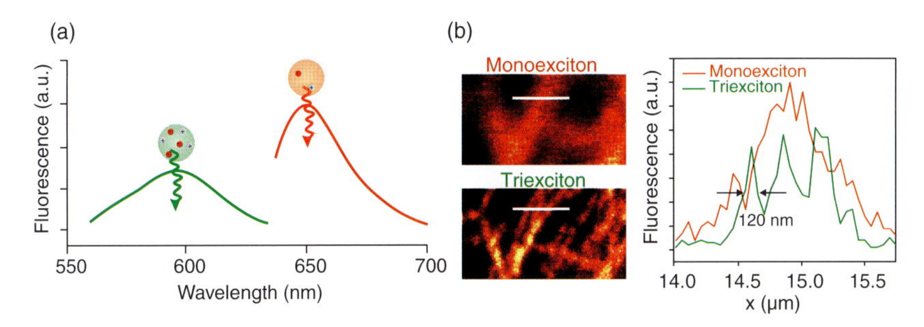

Figure 8.3 QDTI using QDot655 (a) Triexciton fluorescence emission is centered around 590 nm, and is thus well separated from the monoexciton emission centered at 655 nm. (b) Immunofluorescence images of the microtubulin network of COS-7 cells stained with QDot655 labeled secondary antibodies, recorded with a confocal microscope with a resolution of 50 nm per pixel. The monoexciton image was recorded with 20 W cm^{-2} excitation intensity and the triexciton image measured on the short-wavelength detector between 550 and 600 nm applying an excitation intensity of 200 W cm^{-2} at 445 nm (scale bar 5 μm; 1 ms integration time per pixel). The line profiles shown for the monoexciton (red) and triexciton (green) emission channel demonstrate the improvement in resolution.

microscopes available in almost any laboratory. As a consequence, the method is widely applicable, in particular for biological imaging in living cells.

8.3 Stimulated Emission Depletion

The ever-growing demand in biology for improved resolution has stimulated the development of novel fluorescence microscopy techniques that overcome the fundamental diffraction limit and achieve theoretically unlimited resolution. To attain sub-diffraction–resolution fluorescence imaging, methods have been implemented that are based on the fact that the emission of a fluorophore can have a nonlinear dependence on the applied excitation intensities, determined by the excitation wavelength(s) and intensity (e.g., saturation of the emission), and on its position (by applying a spatial intensity distribution) [40, 41]. One way to obtain this is to make use of saturable transitions of a fluorophore between two molecular states, for example, transitions to states with different emission properties, such as intersystem crossing to the triplet state [42]. These strategies have been generalized under the acronym RESOLFT [43], which stands for *re*versible *s*aturable *o*ptical *f*luorescence *t*ransitions. The first realization of RESOLFT with far-field optics was stimulated emission depletion (STED) microscopy [40, 41, 44, 45].

In STED microscopy, the optical resolution is dramatically improved by de-excitation of excited fluorophores via stimulated emission using a red-shifted light beam featuring a local intensity zero in the center. In practice, this is accomplished by overlapping the diffraction-limited spot of the excitation beam of a scanning microscope with a red-shifted donut-shaped beam (generated by a zero-node phase

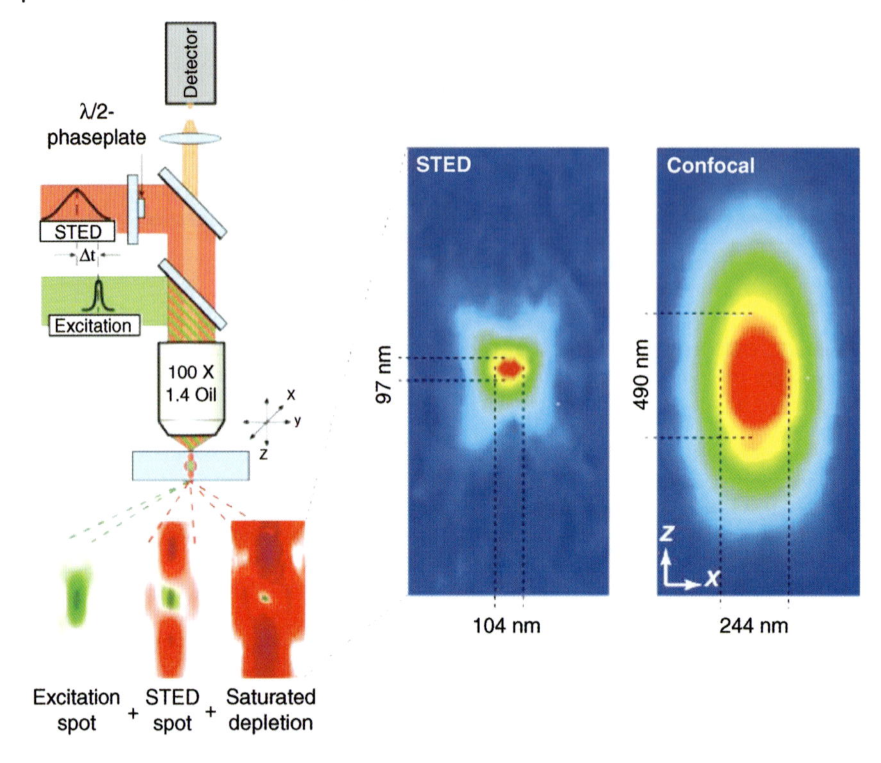

Figure 8.4 Principle of STED microscopy. The fluorophores located in the diffraction-limited spot are excited by a normal mode (Airy disk) laser pulse (green) with a duration of picoseconds or less. A few hundred picoseconds later a second donut-shaped laser pulse (red) irradiates the same region and induces stimulated emission of most fluorophores, except those located at the center of the focus. Because of the depletion of fluorophores in the outer part of the laser focus, the fluorescence is confined spatially to sub-diffraction dimensions leading to increased resolution. Scanning the sample through the laser foci produces a complete image. Experimental set-up used for STED microscopy showing the excitation and depletion pulse, and the resulting reduction of the fluorescent spot size. (With permission from [47].)

mask) for stimulated emission (Figure 8.4). Consequently, excited molecules in the outer parts of the focus are turned off by stimulated emission. Thus, depending on the illumination conditions (intensity and irradiation time) and the quality of the shape of the donut-mode beam (i.e., the stimulated emission beam), theoretically an arbitrarily high resolution can be achieved. The theoretical achievable resolution can be approximated by Equation 8.1 [40].

$$\Delta x \approx \frac{\lambda}{2n \sin \alpha \sqrt{1 + I/I_{sat}}} \tag{8.1}$$

In Equation 8.1, which can be regarded as an extension of Abbe's equation [7], I is the peak intensity (photon flux per unit area) of the donut beam and $I_{sat} = (\sigma \times \tau)^{-1}$

gives the intensity at which the fractional population of the excited state is depleted by stimulated emission to $1/e$. Here σ denotes the cross-section for stimulated emission and τ the fluorescence lifetime of the first excited singlet state. For commonly used fluorophores I_{sat} corresponds to about 30 MW cm^{-2} in the visible range. Thus Δx approaches very small values, ideally $\Delta x \rightarrow 0$, for $I \gg I_{sat}$, that is, using very high excitation intensities for the donut-shaped stimulated emission beam. In combination with fast laser scanners, video-rate (28 Hz) far-field optical imaging of small areas ($\sim 1.8 \times 2.5\ \mu m^2$) of synaptic vesicles by STED with a focal spot size of 62 nm can be performed [46].

Importantly, STED microscopy does not always require pulsed excitation and has been demonstrated by solely applying continuous-wave lasers [48]. With decreasing complexity, STED microscopy can nowadays, likewise, be used for live cell imaging. This is simplified by the fact that STED microscopy is compatible with some photostable variants of fluorescent proteins, reaching a lateral resolution of $\sim$50 nm in a living cell [49]. However, one should be aware of the fact that STED microscopy requires relatively high illumination conditions for efficient stimulated emission, due to the short lifetime of the first excited singlet state, of a few nanoseconds [$I_{sat} = (\sigma \times \tau)^{-1}$]. Therefore, many standard fluorophores are not amenable for use in STED microscopy, and live cell applications have to be carefully controlled with respect to light-induced damage. On the other hand, very photostable fluorescent nitrogen-vacancy defects in diamond can be used as ideal probes to demonstrate the resolving power of STED microscopy. In this context, PDFs of 5.8 nm have been measured for single nitrogen-vacancy defects [50].

It has to be pointed out that in practice any reversible and saturable molecular transition that effectively changes the fluorescence intensity of a fluorophore can be used advantageously to improve resolution. In this context, switching between a dark and a bright state has been identified as an essential element to separate adjacent objects in time. The ultimately appropriate saturable transition represents light-induced switching of fluorescence between two thermally stable states, an off and an on state. These requirements are fulfilled by molecular optical switches, that is, photoswitches that exhibit two stable and selectively addressable states, a fluorescent and a non-fluorescent one, which can be reversibly interconverted upon irradiation with different wavelengths of light under low illumination intensity conditions. Analogous to STED microscopy, optical switches can be selectively switched off upon illumination of the sample with a donut-shaped beam of appropriate wavelength superimposed on the regular beam used to excite fluorescence [40, 51, 52].

Alternatively, the temporal behavior of fluorescence intensity controlled by the local excitation intensity, that is, the nonlinear response of the fluorescence signal with respect to the excitation intensity, can be used for resolution enhancement applying concepts such as, for example, dynamic saturation optical microscopy (DSOM) [53, 54]. Another very promising method that exploits the temporal information of fluorescence fluctuations and mathematically extracts high-resolution spatial information constitutes super-resolution optical fluctuation imaging (SOFI) [55].

8.4
Single-Molecule Based Photoswitching Microscopy

To achieve sub-diffraction–resolution fluorescence imaging, fluorophore emission has to be separated in time. Whereas in the STED/RESOLFT case, the phase mask defines the coordinates of fluorescence emission (predefined in space by the zero-node), alternative methods have emerged that use stochastic activation of individual fluorophores and precise position determination (localization). These approaches randomly separate the emission of stochastically activated individual fluorescent probes in time. Provided the distance between the individual fluorophores enables the analysis of the different emission spots to be unaffected, that is, individual fluorophores are spaced further apart than the distance resolved by the microscope ($>\lambda/2$ on a CCD camera), the standard error of the fitted position is a measure of localization and it can be made arbitrarily small by collecting more photons and minimizing noise factors. The error in the position determination of detected photons can be approximated as Equation 8.2:

$$\left\langle (\Delta x)^2 \right\rangle = \frac{\sigma^2}{N} \tag{8.2}$$

where

Δx is the error in localization
σ is the standard deviation of the point-spread function
N is the number of photons collected [8–10, 56, 57] (Figure 8.5).

Thus, depending on photon statistics fluorescence imaging with one nanometer accuracy (FIONA) can be used, for example, to monitor the migration of single fluorophores conjugated to myosin proteins along actin filaments [58, 59].

If the emission of fluorophores can be separated in time, then the high localization precision can be used for sub-diffraction–resolution fluorescence imaging. One way to achieve this relies on targeting the surface of the object using the diffusion and transient binding of fluorescently labeled probe molecules [62]. Because the diffusion of the fluorophores is too fast to be followed on a CCD camera, the fluorophores only appear when they bind or adhere to the target object. This has two advantages: because the fluorophores, in general, all bind at different time intervals, the localized emission is spread out in time. Moreover, because binding can, ideally, occur anywhere on the target object, the probe molecules will eventually map out the entire structure. A high-resolution image of the target object can then be reconstructed using the localized positions.

With immobilized fluorophores a similar approach is possible if the spatial density of emitting fluorophores can be adjusted in a way that ensures that the majority of the fluorophores resides in a non-fluorescent state. In principle, this can be achieved by exploiting random fluorescence intermittencies of semiconductor quantum dots due to charge separated states [63] or organic fluorophores entering long-lived triplet states in the absence of oxygen [64]. However, for the well defined control of the density of active (fluorescent) fluorophores, molecular optical switches or photo-

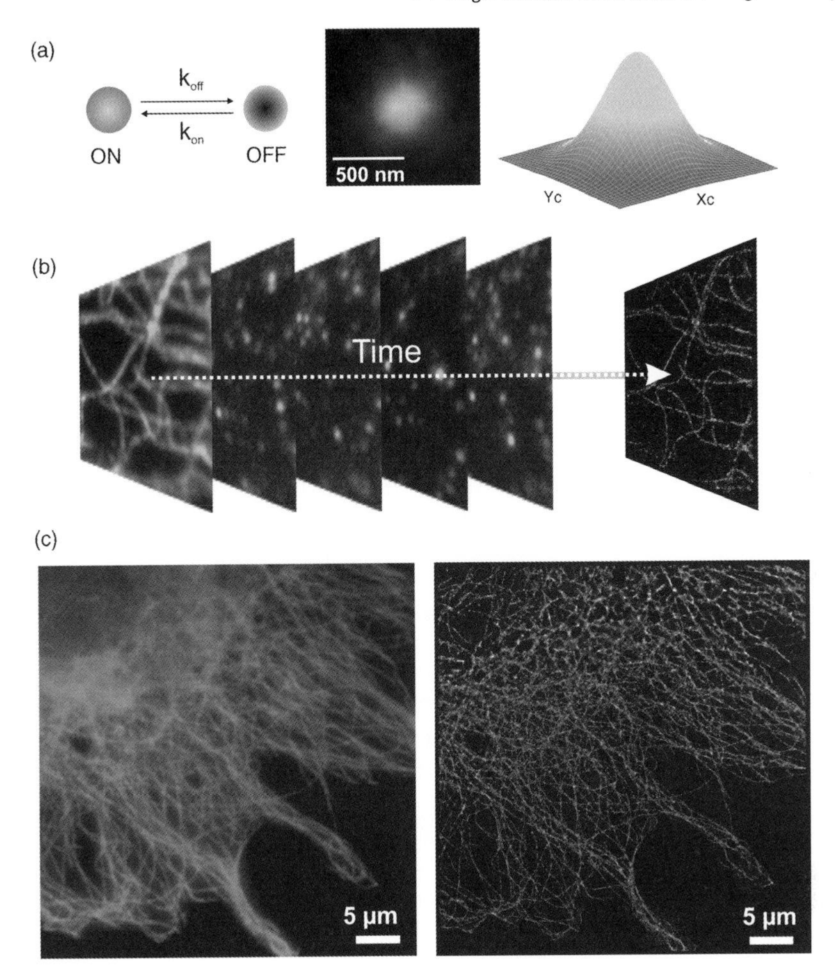

Figure 8.5 (a) Molecular optical switches that can be photoactivated or reversibly photoconverted between two states ("ON" or fluorescent and "OFF" or non-fluorescent) are the key to single-molecule based super-resolution methods. The PSF of a single emitter imaged on a widefield microscope can be approximated by a Gaussian function, which allows determination of the molecule's position with few nanometers precision. (b) By selectively observing only a small subset of all fluorophores in a sample, single emitters can be identified and localized. The localization coordinates determined over many imaging cycles are used to reconstruct an image with higher resolution. (c) Microtubular filaments of a COS-7 cell imaged in both TIRF mode (left) and according to the *d*STORM concept [60, 61].

switches are most promising [65]. Today, many fluorescent probes that exhibit the necessary photoactivation or photoswitching properties are available [65, 66], including conventional organic fluorophores, such as carbocyanine, oxazine, and rhodamine dyes [60, 61, 67–74], caged fluorophores [75], photochromic compounds [76–78] and a large variety of fluorescent proteins [79–83]. All these photoswitchable fluorophores have in common the existence of at least two different states that are distinguishable in their fluorescence emission properties (Figure 8.5).

Photoswitchable fluorophores populate a fluorescent "on"- and a non-fluorescent "off"-state (dark state), and the interconversion between these states can be controlled by light or the chemical nanoenvironment. Photoactivatable fluorophores are initially found in a dark state and require activation to become fluorescent, typically achieved by irradiation with light. All methods that use photoswitchable fluorophores for sub-diffraction fluorescence microscopy employ a temporal confinement of the fluorescence signal. This is achieved by first turning off all fluorophores in a sample. In the next step, only a small subset of fluorophores is transferred into a fluorescent "on"-state. Here one has to make sure that activated fluorophores can be detected as individual emitters that are spaced far enough away from their nearest neighbor. The fluorescence emission signal of a fluorophore is read out, and the position of the fluorophore is determined by approximating the PSF with a Gaussian function (Figure 8.5). This procedure is repeated many times, and the ensemble of coordinates collected from localizing single fluorophores is used to generate ("reconstruct") an artificial image, which provides sub-diffraction or super-resolution information.

Prominent examples of concepts that rely on photoswitchable fluorophores are photoactivated localization microscopy (PALM) [84], fluorescence photoactivation localization microscopy (FPALM) [85], stochastic optical reconstruction microscopy (STORM) [86], PALM with independently running acquisition (PALMIRA) [87], direct STORM (*d*STORM) [60, 61] (Figure 8.5), and other alternative methods [73, 74]. Furthermore, faster variants of PALM have been reported using a stroboscobic illumination scheme [88]. A related method is ground-state depletion followed by individual molecule return (GSDIM), which switches off fluorophores by populating the non-fluorescent triplet state out of which the spontaneous return to the ground state occurs [64].

All the photoswitching-based methods mentioned above are being continuously improved. For example, multi-color imaging with ~20 nm lateral resolution has been demonstrated [69, 79, 89–91], setting the prerequisite to study biomolecular interactions at the molecular level. A variety of concepts that allow 3D-imaging have been developed, either by introducing astigmatism [92] or a helical shape [93] into the beam path, or by recording two imaging planes simultaneously and approximating the PSF to a three-dimensional model function [94]. Alternatively, an interferometric arrangement (iPALM) enables an axial resolution of only a few nanometers (Figure 8.6) [95]. Although the alignment of iPALM is very demanding, it can close the gap between electron tomography and light microscopy.

A challenge for all methods is compatibility with live cell imaging. Whereas methods that rely on the use of fluorescent proteins have the advantage that these probes can easily be implemented in living cells [96, 97], organic fluorophores cannot be used so easily in live cell super-resolution imaging, because they typically require very special buffer conditions for photoswitching [67, 68, 73]. However, a refined understanding of the impact of redox reactivity on photoswitching has paved the way for the use of organic fluorophores even in living cells [61, 70]. Finally, it also has to be mentioned that unsymmetric dimeric cyanine dyes, such as YOYO-1 and others that

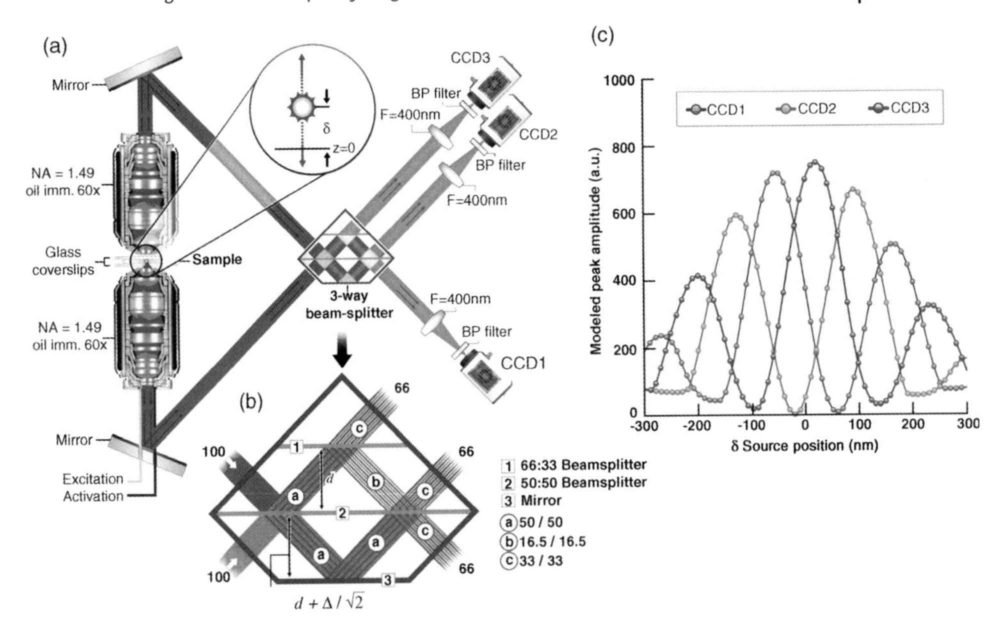

Figure 8.6 Operating principle of iPALM [95]. (a and b) Schematic of the single-photon multiphase fluorescence interferometer. A point source with z-position δ emits a single photon both upwards and downwards. These two beams interfere in a special 3-way beam splitter. (c) The self-interfered photon propagates to the three color-coded CCD cameras with amplitudes that oscillate 120° out of phase, as indicated. The principle takes advantage of the wave-particle duality, which allows a single photon to form its own coherent reference beam. An emitted photon can simultaneously travel two distinct optical paths, which are subsequently recombined so that the photon interferes with itself. The position of the emitter directly determines the difference in the path lengths, hence the relative phase between the two beams. (With permission from [95].)

are known to intercalate into double stranded DNA, can be used for super-resolution imaging, applying conditions as used for "classic" carbocyanine switching [67, 68]. Under these conditions some intercalating fluorophores act as reversible photoswitches and can be used to image DNA with a resolution of better than 40 nm [98].

8.5 Background and Principles of Single-Molecule Based Photoswitching Microscopy Methods

The experimental procedure for all single-molecule based photoswitching methods is similar: a sample has to be densely labeled with a photoswitchable fluorophore, for example, via immunocytochemistry or co-expression of a fluorescent protein, and prepared such that most fluorophores are in their off-state, and only a subset of fluorophores remains fluorescent at any time (Figure 8.5b). As described in the previous section, the emission profiles of single emitters can be localized through the approximation with a Gaussian fit with a precision that is determined by the number of photons detected and the background fluorescence, that is, autofluorescence of the sample and residual fluorescence of surrounding fluorophores in the off-state [10]. Thus, photoswitches with a high intensity contrast between the on- and off-states, together with high extinction coefficients and quantum yields in the on-state, are the key for nanometer accuracy.

For the case of a densely labeled structure, the dark state should exhibit a lifetime τ_{off} that is substantially longer than the lifetime of the fluorescent state τ_{on}. Ideally, the lifetime of the on-state τ_{on} should be very short, with a high photon yield in order to allow precise localization [10]. Upon irradiation the fluorophore is transferred to a metastable dark state at a rate of k_{off}, where it resides until it is converted into the singlet ground state with rate k_{on}. The ratio of these rates is defined as $\lambda = \tau_{off}/\tau_{on} = k_{off}/k_{on}$ [99]. The denser the labeling of a structure, the higher the ratio λ has to be in order to ensure that fluorescence emission of single fluorophores is temporally well separated, to allow unambiguous localization of individual emitters. The on-time or lifetime of the fluorescent state τ_{on} can be shortened by applying high irradiation intensities, thus reaching a higher ratio λ and enabling fast acquisition with commonly available CCD cameras [64, 72, 73, 100]. This has the additional advantage that further effects such as mechanical drifts are reduced.

In this context, the Nyquist–Shannon sampling theorem becomes important, which requires that the average distance between neighboring molecules must be at least half of the desired resolution [101]. Therefore, next to the localization accuracy of single emitters, which is governed by the photons emitted, the achievable resolution is also controlled by the labeling density. In the extreme case, the density of photoswitchable fluorophores can become so high that the photophysical parameters τ_{on} and τ_{off} reach their limit and photobleaching must be applied to achieve the desired fluorophore density. Alternatively, the number of fluorophore labels (e.g., antibodies) can be reduced. However, both strategies fail at the expense of optical resolution according to the Nyquist–Shannon sampling theorem. The importance of λ for super-resolution imaging can be easily demonstrated if we imagine we want to image a PSF area with a typical diameter of 250 nm with a resolution of 20 nm. Hence, according to the Nyquist–Shannon criterion, the sample has to be labeled with photoswitchable fluorophores every 10 nm (Figure 8.7a). In order to achieve sub-diffraction–resolution imaging only one fluorophore is allowed to reside in the on-state at any time within a diffraction-limited area, that is, a ratio $\lambda \geq 600$ is required. Assuming a lifetime of the on-state τ_{on} of only 1 ms, a very stable off-state with $\tau_{off} \geq 600$ ms has to be realized.

Only under these conditions will the majority of isolated fluorophores residing in their fluorescent on-state be appropriately identified and localized (Figure 8.7b). In some instances, it can happen that more than one fluorophore is fluorescent at the same time in a diffraction-limited area because of the stochastic nature of the switching process, or due to inappropriate photoswitching conditions. In this case the shape of the PSF becomes unsymmetrical, which results in "irregular spots" that are usually sorted out during software analysis. However, if two or more fluorophores are very close, the asymmetry in their shape might be too small to be distinguished from single-molecule spots. Consequently they are fitted as one spot whose local-

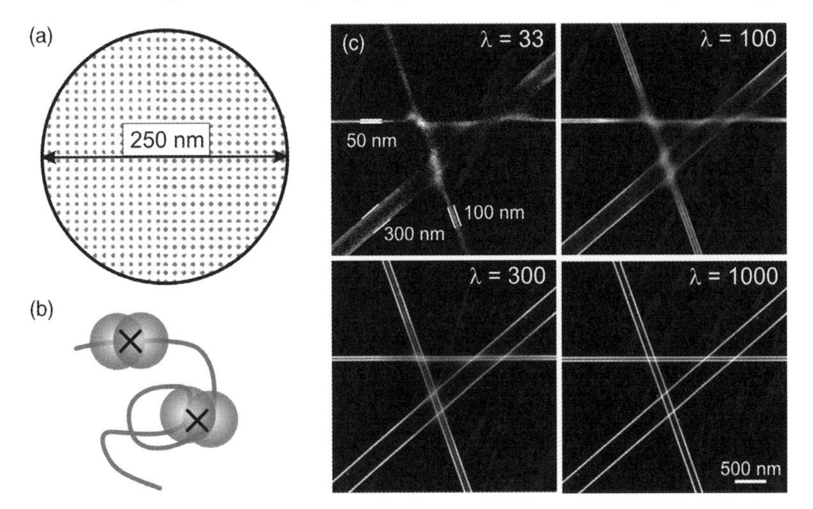

Figure 8.7 (a) In order to image a 2D-area with an optical resolution of 20 nm, the Nyquist–Shannon criterion demands labeling with a photoswitchable fluorophore every 10 nm corresponding to ~600 fluorophores per PSF defined area. Thus, a switching ratio of $\tau_{off}/\tau_{on} \geq 600$ is required. (b) If the photoswitching rates are set inappropriately, two or more fluorophores are fluorescent at the same time and their PSFs may overlap significantly (multiple-fluorophore localizations). The approximation of the PSF of multiple emitters yields a position that does not correspond to the physical position of any of the fluorophores. If the fluorophores are arranged along a single filament (above), the mismatch has no consequences on the resolution enhancement. In the case of two adjacent fluorophores on different, for example, parallel, or crossing filaments (below), a false localization is generated and this will affect the ability to resolve the independent filaments. (c) Simulations on filaments. A network of straight adjacent filament pairs with neighbor distances of 50, 100, 300 nm was simulated. Every filament consists of a line labeled with a fluorophore every 8.5 nm. The simulated photoswitching properties are based on the experimental data of the photoswitchable fluorophore Cy5. The network was resolved for different $\lambda = k_{off}/k_{on}$. Cross-section profiles of three different filament pairs with distances of 50, 100, and 300 nm; within a structure the ratio required to resolve the filaments increases with the complexity of organization, that is, the denser an area is labeled the higher the ratio has to be. A low ratio of $\lambda = 33$ suffices to resolve single filaments with large distances (300 nm), whereas filaments with smaller distances or crossing areas need higher ratios. The image simulated with $\lambda = 100$ is a good demonstration of the influence of false localizations. The two filaments separated by 100 nm show a third artificial filament appearing in the interspace [99].

ization no longer reflects the physical position of a single fluorophore (Figure 8.7b). These localizations are called "multiple-fluorophore localizations" [99].

To demonstrate the power of super-resolution microscopy it is very common to choose cellular polymers such as the microtubule or actin skeleton in fixed cells. Both filaments have diameters well below the diffraction limit and also a large number of binding sites for antibodies or other drug molecules, which can be labeled with photoswitchable fluorophores. However, under realistic conditions, super-resolution imaging should not just show linear filaments that can be easily resolved irrespective of the rate ratio λ. The error rate due to multiple-fluorophore localizations becomes increasingly important, imaging networks of crossing and adjacent filaments. Furthermore, when using fluorophore labeled antibodies as probes, the degree of labeling (DOL) has to be considered because a DOL >1 reduces the off-time of the label. According to a binominal distribution for a sufficiently low (and realistic in terms of antibody labeling) DOL (2–6) and higher rate-ratios ($\lambda > 50$), λ is approximately reduced by λ/DOL.

The issue of resolving straight, single, and well isolated filaments is fairly simple. It does not matter whether a single fluorophore or a bulk is fitted. Each time a multiple-fluorophore localization is made along the filament, it contributes to the artificial reconstructed image in the same way as every regular localization (Figure 8.7c). In this case the localization no longer reflects the position of a single fluorophore, but this does not affect the resolution of the structure. Therefore, the rate-ratio λ can be relatively small, and also short-lived off-states of a standard organic fluorophore can be used under high excitation conditions for super-resolution imaging [64, 73, 102].

The resolution of crossing or adjacent filaments, however, requires a higher λ and a stable non-fluorescent state with a comparably long lifetime τ_{off}, respectively. Localizing more than one fluorophore within a diffraction-limited area affects the resolution of the structure as it yields coordinates where no fluorophore actually resides. In Figure 8.7c, straight filaments were simulated using different λ values with an interfluorophore distance of 8.5 nm, predicting a theoretical resolution of 17 nm [101]. Three crossing pairs of filaments with neighbor distances of 50, 100, and 300 nm were simulated. Photoswitching parameters used in the simulation were derived from experimental values derived from carbocyanine dyes [60, 68, 71, 72, 99]. In case of low switching rates ($\lambda = 33$), adjacent filaments with distances of 50 and 100 nm are not resolvable and crossing filaments are blurred completely ($\lambda = 33$, 100). With an increasing ratio ($\lambda = 300$, 1000) adjacent and also crossing filaments can be resolved clearly. As can be seen for high λ, the bulk of the localizations are made from single molecules, whereas for low rate ratios most of the fluorophores lead to irregular spots. These spots are sorted out automatically due to their asymmetry during software analysis. Hence, for lower ratios the total number of localizations that contribute to the final image is reduced.

After demonstrating the importance of the generation of thermally stable long-lived off-states, the question arises as to how states such as these can be realized in standard organic fluorophores. The triplet state of organic fluorophores can only be used as the off-state when oxygen is efficiently removed in order to prolong the triplet lifetime from micro- to milliseconds in aqueous solvents [64]. This can be achieved by oxygen scavenging systems or embedment in a polymer matrix such as poly(vinly alcohol) (PVA) with low oxygen permeability. Furthermore, relatively high excitation intensities ($>10\,\mathrm{kW\,cm^{-2}}$) have to be applied to shorten τ_{on}. However, besides the triplet states, radical ion states can be efficiently populated by the presence of appropriate redox partners. For example, the triplet state might be reduced by the presence of appropriate reducing substances in the micromolar range. Thus, a semi-reduced radical anion is formed that exhibits a surprising stability in aqueous solvents with a lifetime of up to several seconds, which is dependent on the oxygen concentration. The fluorescent singlet state can be easily recovered by reaction with oxygen or other oxidizing substances; that is, τ_{on} can be controlled by the irradiation intensity and the concentration of the reducer, and τ_{off} by the concentration of the oxidizer. This reducing and oxidizing system (ROXS) was originally introduced to minimize photobleaching and blinking of fluorescent dyes (see Chapter 4) [103] but might, likewise, be used advantageously to control λ [61, 69, 70, 73, 74].

In order to indentify a general mechanism for the generation of reversible long-lived off-states in organic standard fluorophores, the photophysical properties are again in the spotlight. As has been demonstrated in Chapter 7, most rhodamine and oxazine derivatives are efficiently quenched by electron donors, in particular by aromatic amines such as the amino acid tryptophan. This finding can be used as a platform to develop a universal, widely applicable method for super-resolution imaging with small organic fluorophores. With the exception of some carbocyanine dyes such as Cy3, Cy5, and Alexa Fluor 647, most other commercially available fluorophores with absorption maxima between 480 and 700 nm (e.g., Alexa Fluor and ATTO dyes) belong to the class of rhodamine and oxazine derivatives. Because they feature the same basic chromophore structure they exhibit similar redox properties, that is, rhodamine and particularly oxazine derivatives have a high lying one-electron reduction potential, and because of this they are prone to reduction, whereas cyanine dyes exhibit a lower one-electron oxidation potential, and are therefore more easily oxidized than rhodamine and oxazine dyes (see Table 7.1 in Chapter 7).

Following these ideas it was discovered [61] that the triplet states of various Alexa Fluor and ATTO dyes are quenched by thiol containing reducing compounds such as β-mercaptoethylamine (MEA), dithiothreitol (DTT), or glutathione (GSH), that is, substances with slightly lower reduction power compared with aromatic amines [104]. The quenching efficiency depends on the pH of the solvent because most thiols (RSH) have a $pK_{a,SH}$ 8–9 [105, 106] and the reducing species is the thiolate anion (RS^-). Therefore, the reduction efficiency of compounds carrying one thiol group exhibits a plateau at $pH > 9$ with all relevant functional groups ionized, followed by a linear increase in reduction potential with pH decrease. Thus, the reducing power can be easily and sensitively controlled by the pH of the solvent.

Importantly, the first excited singlet state of dyes such as ATTO655 are quenched by electron transfer from various thiol compounds (RSH/RS^-) at pH values >9. The triplet state, however, has a longer lifetime and thus is quenched even at pH 7–8, that is, at lower RS^- concentrations. Furthermore, the thiolate (RS^-) competes with oxygen, which is known to efficiently quench triplet states and is present at a concentration of $\sim$250 μM under standard conditions in aqueous solution. As a consequence, the pH value of the solvent and the concentration of the thiol compound are crucial experimental parameters, and efficient quenching of the triplet state under physiological conditions (pH 7–8) requires a concentration of 10–100 mM of the thiol compound. The reaction scheme for reversible radical anion formation by thiol compounds is presented in detail in Figure 8.8.

The semi-reduced radical anions formed in the first reaction step can be re-oxidized by molecular oxygen in a photoinduced process or can be further reduced to very stable non-fluorescent species that exhibits a lifetime of several seconds, even in

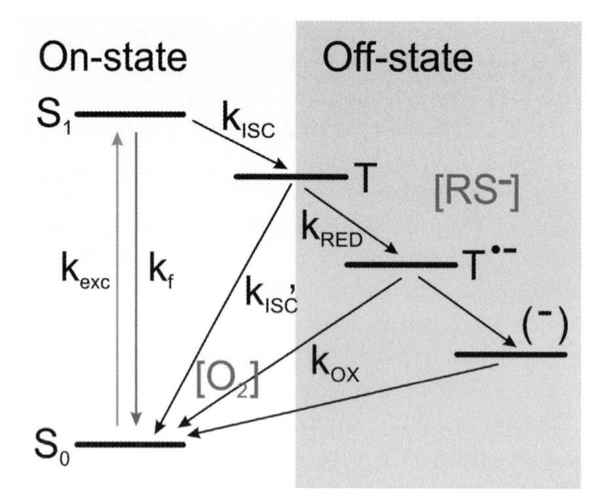

Figure 8.8 Underlying photophysical processes of reversible photoswitching of Alexa Fluor and ATTO dyes. Following excitation of the fluorophores (k_{exc}) into their first excited singlet state, S_1, the excited-state energy is either released via fluorescence emission with rate k_f, or the triplet state is occupied via intersystem crossing (k_{ISC}). The triplet state (T) is depopulated either via k_{ISC}' by oxygen or reduced by thiolate (RS^-) with rate k_{RED} to form radical anions ($T^{\bullet -}$). The semi-reduced radical anions formed in the first reaction can be further reduced to yield the fully reduced species. Both the semi-reduced and the fully reduced form can be re-oxidized by molecular oxygen with rate k_{OX}. Thus, molecular oxygen plays a crucial role in single-molecule based photoswitching methods according to the *d*STORM principle [61]. In other words, the concentration of molecular oxygen and thiol have to be carefully balanced to adjust the desired switching ratio λ.

the presence of oxygen (Figure 8.8). The role of oxygen is twofold: (i) it quenches the triplet state in competition with the reducing thiolate and (ii) oxygen is responsible for generation of the on-state by oxidizing the reduced dyes produced. The stability of the reduced off-state is further corroborated by the fact that the fluorescent form cannot be recovered efficiently when oxygen is removed from the solution by purging with nitrogen [61].

According to these results, in the first step, the reaction of RS^- with the triplet states of fluorophores produces radical anions and thiyl radicals ($T + RS^- \rightarrow T^{\bullet -} + RS^{\bullet}$). The main reactions of thiyl radicals in aqueous solution are conjugation with thiols or thiolates, or dimerization to the corresponding disulfides or reaction with molecular oxygen [107, 108]. The free radical reactions can generate superoxide radicals and hydrogen peroxide. As the formation of superoxide radicals and hydrogen peroxide is less efficient, the main consequence of the reaction mechanism is oxygen consumption. Both thiol oxidation and oxygen consumption have been shown to increase with pH as a consequence of the increasing fraction of thiolate in the reaction medium [107, 108].

Consequently, direct stochastic reconstruction microscopy (*d*STORM) using standard fluorophores in the absence of activator fluorophores has become feasible and allows super-resolution imaging with standard fluorophores under identical

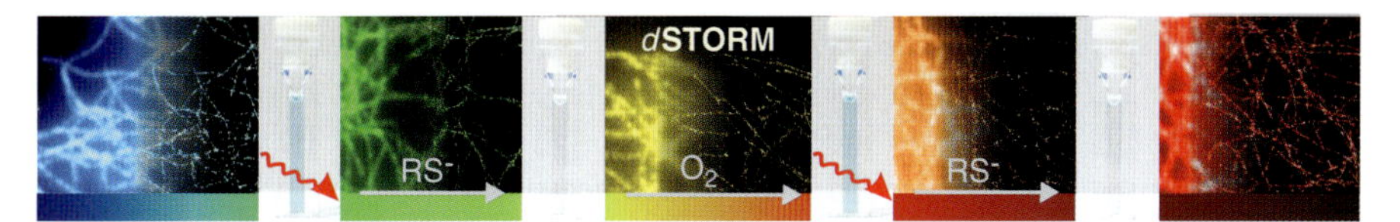

Figure 8.9 Super-resolution imaging of the cytoskeletal network of mammalian cells with the five spectrally different Alexa Fluor and ATTO dyes Alexa Fluor 488, ATTO520, ATTO565, and ATTO655, spanning the visible wavelength range according to the *d*STORM principle [60, 61]. The immunofluorescence images were recorded using conventional TIRF microscopy of microtubules in COS-7 cells. The left-hand sides show the TIRF images superimposed by the reconstructed *d*STORM images on the right-hand sides. Experiments were performed in PBS, 10–200 mM MEA, with a frame rate of 10–33 Hz and excitation intensities between 1 and 4 kW cm^{-2} matched to balance the different extinction coefficients of the fluorophores at the excitation wavelength (Alexa Fluor 488: 488 nm, 3 kW cm^{-2}, 100 mM MEA; ATTO520: 514 nm, 3 kW cm^{-2}, 100 mM MEA; ATTO565: 568 nm, 1.5 kW cm^{-2}, 100 mM MEA; ATTO655: 647 nm, 1.5 kW cm^{-2}, 10 mM MEA). Between the fluorescence images, cuvette ensemble experiments demonstrate the efficiency of reversible photoswitching. An aqueous solution of ATTO655 (10–6 M) in PBS, pH 8.3 in the presence of 100 mM MEA can be switched between a fluorescent and uncolored non-fluorescent form upon irradiation with light. The colored fluorescent from is recovered upon delivery of fresh oxygen and agitation of the cuvette.

experimental conditions by simply adding millimolar concentrations of thiols (Figure 8.9). Thus, the list of organic fluorophores that can be used for super-resolution imaging can be extended considerably by all common rhodamine and oxazine derivatives from the blue to the red part of the electromagnetic spectrum. The underlying mechanism can be described as a remarkably efficient cycling between a fluorescent and non-fluorescenct state of the fluorophores in the presence of millimolar concentrations of thiols. Hence, the lifetime of the on-state can be adjusted by the excitation intensity, provided that the concentration of the reducing thiolate species ensures efficient quenching of the triplet state via formation of radical anions and subsequent secondary reactions to a stable off-state. The lifetime of the off-state, that is, the time it takes until the reduced species are oxidized by oxygen to repopulate the singlet state, is determined by the oxygen concentration. In the *d*STORM experiments shown in Figure 8.9, the-laser intensities were adjusted to ensure that the lifetime of the off-state is substantially longer than the lifetime of the on-state. Thus, only a subset of fluorophores is activated at any time in the field of view. In order to reconstruct super-resolution images as shown in Figure 8.9. typically 10 000–20 000 frames at frame rates of 10–33 Hz were recorded to achieve an optical resolution of ~20 nm. Higher frame rates can be achieved using faster EMCCD cameras in combination with higher excitation intensities. Under the applied experimental conditions, all fluorophores tested exhibit fluorescence count rates of 10–30 kHz. Thus, 500–3000 photons can be used to calculate fluorophore localizations with a theoretical precision of 5–15 nm [10].

Interestingly, *d*STORM is not restricted to the use of MEA but works similarly with other thiols, such as GSH, under physiological conditions. The tripeptide GSH is the most abundant low molecular weight thiol protectant and antioxidant in mammalian biology. The thiol groups are kept in a reduced state at millimolar concentrations in animal cells [109]. Thus, super-resolution imaging in living cells is also possible with small organic fluorophores such as ATTO655, ATTO680, ATTO700, and ATTO520, which exhibit the most pronounced electron accepting properties and require only low millimolar concentrations of thiols to be efficiently transferred to the long-lived off-state [70]. Although the pH value and GSH concentration of cells varies considerably between the various cell types and conditions, the method can offer live cell experiments using selected fluorophores [61]. Importantly, the method enables screening for suitable live cell fluorophores by simple ensemble cuvette experiments (see Figure 8.9). The intriguing simplicity of the method facilitates its application and opens avenues for multicolor super-resolution imaging with combinations of small organic fluorophores. The results indicate that the development of new methods combining the genetic labeling approach with small, bright and photostable organic fluorophores represents an elegant method for super-resolution imaging and precision colocalization experiments. Therefore, *d*STORM is ideally suited to the study of subcellular structures and cluster analysis of protein heterogeneity (distribution) in fixed and living cells with, so far, unmatched resolution [110–112].

8.6
Temporal Resolution of Super-Resolution Imaging Methods

A major drawback to all photoswitching methods is that a large set of individual images has to be recorded, typically several thousands, which drastically reduces the temporal resolution. Even though the lifetime of the on- and off-states can be reduced in combination with higher excitation intensities, to enable higher imaging speeds, one should be aware of the fact that higher excitation intensities likewise promote light induced cell damage. Temporal resolution in STED microscopy, for example, is limited by the scanning process. The use of fast beam scanners made STED microscopy at video rate possible, fast enough to observe the dynamics of synaptic vesicles inside the axons of cultured neurons [46]. However, even the fastest beam scanner restricts the observable area to a rather small area of $1.8 \times 2.5\,\mu m^2$. Furthermore, fast scanning limits the achievable resolution, which was determined to about 62 nm. In contrast to sequential spot-scanning in STED microscopy, SIM, as a parallelized imaging approach, allows fast imaging of larger areas of a size that is essentially determined by the imaging optics and the camera chip. SIM has been used to study tubulin and kinesin dynamics in living cells with a lateral resolution of 100 nm and a frame rate of up to 11 Hz [36]. Here the constraint is the resolution limit of about 120 nm, which is inherent to linear SIM, and the number of images that have to be recorded at different experimental settings.

Very different constraints have to be considered in single-molecuel based photoswitching methods. First of all, images with sub-diffraction–resolution are obtained by reconstruction from individual localizations that were determined from thousands of individual imaging frames. In other words, the temporal resolution is, at first order, determined by the number of imaging frames that are required to obtain a satisfactorily reconstructed image with sub-diffraction–resolution. Typically, thousands of images are required, recorded in an experimental time of tens of seconds to minutes. Other constraints lie in the nature of the photoswitchable fluorescent probes themselves. On the one hand, the fluorescent probes need to exhibit photoswitching under the particular experimental conditions, for example, in living cells, and on the other hand, the kinetics of the photoswitching process determines the temporal resolution. Fluorescent proteins can readily be handled in living cells, but exhibit slow photoswitching kinetics. Live-cell imaging with sub-diffraction–resolution using fluorescent proteins has therefore only been demonstrated for relatively slow processes, such as the dynamics of adhesion complexes [97]. Organic fluorophores are brighter than fluorescent proteins and at the same time are less prone to photobleaching, but their photoswitching requires specific chemical conditions. However, as has been demonstrated by the *d*STORM method, the refined understanding of photophysical and photochemical processes that drive the transition of a fluorophore between a fluorescent and a dark state enables the identification of suitable fluorophores, even for live-cell imaging using the natural reducing redox buffer present in all cells.

An important advantage of organic fluorophores is that very fast photoswitching cycles can be achieved. Taking the example of carbocyanine fluorophores, such as the commercial derivatives Cy5 and Alexa Fluor 647, both the on- and the off-switching of the fluorophores is controlled by the irradiation intensity of a green and a red laser, respectively [60]. Using these carbocyanine fluorophores, rapid photoswitching with an imaging frame rate up to 1 kHz and a lateral resolution of ~30 nm has been demonstrated [72]. As such, switching cycles of ~1 ms can be realized with organic fluorophores that are about one hundred times faster than those reported for live cell PALM [97] or FPALM [96] experiments. Sliding window analysis of *d*STORM data taken with frame rates of ~100 Hz allow the generation of video-like (~10 Hz) super-resolution movies [100]. Here just 100 consecutive frames are sufficient to generate a single high-resolution image with a lateral resolution of ~30 nm.

References

1 Stephens, D.J. and Allan, V.J. (2003) *Science*, **300**, 82–86.

2 Weiss, S. (1999) *Science*, **283**, 1676–1683.

3 Tinnefeld, P. and Sauer, M. (2005) *Angew. Chem. Int. Ed.*, **44**, 2642–2671.

4 Lin, M.Z. and Wang, L. (2008) *Physiology*, **23**, 131–141.

5 Wilt, B.A., Burns, L.D., Ho, E.T.W., Gosh, K.K., Mukamel, E.A., and Schnitzer, M.J. (2009) *Annu. Rev. Neurosci.*, **32**, 435–506.

6 Pawley, J. (2006) *Handbook of Biological Confocal Microscopy*, Plenum Press, New York.

7 Abbe, E. (1873) *Arch. Mikrosk. Anat.*, **9**, 413–420.

8 Schmidt, T., Schütz, G.J., Baumgartner, W., Gruber, H.J., and Schindler, H. (1996) *Proc. Natl. Acad. Sci. USA*, **93**, 2926–2929.

9 Kubitscheck, U., Kuckmann, O., Kues, T., and Peters, R. (2000) *Biophys. J.*, **78**, 2170–2179.

10 Thompson, R.E., Larson, D.R., and Webb, W.W. (2002) *Biophys. J.*, **82**, 2775–2783.

11 Betzig, E. (1995) *Opt. Lett.*, **1995**, 237–239.

12 van Oijen, A., Köhler, J., Schmidt, J., Müller, M., and Brakenhoff, G.J. (1999) *J. Opt. Soc. Am.*, **16**, 909–915.
13 Lacoste, T.D., Michalet, X., Pinaud, F., Chemla, D.S., Alivisatos, A.P., and Weiss, S. (2000) *Proc. Natl. Acad. Sci. USA*, **97**, 9461–9466.
14 Churchman, L.S., Okten, Z., Rock, R.S., Dawson, J.F., and Spudich, J.A. (2005) *Proc. Natl. Acad. Sci. USA*, **102**, 1419–1423.
15 Heilemann, M., Herten, D.P., Heintzmann, R., Cremer, C., Weston, K.D., Wolfrum, J., and Sauer, M. (2002) *Anal. Chem.*, **74**, 3511–3517.
16 Heinlein, T., Biebricher, A., Schlüter, P., Herten, D.P., Wolfrum, J., Heilemann, M., Müller, C., Tinnefeld, P., and Sauer, M. (2005) *ChemPhysChem*, **6**, 949–955.
17 Gordon, M.P., Ha, T., and Selvin, P.R. (2004) *Proc. Natl. Acad. Sci. USA*, **101**, 6462–6465.
18 Qu, X., Wu, D., Mets, L., and Scherer, N.F. (2004) *Proc. Natl. Acad. Sci. USA*, **101**, 11298–11303.
19 Spector, D.L. (1993) *Annu. Rev. Cell Biol.*, **9**, 265–315.
20 Lamond, A.I. and Earnshaw, W.C. (1998) *Science*, **280**, 547–553.
21 Cook, P.R. (1999) *Science*, **284**, 1790–1795.
22 Synge, E.A. (1928) *Philos. Mag. J. Sci.*, **6**, 356–362.
23 Novotny, L. and Hecht, B. (2006) *Principles of Nano-Optics*, Cambridge University Press, Cambridge.
24 Denk, W., Strickler, J.H., and Webb, W.W. (1990) *Science*, **248**, 73–76.
25 Helmchen, F. and Denk, W. (2002) *Curr. Opin. Neurobiol.*, **12**, 593–601.
26 Hell, S.W. and Stelzer, E.H.K. (1992) *Opt. Commun.*, **93**, 277–282.
27 Gustafson, M.G. (1999) *Curr. Opin. Struct. Biol.*, **9**, 627–634.
28 Gugel, H., Bewersdorf, J., Jakobs, S., Engelhardt, J., Storz, R., and Hell, S.W. (2004) *Biophys. J.*, **87**, 4146–4152.
29 Gustafson, M.G. (2005) *Proc. Natl. Acad. Sci. USA*, **102**, 13081–13086.
30 Gustafson, M.G., Agard, D.A., and Sedat, J.W. (1999) *J. Microsc.*, **195**, 10–16.
31 Egner, A., Verrier, S., Goroshkov, A., Soling, H.D., and Well, S.W. (2004) *J. Struct. Biol.*, **147**, 70–76.
32 Medda, R., Jakobs, S., Hell, S.W., and Bewersdorf, J. (2006) *J. Struct. Biol.*, **156**, 517–523.
33 Bewersdorf, J., Schmidt, R., and Hell, S.W. (2006) *J. Microsc.*, **222**, 105–117.
34 Gustafson, M.G. (2000) *J. Microsc.*, **198**, 82–87.
35 Schermelleh, L., Carlton, P.M., Haase, S., Shao, L., Winoto, L., Kner, P., Burke, B., Cardoso, M.C., Agard, D.A., Gustafson, M.G., Leonhardt, H., and Sedat, J.W. (2008) *Science*, **320**, 1332–1336.
36 Kner, P., Chhun, B.B., Griffis, E.R., Winoto, L., and Gustafson, M.G. (2009) *Nat. Methods*, **6**, 339–342.
37 Gustafson, M.G., Shao, L., Carlton, P.M., Wang, C.J., Golubovskaya, I.N., Cande, W.Z., Agard, D.A., and Sedat, J.W., *Biophys. J* (2008) **94**, 4957–4970.
38 Ben-Haim, N.R. and Oron, D. (2008) *Opt. Lett.*, **33**, 2089–2091.
39 Hennig, S., van de Linde, S., Heilemann, M., and Sauer, M. (2009) *Nano Lett.*, **9**, 2466–2470.
40 Hell, S.W. (2007) *Science*, **316**, 1153–1158.
41 Hell, S.W. and Wichmann, J. (1994) *Opt. Lett.*, **19**, 780–783.
42 Hell, S.W. and Kroug, M. (1995) *Appl. Phys. B*, **60**, 495–497.
43 Hofmann, M., Eggeling, C., Jakobs, S., and Hell, S.W. (2005) *Proc. Natl. Acad. Sci. USA*, **102**, 17565–17569.
44 Eggeling, C., Ringemann, C., Medda, R., Schwarzmann, G., Sandhoff, K., Polyakova, S., Belov, V.N., Hein, B., von Middendorff, C., Schönle, A., and Hell, S.W. (2009) *Nature*, **457**, 1159–1162.
45 Schmidt, R., Wurm, C.A., Punge, A., Egner, A., Jakobs, S., and Hell, S.W. (2009) *Nano Lett.*, **9**, 2508–2510.
46 Westphal, V., Rizzoli, S.O., Lauterbach, M.A., Kamin, D., Jahn, R., and Hell, S.W. (2008) *Science*, **320**, 246–249.
47 Hell, S.W. (2003) *Nat. Biotechnol.*, **21**, 1347–1355.
48 Willig, K.I., Harke, B., Medda, R., and Hell, S.W. (2007) *Nat. Methods*, **4**, 915–918.
49 Hein, B., Willig, K.I., and Hell, S.W. (2008) *Proc. Natl. Acad. Sci. USA*, **105**, 14271–14276.

50 Rittweger, E., Han, K.Y., Irvine, S.E., Eggeling, C., and Hell, S.W. (2009) *Nat. Photonics*, **3**, 144–147.
51 Hell, S.W., Dyba, M., and Jakobs, S. (2004) *Curr. Opin. Neurobiol.*, **14**, 599–609.
52 Dedecker, P., Hotta, J.I., Flors, C., Sliwa, M., Uji-I, H., Roeffaers, M.B.J., Ando, R., Mizuno, H., Miyawaki, A., and Hofkens, J. (2007) *J. Am. Chem. Soc.*, **129**, 16132–16141.
53 Enderlein, J. (2005) *Appl. Phys. Lett.*, **87**, 094105.
54 Fujita, K., Kobayashi, M., Kawano, S., Yamanaka, M., and Kawata, S. (2007) *Phys. Rev. Lett.*, **99**, 228105.
55 Dertinger, T., Colyer, R., Iyer, G., Weiss, S., and Enderlein, J. (2010) *Proc. Natl. Acad. Sci. USA*. doi: 10.1073/pnas.0907866106.
56 Cheezum, M.K., Walker, W.F., and Guilford, W.H. (2001) *Biophys. J.*, **81**, 2378–2388.
57 Ram, S., Ward, E.S., and Ober, R.J. (2006) *Proc. Natl. Acad. Sci. USA*, **103**, 4457–4462.
58 Yildiz, A., Forkey, J.N., McKinney, S.A., Ha, T., Goldman, Y.E., and Selvin, P.R. (2003) *Science*, **300**, 2061–2065.
59 Yildiz, A. and Selvin, P.R. (2005) *Acc. Chem. Res.*, **38**, 574–582.
60 Heilemann, M., van de Linde, S., Schüttplez, M., Kasper, R., Seefeldt, B., Mukherjee, A., Tinnefeld, P., and Sauer, M. (2008) *Angew. Chem. Int. Ed.*, **47**, 6172–6176.
61 Heilemann, M., van de Linde, S., Mukherjee, A., and Sauer, M. (2009) *Angew. Chem. Int. Ed.*, **48**, 6903–6908.
62 Sharonov, A. and Hochstrasser, R.M. (2006) *Proc. Natl. Acad. Sci. USA*, **103**, 18911–18916.
63 Lidke, K.A., Rieger, B., Jovin, T.M., and Heintzmann, R. (2005) *Opt. Express*, **13**, 7052–7062.
64 Foelling, J., Bossi, M., Bock, H., Medda, R., Wurm, C.A., Hein, B., Jakobs, S., Eggeling, C., and Hell, S.W. (2008) *Nat. Methods*, **5**, 943–945.
65 Heilemann, M., Dedecker, P., Hofkens, J., and Sauer, M. (2009) *Laser & Photon. Rev.*, **3**, 180–202.
66 Fernandez-Suarez, M. and Ting, A.Y. (2008) *Nat. Rev. Mol. Cell Biol.*, **9**, 929–943.
67 Bates, M., Blosser, T.R., and Zhuang, X.W. (2005) *Phys. Rev. Lett.*, **94**, 101108.
68 Heilemann, M., Margeat, E., Kasper, R., Sauer, M., and Tinnefeld, P. (2005) *J. Am. Chem. Soc.*, **127**, 3801–3806.
69 van de Linde, S., Endesfelder, U., Mukherjee, A., Schüttplez, M., Wiebusch, G., Wolter, S., Heilemann, M., and Sauer, M. (2008) *Photochem. Photobiol. Sci.*, **8**, 465–469.
70 van de Linde, S., Kasper, R., Heilemann, M., and Sauer, M. (2008) *Appl. Phys. B*, **93**, 725–731.
71 van de Linde, S., Sauer, M., and Heilemann, M. (2008) *J. Struct. Biol.*, **164**, 250–254.
72 Wolter, S., Schüttpelz, M., Tscherepanow, S., van de Linde, S., Heilemann, M., and Sauer, M. (2010) *J. Microsc.*, **237**, 12–22.
73 Steinhauer, C., Forthmann, C., Vogelsang, J., and Tinnefeld, P. (2008) *J. Am. Chem. Soc.*, **130**, 16840–16841.
74 Vogelsang, J., Cordes, T., Forthmann, C., Steinhauer, C., and Tinnefeld, P. (2009) *Proc. Natl. Acad. Sci. USA*, **106**, 8107–8112.
75 Belov, V.N., Bossi, M.L., Foelling, J., Boyarskiy, V.P., and Hell, S.W. (2009) *Chemistry*, **15**, 10762–10776.
76 Irie, M. (2000) *Chem. Rev.*, **100**, 1685–1716.
77 Irie, M., Fukaminato, T., Sasaki, T., Tamai, N., and Kawai, T. (2002) *Nature*, **420**, 759–760.
78 Seefeldt, B., Kasper, R., Beining, M., Arden-Jacobm, J., Kemnitzer, N., Mattay, J., Drexhage, K.H., Heilemann, M., and Sauer, M. (2010) *Photochem. Photobiol. Sci.* doi: 10.1039/B9PP00118B.
79 Andresen, M., Stiel, A.C., Foelling, J., Wenzel, D., Schönle, A., Egner, A., Eggeling, C., and Hell, S.W. (2008) *Nat. Biotechnol.*, **26**, 1035–1040.
80 Schönle, A. and Hell, S.W. (2007) *Nat. Biotechnol.*, **25**, 1234–1235.
81 Patterson, G.H. and Lippincott-Schwartz, J. (2002) *Science*, **297**, 1873–1877.
82 Wiedenmann, J., Ivanchenko, S., Oswald, F., Schmitt, F., Rocker, C., Salih, A., Spindler, K., and Nienhaus, G.U. (2004) *Proc. Natl. Acad. Sci. USA*, **101**, 15905–15910.

83 Lukyanov, K.A., Chudakov, D.M., Lukyanov, S., and Verkhusha, V.V. (2005) *Nat. Rev.*, **6**, 885–891.

84 Betzig, E., Patterson, G.H., Sougrat, R., Lindwasser, O.W., Olenych, S., Bonofacino, J.S., Davidson, M.W., Lippincott-Schwartz, J., and Hess, H.F. (2006) *Science*, **313**, 1642–1645.

85 Hess, S.T., Girirajan, T.P., and Mason, M.D. (2006) *Biophys. J.*, **91**, 4258–4272.

86 Rust, M.J., Bates, M., and Zhuang, X.W. (2006) *Science*, **3**, 793–795.

87 Egner, A., Geisler, C., Middendorff, C., Bock, H., Wenzel, D., Medda, R., Andresen, M., Stiel, A.C., Jakobs, S., Eggeling, C., Schönle, A., and Hell, S.W. (2007) *Biophys. J.*, **93**, 3285–3290.

88 Flors, C., Hotta, J., Uji-I, H., Dedecker, P., Ando, R., Mizuno, H., Miyawaki, A., and Hofkens, J. (2007) *J. Am. Chem. Soc.*, **129**, 13970–13977.

89 Bates, M., Huang, B., Dempsey, G.T., and Zhuang, X.W. (2007) *Science*, **317**, 1749–1753.

90 Bock, H., Geisler, C., Wurm, C.A., von Middendorff, C., Jakobs, S., Schönle, A., Egner, A., Hell, S.W., and Eggeling, C. (2007) *Appl. Phys. B*, **88**, 161–165.

91 Shroff, H., Galbraith, C.G., Galbraith, J.A., White, H., Gillette, J., Olenych, S., and Betzig, M.W. (2007) *Proc. Natl. Acad. Sci. USA*, **104**, 20308–20313.

92 Huang, B., Wang, W.Q., Bates, M., and Zhuang, X.W. (2008) *Science*, **319**, 810–813.

93 Pavani, S.R., Thompson, M.A., Biteen, J.S., Lord, S.J., Liu, N., Twieg, R.J., Piestun, R., and Moerner, W.E. (2009) *Proc. Natl. Acad. Sci. USA*, **106**, 2995–2999.

94 Juette, M.F., Gould, T.J., Lessard, M.D., Mlodzianoski, M.J., Nagpure, B.S., Bennett, B.T., and Bewersdorf, S.T. (2008) *Nat. Methods*, **5**, 527–529.

95 Shtengel, G., Galbraith, J.A., Galbraith, C.G., Lippincott-Schwartz, J., Gillette, J.M., Manley, S., Sougrat, R., Waterman, C.M., Kanchanawong, P., Davidson, M.W., Fetter, R.D., and Hess, H.F. (2009) *Proc. Natl. Acad. Sci. USA*, **106**, 3125–3130.

96 Hess, S.T., Gould, T.J., Gudheti, M.V., Maas, S.A., Mills, K.D., and Zimmerberg, J. (2007) *Proc. Natl. Acad. Sci. USA*, **104**, 17370–17275.

97 Shroff, H., Galbraith, C.G., Galbraith, J.A., and Betzig, E. (2008) *Nat. Methods*, **5**, 417–423.

98 Flors, C., Ravarani, C.N., and Dryden, D.T. (2009) *ChemPhysChem*, **10**, 2201–2204.

99 van de Linde, S., Wolter, S., Heilemann, M., and Sauer, M. (2010) *J. Biotechnol.*, **149**, 260–266.

100 Endesfelder, U., van de Linde, S., Wolter, S., Sauer, M., and Heilemann, M. (2010) *ChemPhysChem*, **11**, 836–840.

101 Shannon, C.E. (1949) *Proc. IRE*, **37**, 10–21.

102 Baddeley, D., Jayasinghe, I.D., Cremer, C., Cannell, M.B., and Soeller, C. (2009) *Biophys. J.*, **96**, 22–24.

103 Vogelsang, J., Kasper, R., Steinauer, C., Person, B., Heilemann, M., Sauer, M., and Tinnefeld, P. (2008) *Angew. Chem. Int. Ed.*, **47**, 5465–5469.

104 Wardman, P. (1989) *J. Phys. Chem. Ref. Data*, **18**, 1639–1755.

105 Cleland, W.W. (1964) *Biochemistry*, **3**, 480–482.

106 Madej, E. and Wardman, P. (2007) *Arch. Biochem. Biophys.*, **462**, 94–102.

107 Burner, U., Jantschko, W., and Obinger, C. (1999) *FEBS Lett.*, **443**, 290–296.

108 Burner, U. and Obinger, C. (1997) *FEBS Lett.*, **411**, 269–274.

109 Sies, H. (1999) *Free Radical Biol. Med.*, **27**, 916–921.

110 Wombacher, R., Heidbreder, M., van de Linde, S., Sheetz, M. P., Heilemann, M., Cornish, V. W., and Sauer, M. (2010) *Nat. Methods*, **7**, 717–719.

111 Owen, D.M., Rentero, C., Rossy, J., Magenau, A., Williamson, D., Rodriguez, M., and Gaus, K. (2010) *J. Biophotonics*, **3**, 446–454.

112 Baddeley, D., Jayasinghe, I.D., Lam, L., Rossberger, S., Cannell, M.B., and Soeller, C. (2010) *Proc. Natl. Acad. Sci. USA*, **106**, 22275–22280.

9
Single-Molecule Enzymatics

9.1
Introduction: Why Study Enzymes on a Single-Molecule Level?

Enzymes, the protein workhorses of almost all biochemical reactions, have been studied extensively because of their exceptional catalytic activity and selectivity towards their natural substrates. Although proteins consist mainly of a linear arrangement of a limited number of different amino acids, they show a very complex behavior from a conformational point of view. The linear chain tends to fold into a very specific three-dimensional structure due to electrostatic and van der Waals interactions between the side chains of the amino acids and due to the formation of covalent disulfide bridges between cystein residues. The enzymatic activity is, to a large extent, determined by this three-dimensional structure and is very sensitive to small structural fluctuations. On the other hand, some degree of conformational freedom is necessary for the substrate to be able to reach the active site and to be converted by the enzyme. For instance, an "induced fit" mechanism is only possible if the protein backbone of the biocatalyst is sufficiently flexible. This knowledge led to the insight that enzymes cannot be regarded as static entities, but are fairly dynamic with respect to conformation and also activity [1].

However, classical bulk experiments return to an activity value that is averaged over 10^{12}–10^{18} enzymes, and can therefore never reveal such activity fluctuations. In order to unravel the hidden heterogeneities in enzyme populations, a growing interest in the study of individual enzymes arose. By 1961, Rotman had already published data on the denaturation behavior and the time-averaged activity of individual β-D-galactosidase enzymes, as will be highlighted later in this chapter [2]. Thanks to the advances in ultrasensitive fluorescence microscopy – since the late 1980s it has been possible to detect single fluorophores at room temperature – the first results on single-enzyme activity with a single-turnover time resolution were published halfway through the 1990s [3]. Today single-molecule enzymatic research is one of the hot-topics in biochemistry. After a short introduction to the basic biochemical principles that are involved in enzymatic processes, we will first take a closer look at the different

Handbook of Fluorescence Spectroscopy and Imaging. M. Sauer, J. Hofkens, and J. Enderlein

ISBN: 978-3-527-31669-4

strategies one can use to study single-enzyme activity and what information can be obtained. Secondly a few examples will be given of how single enzyme experiments can reveal some interesting enzymatic mechanisms.

9.2
Biochemical Principles of Enzymatic Activity: the Michaelis–Menten Model

In 1913 Leonor Michaelis and Maud Menten developed a simple kinetic scheme that models the rate of almost all enzyme-catalyzed reactions. This model describes the enzymatic action as a two-step reaction in which a reversible complex formation between enzyme (E) and substrate (S) is followed by the conversion of the complex (ES) into the reaction product (P) and the free enzyme.

$$E + S \underset{k_{-1}}{\overset{k_1}{\rightleftharpoons}} ES \xrightarrow{k_2} E + P \tag{9.1}$$

Initially, no product P is present and the rate of product formation is given by

$$V_0 = k_2[ES] \tag{9.2}$$

Under steady-state conditions of constant $[ES]$ this model was mathematically elaborated by Briggs and Haldane [4].

$$V_0 = \frac{V_{max} \cdot [S]}{[S] + K_M} \tag{9.3}$$

$$K_M = \frac{k_{-1} + k_2}{k_1} = \frac{[E][S]}{[ES]} \tag{9.4}$$

$$V_{max} = k_2[E]_{total} \tag{9.5}$$

In this instance k_{cat} equals k_2. As can be seen from the formula, there are two limiting cases. At low substrate concentrations ($[S] \ll K_M$) the complex formation is rate limiting and the initial reaction rate is directly proportional to the substrate concentration. At high substrate concentrations ($[S] \gg K_M$), k_2 becomes rate limiting and V_0 is independent of $[S]$ and is given by V_{max}. The enzyme is then saturated with substrate. The apparent rate constant at these saturating conditions, k_{cat}, then equals k_2. K_M is the substrate concentration at which the initial rate is half the maximum rate (Figure 9.1). This very simplified kinetic model also accounts for the activity of more complex enzymatic systems. For instance, in many protease and lipase catalyzed hydrolysis reactions, the enzymatic cleavage proceeds in two steps. Firstly, the acyl group is transferred to the enzyme resulting in an acyl-enzyme intermediate (AE) and a free alcohol or amine (P_1). In the second step this acyl-enzyme intermediate is hydrolyzed, resulting in a free carboxylic acid (P_2) and free enzyme.

$$E + S \underset{k_{-1}}{\overset{k_1}{\rightleftharpoons}} ES \xrightarrow{k_2} P_1 + AE \xrightarrow{k_3} E + P_2 \tag{9.6}$$

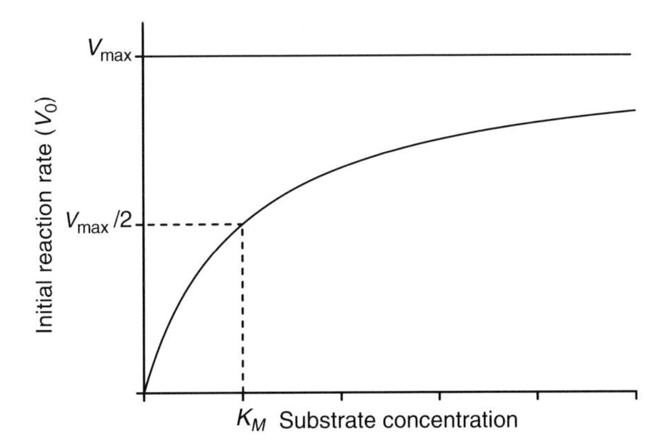

Figure 9.1 Michaelis–Menten kinetics. V_{max} is the reaction rate at saturating substrate concentrations. K_M represents the substrate concentration at which the rate equals $V_{max}/2$.

In this case the MM (Michaelis–Menten) parameters are given by:

$$k_{cat} = \frac{k_2 k_3}{k_2 + k_3} \quad \text{and} \quad K_M = \frac{(k_{-1} + k_2)k_3}{k_1(k_2 + k_3)} \tag{9.7}$$

In general, the deacylation step (k_3) is rate limiting over k_2, so k_{cat} approaches k_3. Effects of inhibition by substrates or products can also be incorporated in the MM equation, but this is beyond the scope of the present text.

The key issue in single-molecule enzymology is the validation of this Michaelis–Menten theory in single-enzyme conditions. Does a single enzyme still obey the same kinetic laws compared with the averaged activity of a huge ensemble of seemingly identical enzymes?

9.3 “Looking” at Individual Enzymes

The very high sensitivity of fluorescence measurements allows the detection of low concentrations of fluorophores produced by the action of one individual enzyme. The crucial factor in this type of time-averaged measurement is that the individual enzyme molecules need to be spatially resolved.

The possibility of detecting single fluorophores with fluorescence microscopes (wide-field or confocal) opened up a new area of enzymatic assays. Several strategies were developed to study individual enzymes with a single turnover precision, thereby revealing even more detailed information on the time-resolved activity and mechanisms of several enzymatic systems.

This section is divided in two main parts: first we will take a closer look at what single enzymes can teach us about catalysis kinetics on the scale of individual molecules. In the second part, a short overview is given about how single-enzyme research can unravel some interesting enzymatic mechanisms. More detailed information can be found in the recent reviews by Blank *et al.* and by Chen *et al.* [5, 6].

9.3.1 Single-Enzyme Studies and Kinetics

9.3.1.1 Space-Resolved, but Time-Averaged Single Enzyme Assays

Before the emergence of single-molecule fluorescence spectroscopy (SMFS), scientists invented some ingenious techniques for studying single enzymes, by detecting the accumulated fluorescent products from the action of an individual enzyme. These spatially-, but not time-resolved experiments revealed useful information on the distribution of enzymatic properties along a population. A few strategies will be highlighted in this section.

Water-in-Oil Emulsions Spatial resolution of individual enzymes can be achieved by confining the enzymes in the small water droplets of a water-in-oil emulsion. By starting from a very dilute aqueous enzyme solution ($\pm 10^{-10}$ M or $\pm 10^{-2}$ mol μm^{-3}), one can statistically assume that almost none of the μm-sized water droplets will contain more than one enzyme molecule; most of them will contain no enzyme at all, while some will contain exactly one enzyme. If the aqueous solution also contains a profluorescent substrate, the time-averaged activity of the individual enzymes can be monitored by measuring the gradual increase in the fluorescence of the "active" droplets with a wide-field fluorescence microscope (Figure 9.2).

The first real single-enzyme experiment was carried out by Rotman in 1961, who used this technique to analyze the activity, and in particular the effect of thermal denaturation on the activity, of individual β-D-galactosidase enzymes catalyzing the hydrolysis of 6-hydroxyfluoran-β-D-galactose (6HFG) to the strong emitting 6-hydroxyfluoran (6HF) [2]. The results suggested that the decrease in activity upon heating is a

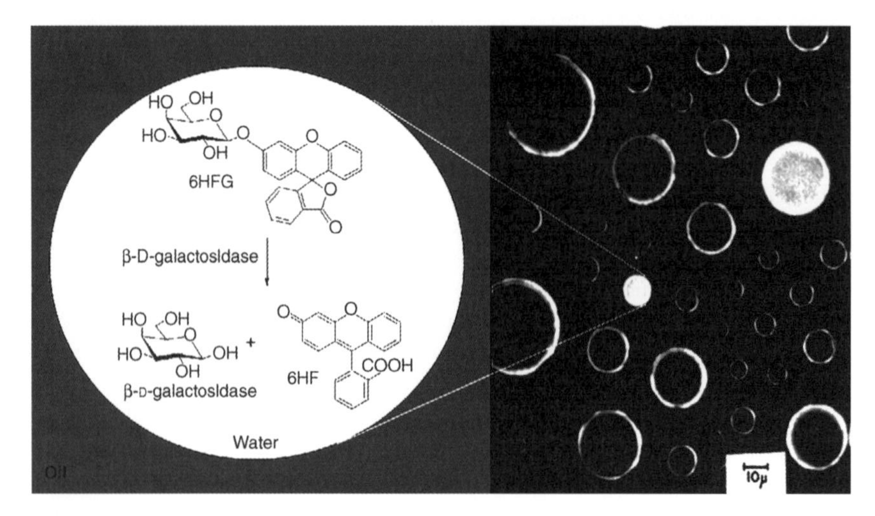

Figure 9.2 Studying individual enzymes using water-in-oil emulsions. Upon enzymatic action a fluorescent product is formed that accumulates in the water droplets. This is visualized as an increase in intensity of the droplets containing the enzymes. (Copyright Rotman (1961) (*Proc. Natl. Acad. Sci. USA*, **47**, 1981–1991 [2].)

consequence of the complete loss of activity of a few enzymes, rather than the gradual decrease of activity of all enzymes. The discovery that thermal denaturation is an "all-or-nothing" process immediately proved the power of single-enzyme experiments.

Lee and Brody picked re-visited this experimental scheme to study the protease activity of α-chymotrypsin [7]. Upon enzyme action a non-fluorescent peptide-derivatized rhodamine dye is converted into the very fluorescent free rhodamine. They compared several individual enzymes and discovered that there is a fairly large heterogeneity or static disorder among the enzyme population. Although the time resolution and sensitivity of these experiments does not allow the detection of single-turnovers, they saw a nonlinear increase of fluorescence intensity over time. The enzyme seemed to fluctuate between states with different activities, a phenomenon that is termed dynamic disorder. The observation of these fluctuation phenomena gave rise to much discussion concerning the validity of the traditional Michaelis–Menten model at the individual enzyme level.

Capillary Electrophoresis Spatial separation of individual enzymes can also be achieved by loading an extremely dilute enzyme solution together with the pro-fluorescent substrate into a capillary tube, with the size of a few microns in the radial dimension and a few centimeters in the axial dimension. After incubation, enzymes, substrate, and product are led to a fluorescence detector by applying an electrical field of about 400 kV cm^{-1}. The difference in electrophoretic mobility allows the separation of substrate and product, thereby preventing interference of substrate fluorescence in the detection of the product (Figure 9.3).

An additional advantage of this technique is the relatively high electrophoretic mobility of the enzyme compared with that of the substrate and product. This implies that the enzyme can be moved from one substrate zone to another by applying smaller electrical fields, while for instance the temperature is altered. Subsequently, the various substrate and product zones can be moved to the detector, and analysis of the temperature dependence of one single enzyme is possible.

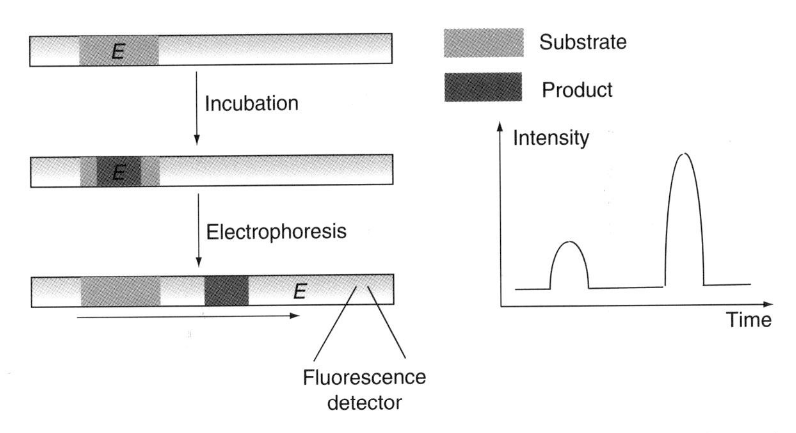

Figure 9.3 General scheme for electrophoretic assays of single-enzymes (*E*). After incubation of the enzyme in its substrate environment, a voltage is applied to lead the enzyme, substrate, and product separately to a fluorescence detector.

Xue and Yeung became well known in this field after the publication of their results on the activity of single lactate dehydrogenase enzymes towards the oxidation of lactic acid by NAD^+ [8]. Using capillary electrophoresis, they measured the activity by detecting the typical fluorescence of the NADH formed upon enzymatic reaction. These researchers were actually the first to prove static disorder in enzymatic activity: the activities of the individual enzymes differed by a factor of up to four and this heterogeneity remained stable over a period of at least two hours.

Alkaline phosphatase even showed a tenfold activity difference in the dephosphorylation of a hydroxybenzothiazol phosphate according to Craig *et al.* [9]. However, these workers attributed this disorder to a difference in the degree of post-translational glycosylations, rather than the equilibrium between several conformers.

Femtoliter Chambers Via photolitographical procedures it is possible to produce materials with extremely small – down to femtoliter volumes – and well defined holes. These femtoliter chambers are very well suited for use as minireactors for individual catalysts. Wide-field fluorescence microscopy applications are possible by using a transparant matrix such as polydimethylsiloxane (PDMS), as shown by Rondelez *et al.* (Figure 9.4) [10]. Loading the chambers with substrate and one individual enzyme is possible by sandwiching a droplet containing the substrate and extremely dilute enzyme concentrations between a glass coverslip and the PDMS matrix, with

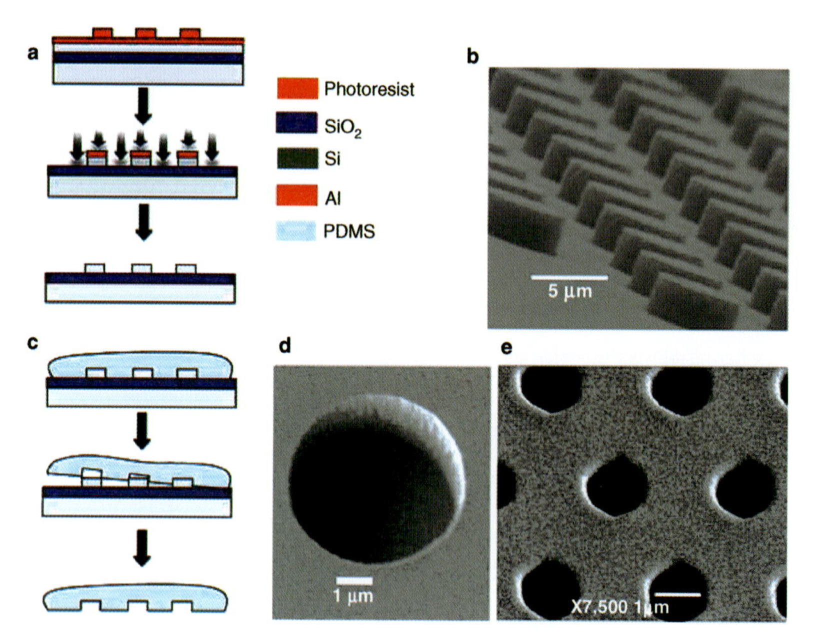

Figure 9.4 Fabrication process of femtoliter chambers. (a) Using photolithographic processes and chemical etching the mold is made. (b) SEM picture of the microstructure of the mold. (c) Preparation of the PDMS with femtoliter chambers is carried out by pouring the PDMS on top of the mold, followed by curing. (d) SEM images of the resulting PDMS polymer. (Copyright Rondelez *et al.* (2005) *Nat. Biotechnol.*, **23** (3), 361–365[10]).)

Figure 9.5 Loading of the femtoliter chambers. A droplet containing substrate and dilute enzyme concentration is sandwiched between a glass coverslip and the PDMS matrix, with the chambers facing towards the coverslip. (Copyright Rondelez *et al.* (2005) *Nat. Biotechnol.*, **23** (3), 361–365[10]).)

the chambers facing towards the coverslip (Figure 9.5). As the amounts of enzyme in the chambers are Poissonian distributed, most of the chambers will contain no enzyme, some have one enzyme and only a few contain more than one enzyme molecule. Wide-field fluorescence microscopy then monitors the fluorescence increase of each of these chambers.

As in the abovementioned methods, this technique also requires the use of a profluorescent substrate in order the measure the enzymatic activity as an increase in fluorescence intensity. To prove the usefulness of their technique, Rondelez and coworkers studied the activity of single β-D-galactosidase molecules, the same enzyme that Rotman used in 1961 (Figure 9.6). To prevent enzyme adsorption and denaturation on the hydrophobic PDMS-polymer, the walls were coated with a bovine serum albumine (BSA) layer. They could prove that more than 70% of the enzymes remained active under these conditions.

Femtoliter Wells Etched in Optical Fiber Bundles Another elegant approach to confine individual enzymes in small volumes is to use an optical fiber bundle, consisting of a few tens of thousands of individual fibers of about 5 μm diameter, in which the distal ends of the fiber cores were etched away, creating small wells of homogeneous size. Upon enclosure of the reaction mixture in the individual wells, by sealing the distal end with, for instance, a silicon gasket, the reaction can be monitored by illumination

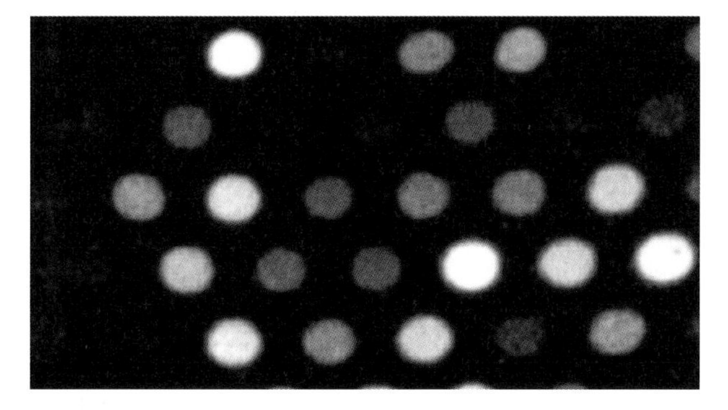

Figure 9.6 Wide-field image of β-D-galactosidase activity in femtoliter chambers. Upon enzymatic action strong fluorophores accumulate in the chambers, yielding bright emission. (Copyright Rondelez *et al.* (2005) *Nat. Biotechnol.*, **23** (3), 361–365 [10]).)

Table 9.1 Selection of publications highlighting various approaches to single-enzyme activity assays without single-turnover sensitivity.

Enzyme	Substrate	Approach	Ref.
β-Galactosidase	6-Hydroxyfluoran β-D- galactopyranoside	Emulsion droplets	[2]
Chymotryspin	$(sucAAPF)_2$-rhodamine 110	Emulsion droplets	[7]
Lactate dehydrogenase	Lactate + NAD	Capillary electrophoresis	[8]
Alkaline phosphatase	2′-(2-Benzothiazolyl)-6′-hydroxybenzothiazole phosphate (AttoPhos)	Capillary electrophoresis	[9]
β-Galactosidase	Resorufin β-D-galactopyranoside	Capillary electrophoresis	[12]
Lactate dehydrogenase	Lactate + NAD	Microfabricated wells in quartz	[13]
β-Galactosidase	Fluorescein di-β-D-galactopyranoside	Microfabricated wells in PDMS	[10]
β-Galactosidase	Resorufin β-D-galactopyranoside	Wells etched in optical fiber bundles	[11]

of the wells via the proximal side of the fiber bundle. The elegance of this approach thus lies in the fact that the fibers play a bifunctional role in the experiment. Firstly, they separate the individual enzymes and provide a confined reaction environment for each individual enzyme. Secondly, they provide a way of illuminating the individual chambers homogeneously. This approach was pioneered by the group working with Walt, who investigated stochastic inhibitor binding and inhibitor release in individual β-galactosidase enzymes. For the D-galactal inhibitor, cooperative binding and release on the four subunits of the tetrameric enzyme was found [11].

A summary of landmark papers on single-enzyme activity assays without single-turnover sensitivity is given in Table 9.1.

9.3.1.2 Single-Turnover Experiments: Space- and Time-Resolved Enzyme Assays

Although time-averaged single-enzyme experiments yielded some very important new insights into the functioning of enzymes, researchers still wanted to go a step further to reveal the secrets of enzyme mechanisms. As fluorescence detectors and optics became more and more sensitive and single-molecule detection of strong emitters under ambient conditions was no longer utopia, enzymologists took the opportunity to study *in situ* individual enzymatic turnovers.

Interestingly, the concept of "kinetics" takes on a slightly different mathematical meaning on moving down from the "ensemble" level to the level of individual turnovers. In bulk experiments, kinetics are expressed as a change in concentration of reactants and products over time. However, when we look at single-product molecules formed by an individual enzyme, concentration is an irrelevant concept. In this case, kinetics are expressed as the *probability* that one product molecule is formed from a reactant by the single catalyst. As turnovers become stochastic events, it is clear that the classical Michaelis–Menten model needs to be revised to a single-enzyme MM-model. After a non-exhaustive overview of the most important and original techniques for single-turnover experiments, we will go into more detail on the implications of this research on the kinetic model for enzymes.

In the first real single-turnover study, dating from 1995, individual catalytic cycles of ATP-hydrolysis by the molecular motor myosin were recorded, using wide-field fluorescence microscopy [3]. For this purpose, extremely small amounts of myosin, marked with a Cy5 fluorescent dye, were immobilized on a quartz coverslip, followed by addition of a solution containing the substrate, Cy3-labeled ATP. As Cy5 has a distinct red-shifted absorption and emission spectrum compared with Cy3, fluorescence from the enzyme and substrate can be distinguished using two different lasers and two detection channels, with the appropriate mirrors and filters. Labeling of the enzyme allows the exact localization of individual enzyme molecules. By using the total internal reflection technique (TIR), only emission originating close to the coverslip is collected. In this way, emission from the substrate is only detected upon binding to the enzyme (Figure 9.7). This results in a blinking pattern of the emission in the substrate detection channel, originating from the binding of the substrate and subsequently unbinding of the product from the enzyme. Despite the original approach, this technique is hampered by some assumptions. For instance, these workers assumed that every substrate-binding leads to enzymatic hydrolysis. Of course this is not a priori correct. In general, substrate binding or formation of the enzyme–substrate complex is considered to be under the control of dynamic equilibrium. As a result, the spikes in the recorded trace can also originate from substrate-binding and unbinding without enzymatic conversion.

A few years later, Lu *et al.* published their results on the kinetics of cholesterol oxidase [14]. Here the catalytic turnovers were probed by the fluorescent FAD cofactor used by the enzyme as an electron shuttle. During a catalytic cycle, FAD is reduced by

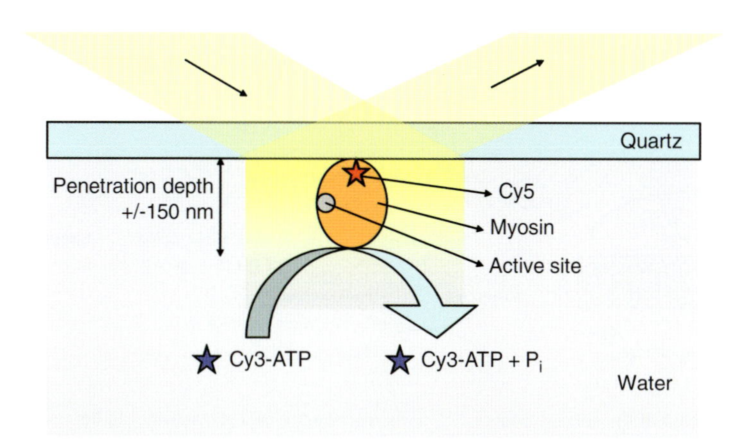

Figure 9.7 Scheme of the single-turnover study of myosin, using wide-field TIRFM. Upon binding of the fluorescent labeled ATP-substrate, a fluorescence signal is detected at the enzyme's location.

cholesterol or cholesterol derivatives, resulting in a non-fluorescent $FADH_2$. After regeneration by O_2 the fluorescence is recovered. As a result, catalytic turnovers are represented by off-times in the fluorescence time trace. For this experiment, spatial resolution and immobilization of the enzymes was achieved by entrapping them in an agarose polymer. The large diameter of the enzyme molecules restricts translational movements, while the smaller substrate and product molecules can freely diffuse through the porous polymer layer (Figure 9.8). The main advantage of this immobilization method is that the enzyme is not restricted in its conformational dynamics and therefore it gives a more representative impression of the activity of "free" enzyme molecules.

Probing a fluorescent cofactor of the enzyme of interest also has some drawbacks. Firstly, blinking phenomena, not related to enzymatic activity, might occur and are not always distinguishable from enzymatic on–off behavior. Secondly, the photostability of the probe determines the time span over which the activity can be followed. Luckily, the photostability of the FAD-cofactor is strongly enhanced by the presence of the protective protein shell.

This research led to some interesting conclusions on the dynamic behavior of enzyme conformations, which will be discussed further. Inspired by the pioneering work of Lu, other single-molecule groups used this elegant method to study conformational dynamics of some other flavo-enzymes, such as dihydroorotate dehydrogenase and *p*-hydroxybenzoate hydroxylase [15, 16].

The most common way to visualize individual turnovers is to make use of fluorogenic substrates that are converted into strong fluorophores upon enzyme action. The big advantage of this technique is that photostability of the dye is not a limit for the time over which the activity can be followed, because by every turnover a new dye molecule is formed. However, this technique also has its drawbacks: as fluorogenic substrates are generally large aromatic compounds, only enzymes with large and flexible active sites can be examined. In addition, fluorogenic substrates are often chemically fairly different compared with the enzyme's natural substrates, and therefore the observed mechanism is not necessarily representative of the "natural" mechanism.

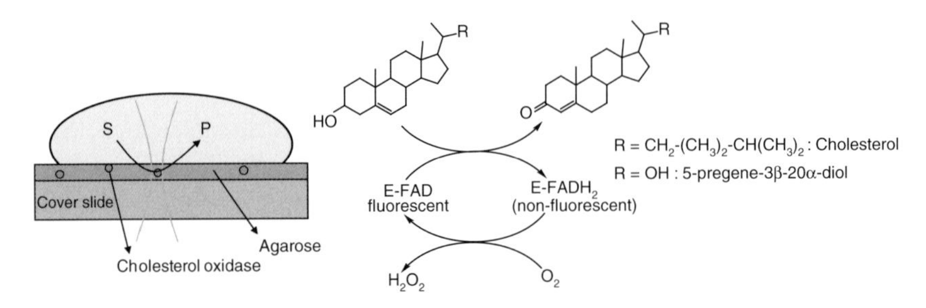

Figure 9.8 Scheme of the single-turnover study of the oxidation of cholesterol derivatives by cholesterol oxidase. Using a confocal fluorescence microscope the catalytic activity can be monitored, as the fluorescence of the enzyme's cofactor switches on and off during the enzymatic cycle.

Figure 9.9 ON-OFF representation of the kinetic scheme in single-turnover experiments when using a fluorogenic substrate. An on-state arises upon formation of the fluorescent product. After diffusion of this product the intensity falls back to its background level.

In the recorded trace, the kinetic scheme of the enzyme activity is simplified to a two-state-system: the off-state represents the enzyme in its free form and the enzyme–substrate complex. The on-state corresponds to the enzymatic reaction and the diffusion of the fluorescent product out of the active site (Figure 9.9). For example, in 1999, Edman *et al.* published their results on the oxidation of the fluorogenic dihydrorhodamine 6G to the strong fluorescent dye rhodamine 6G by horseradish peroxidase, covalently immobilized by a biotin–streptavidin bonding [17]. Analysis of these data is more complex as each catalytic cycle consists of the oxidation of two dihydrorhodamine substrates. The reduced enzyme is subsequently regenerated in one step by hydrogen peroxide (Figure 9.10).

Until now, it has mainly been hydrolysis reactions that have been studied by this technique. For this purpose, rhodamine and fluorescein dyes are particularly suitable: the amino and phenol groups are easily derivatized towards amides, respectively, esters or ethers, resulting in a complete quenching of the fluorescence. For instance, Velonia and coworkers used a fluorescein-ester to study the hydrolysis activity of the lipase CALB (Figure 9.11) [18, 19], thereby confirming the fluctuating enzyme model, as will be discussed later on in this chapter.

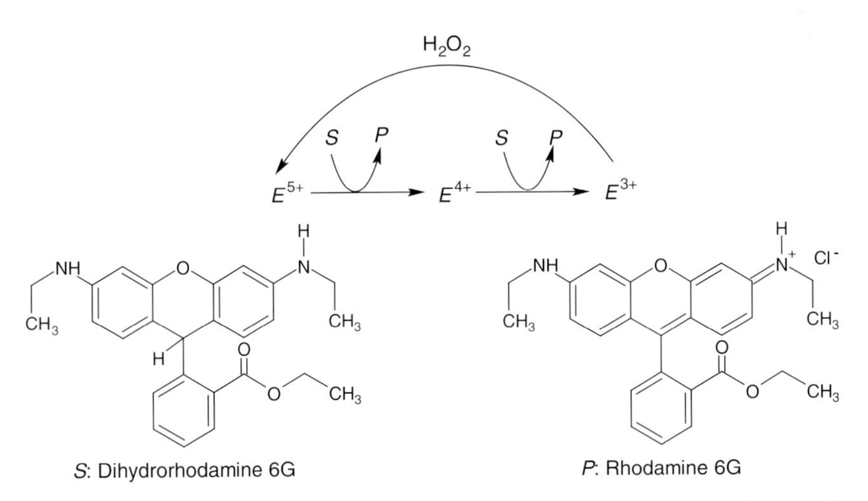

Figure 9.10 Catalytic cycle of horseradish peroxidase in the oxidation of dihydrorhodamine. The non-fluorescent dihydrorhodamine 6G is converted by the enzyme into the strong fluorophore rhodamine 6G.

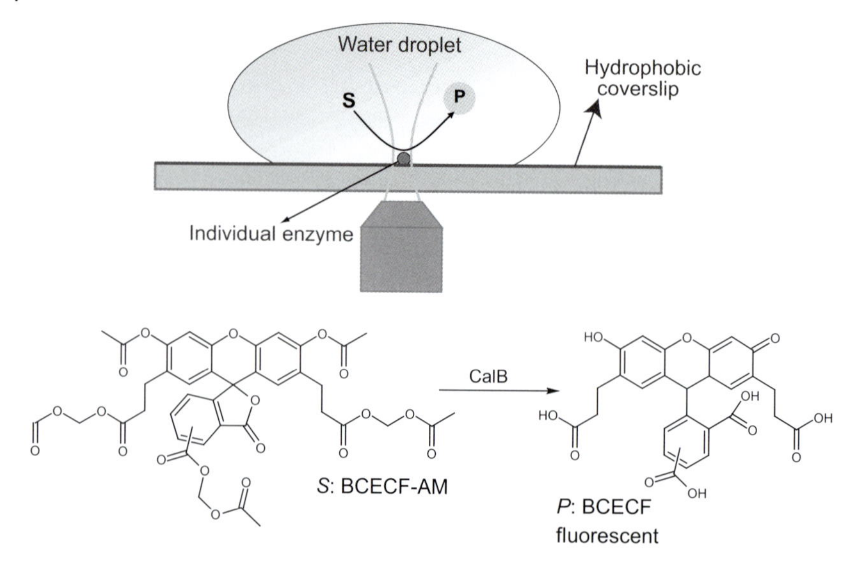

Figure 9.11 Scheme of the single-turnover study of the CALB-catalyzed hydrolysis of BCECF-AM. By using a confocal fluorescence microscope the formation of individual fluorophores from a non-fluorescent substrate in the enzyme can be monitored.

The group working with Xie studied β-D-galactosidase through the hydrolysis of the galactose-ether of resorufin (RGP) (Figure 9.12) [20]. As illustrated in the figure, these workers used an innovative set-up for reducing the background. By shining a broad high-intensity bleaching beam over the focal area, freely diffusing fluorescent product molecules are photobleached before reaching the small confocal volume.

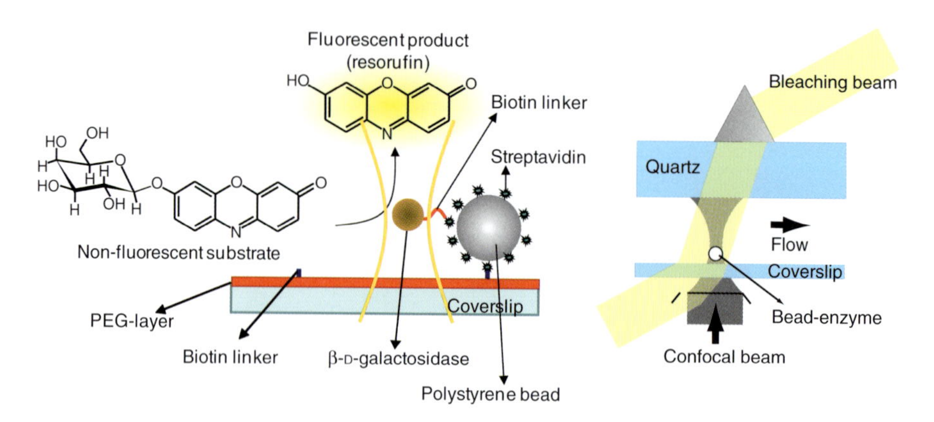

Figure 9.12 Scheme of the set-up for the single-turnover study of β-D-galactosidase by English *et al.* [20]. The enzyme is immobilized on a large polymeric bead that can be easily localized by optical microscopy. Confocal fluorescence microscopy allows the detection of single fluorescent product molecules formed inside the enzyme from a non-fluorescent substrate. The right-hand panel shows the principle of the bleaching beam: a second strong excitation beam overlaps the confocal volume and bleaches freely diffusing fluorophores before they can reach the detection volume.

The results showed that activity fluctuations only occur when the enzyme is nearly saturated with substrate, indicating that interconversion between the different conformers mainly takes place in the *ES*-complex or that substrate binding is hardly affected by small conformational changes.

Single-turnover counting can reveal more than just conformational dynamics. By studying the activity of individual chymotrypsin enzymes, a spontaneous deactivation pathway was observed [21]. Surprisingly, rather than a sudden one-step deactivation due to an abrupt denaturation, a long transient phase of deactivation was observed. Throughout this transient phase a reversible conformational change causes the enzyme to switch between active and inactive states (see Figure 9.13). During this equilibrium, stepwise inactivation occurs before the enzyme deactivates irreversibly. These results provide important new insights into how transitions occur during deactivation.

Also, the localization of a secondary reaction in enzymatic systems can be investigated by single-molecule fluorescence microscopy. This has recently been illustrated for the *Curvularia verruculosa* bromoperoxidase [22]. In the presence of

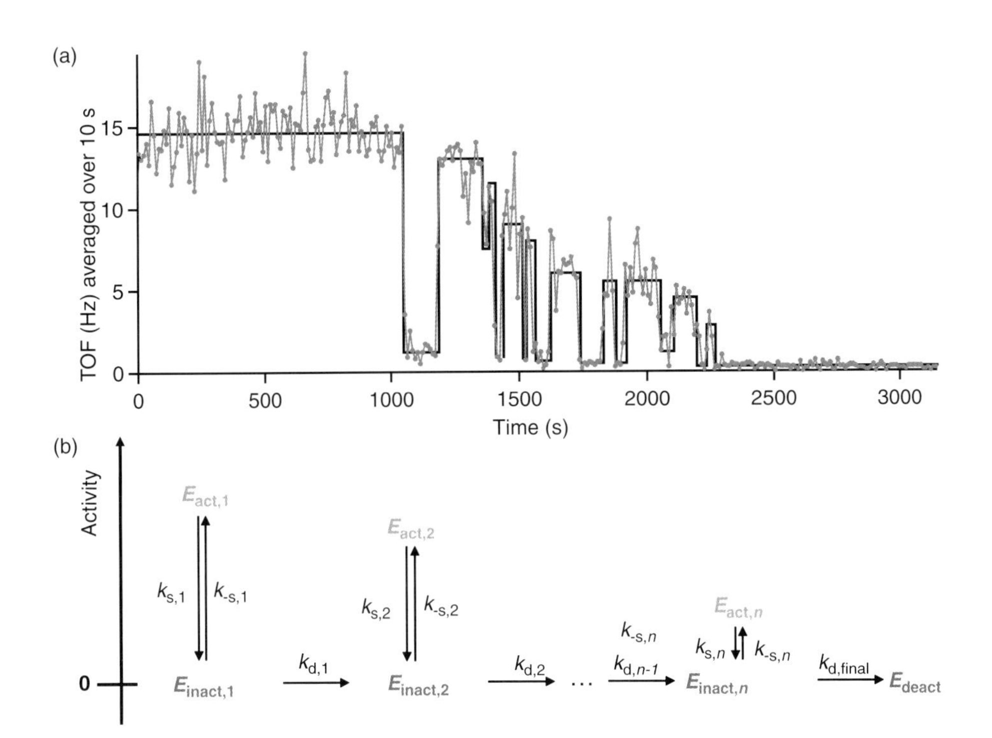

Figure 9.13 Spontaneous deactivation of a single α-chymotrypsin enzyme. (a) Time transient of the turnover frequency (TOF, averaged over 10 s) of a single deactivating chymotrypsin enzyme. The solid black line indicates the different states. (b) The extended single-molecule deactivation model for enzymes: a reversible conformational change causes the enzyme to switch between active and inactive states. During this equilibrium, stepwise inactivation occurs before the enzyme deactivates irreversibly.

hydrogen peroxide this enzyme converts bromide salts into the reactive hypobromite. The formed hypobromite will in turn oxidize or brominate organic compounds in a secondary reaction. However, there is currently a lively debate on where this secondary reaction takes place. The presence of regio- and stereoselectively brominated compounds in organisms that produce haloperoxidase enzymes, raised the idea that this very enzyme has two binding places: one for the conversion of bromide into hypobromite, and one providing a chiral environment for the secondary reaction [23–25]. To monitor the secondary reaction with single-molecule sensitivity, the fluorogenic probe aminophenyl fluorescein was applied. This probe selectively reacts with hypohalites, yielding the strongly fluorescent fluorescein. By positioning the laser focus of a confocal fluorescence microscope at various distances with respect to the immobilized enzyme, the distance profiles of HOBr reactivity towards aminophenyl fluorescein can be constructed by counting the number of single-reaction events as a function of time for each position. It was found that for the studied enzyme, hypobromite is released into the solution where it can freely oxidize the organic probe. In this way, the active oxygen atom is transported in the form of a small molecular vehicle (HOBr) from the catalyst towards the substrate, enabling the oxidation or bromination of very bulky substrates.

Other approaches for visualizing individual turnover events rely on changes in energy transfer efficiencies during the catalytic cycle. Electron transfer [26, 27] and also fluorescence resonance energy transfer (FRET) [15, 28, 29] have been applied for single-enzyme activity assays. For instance, Kuznetsova *et al.* studied the enzyme copper nitrite reductase (NiR) by labeling the enzyme with a fluorescent donor dye [30]. NiR is a homotrimeric enzyme with two copper cofactors per monomer, which catalyze the reduction of nitrite. This reduction involves the transfer of an electron from one copper to the second copper, before it is finally transferred to nitrite, yielding nitric oxide. The first copper absorbs light in the visible range, only in its oxidized and not in its reduced state. Thus, in its oxidized state it can quench the fluorescence of the attached label. One cycle of quenching/emitting then corresponds to one catalytic cycle. Of course this approach has similar limitations to that where a fluorescent cofactor is used as reporter system. In particular, the limited photostability of the dye puts an upper limit to the time range over which the activity of a single enzyme can be monitored.

An overview of reports dealing with fluorescence-based single turnover counting in enzymes is given in Table 9.2.

For some particular enzymatic systems time-resolved information on the activity of individual enzymes can also be obtained by non-fluorescence based techniques. In 2003 the first of a series of papers appeared on the activity dynamics of the λ-exonuclease enzyme toward the hydrolysis of double-strand DNA (ds-DNA) to single-strand DNA (ss-DNA) [37–39]. The activity assay was based on the fact that ss-DNA, but not ds-DNA, tends to coil into a very compact structure. After binding of the 5′-terminal of the ds-DNA substrate through a biotin–streptavidin linker on a glass slide and attaching a large polystyrene particle to the other end of the DNA strand, a laminar flow containing the nuclease enzymes is applied, causing the DNA-substrate to stretch along the flow (Figure 9.14). The exact position of the polystyrene particle

Table 9.2 Selection of publications dealing with single-turnover counting in individual enzymes.

Enzyme	Substrate	Reporter system	Ref.
Lysozyme	*E. coli* cell wall particles	FRET	[28]
Dihydrofolate reductase	7,8-Dihydrofolate + NADPH	Electron transfer	[26, 27]
Dihydrofolate reductase	7,8-Dihydrofolate + NADPH	FRET	[15]
Staphylococcal nuclease	+/− Single stranded DNA	FRET	[29]
Cholesterol oxidase	Cholesterol or 5-pregene3β-20α-diol	Fluorescent cofactor	[14]
Dihydroorotate dehydrogenase A	Dihydroorotate + dichlorophenol indophenol	Fluorescent cofactor	[31]
Dihydroorotate dehydrogenase A	Dihydroorotate + fumarate	Fluorescent cofactor	[32]
Nitrite reductase	Nitrite	FRET	[30]
Horseradish peroxidase	Dihydrorhodamine 6G	Fluorogenic substrate	[17, 33, 34]
Horseradish peroxidase	Dihydrorhodamine 123	Fluorogenic substrate	[35]
Lipase (CalB)	2′,7′-Bis-(2-carboxy-ethyl)-5/6-carboxyfluorescein, acetoxymethylester	Fluorogenic substrate	[18, 19]
β-Galactosidase	Resorufin β-D-galactopyranoside	Fluorogenic substrate	[20]
Lipase (TLL)	Carboxyfluorescein diacetate	Fluorogenic substrate	[36]
Chymotrypsin	$(sucAAPF)_2$-rhodamine 110	Fluorogenic substrate	[21]
Haloperoxidase	Aminophenyl fluorescein	Fluorogenic substrate	[22]

can easily be tracked by optical microscopy. λ-Exonuclease can dock to the substrate only at the 5′-terminal and from this position it converts the ds-DNA into ss-DNA by progressive hydrolysis of the complementary strand. At the applied extension forces through the laminar flow, the ds-DNA is almost completely stretched, whereas the ss-DNA coils into a compact structure, causing a displacement of the polymeric bead in the direction opposite to the flow direction. This displacement can be tracked as a function of time and directly relates to the enzyme activity.

Although the resolution in this experiment was restricted to approximately 500 turnovers, dynamic disorder in the activity was observed. In this case the activity fluctuations were not exclusively assigned to enzyme dynamics, as the free energy of base melting – a crucial step in the enzymatic reaction – also differed along the DNA substrate, depending on the exact nucleotide sequence. By estimating this melting energy as a superposition of hydrogen-bonding and base-stacking interactions, these workers found a good correlation between the digestion speed of the enzyme and the free energy of base melting.

9.3.1.3 **Results: Revision of the Classical Michaelis–Menten Model**

Although from a mathematical point of view, it is not that difficult to translate the classical MM-equation into its single-enzyme analog, a detailed deduction of this equation is beyond the scope of this textbook. As single-enzyme turnovers are stochastic events, the reaction rate is evaluated by the inverse of the mean waiting

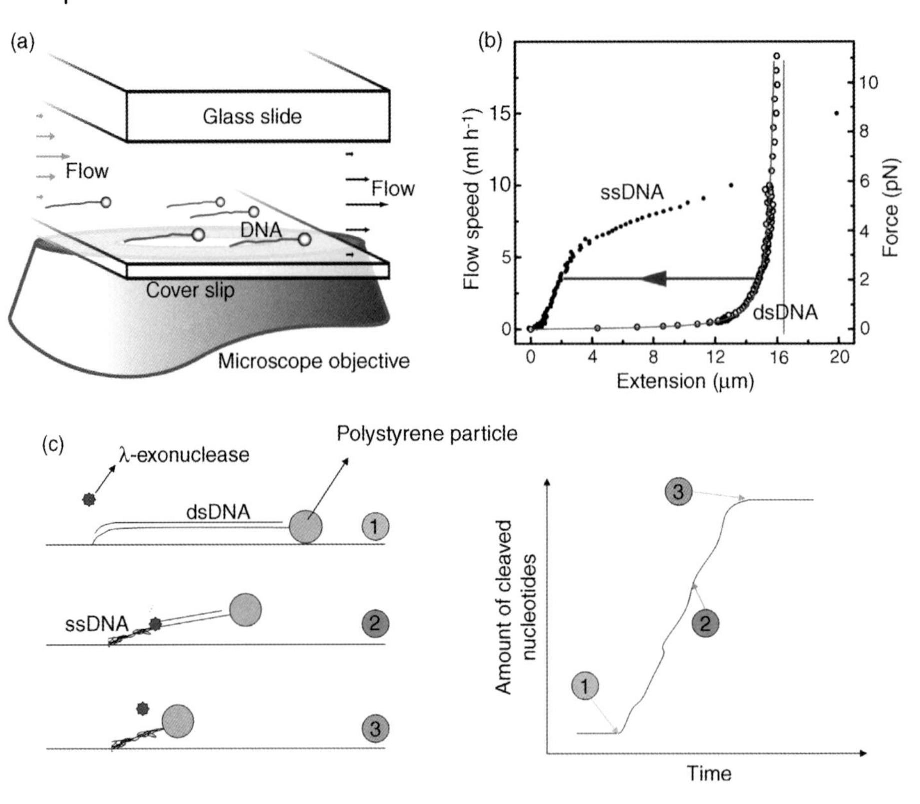

Figure 9.14 Time-resolved activity study of γ-exonuclease. (a) By attaching one end of the ds-DNA substrate to the glass coverslip and the other end to a large polymeric bead and by applying a laminar flow, the DNA-substrate is stretched along the flow. (b) and (c) As ds-DNA is more easily stretched in the flow than ss-DNA, the enzymatic conversion of ds- into ss-DNA will cause the bead to move in a direction opposite to the flow direction. By measuring the displacement as a function of time the time-dependent activity can be monitored. (Copyright Van Oijen *et al.* (2003) *Science*, **301** (5637), 1235–1238 [37].)

time, $\langle\tau\rangle$ between two successive turnovers [40]. The single-molecule MM-equation then becomes:

$$\frac{1}{\langle\tau\rangle} = \frac{k_2 \cdot [S]}{[S] + K_M} \tag{9.8}$$

Because in the absence of static disorder, time-averaging of one individual is equivalent to averaging out over a large number of individuals at a certain point in time – a principle that is known as ergodicity – one can compare this equation with the traditional MM-equation (Equation 9.3) and see that $1/\langle\tau\rangle$ equals $V_0/[E]_{total}$.

As the successive steps in the traditional MM-model are supposed to obey first-order kinetics, a histogram of the waiting times between two turnovers is expected to show a single-exponential decay in the case of static (non-fluctuating) enzymes in a homogeneous population with k_1 or k_2 (see Equation 9.1) as the rate-limiting factor.

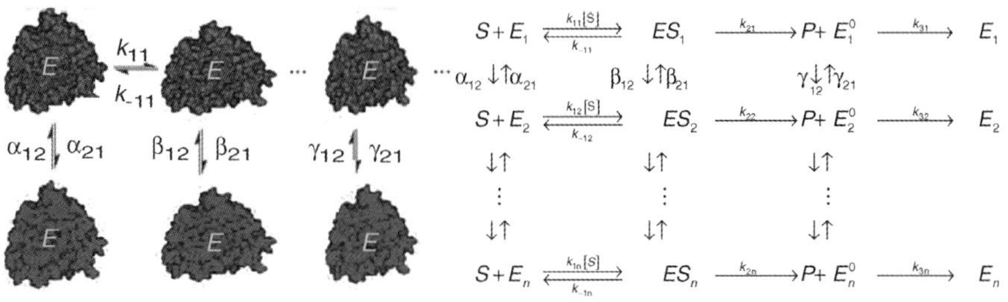

Figure 9.15 The fluctuating enzyme model. The enzyme is in thermodynamic equilibrium with several conformational substates, each having their own catalytic parameters. Interconversion between those substates gives rise to dynamic disorder in the catalytic activity. (Copyright Velonia (2005) *Angewandte Chemie*.)

However, the truth is more complex than this. As single-turnover experiments proved that static and dynamic disorder seems to be the rule, more than the exception, in enzymology, the MM-model needs to be extended to include conformational fluctuations and its influence on the catalytic kinetics [14, 18, 20]. In the fluctuating-enzyme model, enzymes are considered to switch between several conformational states, each with their own characteristic kinetic parameters. This switching is taken into account in the MM-model by adding a thermodynamic “dimension” to the classical MM-model, as is shown in Figure 9.15. Every enzyme intermediate (*E*, *ES* . . .) of the kinetic component has its own thermodynamic equilibrium between the different conformational states, and in addition, memory effects from substrate imprinting come into play. This means that the binding of an appropriate substrate can induce a conformational change that is “remembered” over a certain time, thereby allowing easier substrate binding for the next turnovers. As a result, enzymes show non-Markovian behavior: the activity at a certain point in time is strongly dependent on what happened before.

All this has some important implications on the distribution of the waiting times between successive turnovers. These waiting times can be derived after thresholding the intensity time transient recorded for a single enzyme (see Section 9.4). For a non-fluctuating, static enzyme, a single exponential decay of the probability density function (pdf) of waiting times is theoretically expected (under conditions of a first-order rate limiting step). The pdf or probability density function $f(x)$ of the waiting times is the continuous form of the waiting time histogram, normalized such that $\int_{-\infty}^{+\infty} f(x)dx = 1$. The probability that a waiting time has a value between a and b is then given by $\int_a^b f(x)dx$.

If the enzyme is switching between two possible conformational states, the waiting time pdf will show double exponential behavior. In the general case of a quasi-continuum of conformers, one would expect an infinite sum of exponential terms, which can be represented by the following integral:

$$P_{\text{off-times}} = \int_0^\infty A(\tau)\exp\left(-\frac{t}{\tau}\right)d\tau \qquad (9.9)$$

Mathematically, this situation can be modeled by a stretched exponential decay of the pdf:

$$P_{\text{off-times}} = A \cdot \exp\left[-\left(\frac{t}{\tau_{\text{off}}}\right)^{\alpha}\right] \tag{9.10}$$

The equation is fairly similar to a mono-exponential function, except for the exponent α, which represents a value between 0 and 1. The more α approaches zero, the more the decay is stretched over several orders of magnitude (Figure 9.16). For instance, Velonia *et al.* proved a stretched exponential decay for the waiting time pdf of single CALB enzymes with an α-factor of 0.15 [18]. β-D-galactosidase also shows a significant tailing of the waiting time pdf, fitted by English *et al.* with a stretching parameter of 0.4 [20].

The drawback of analyzing the pdf waiting times is that, in the time-scrambled histograms, all time-correlated information on the dynamics is lost. This problem can be overcome by plotting the waiting times as a function of their event index. As illustrated in Figure 9.17 one can distinguish groups of correlated events, originating from conformational dynamics of the enzyme. Another way of analyzing such correlation effects in the waiting times is the construction of 2D-autocorrelation graphs of successive turnovers. Such plots are a graphical representation of all waiting-time couples separated by n turnovers. Starting from an array of the successive waiting times of an individual enzyme, the x-axis represents the xth waiting time, while the y-axis represents waiting time $x+n$. A high density of points among the diagonal of the plot indicates a high degree of correlation. This means that waiting time $x+n$ has a high probability of being similar to waiting time x, because the enzyme is still in the same conformation over this time period. After longer times, conformational fluctuations will have happened and the kinetic properties will have changed, so less correlation is expected when n increases (Figure 9.18).

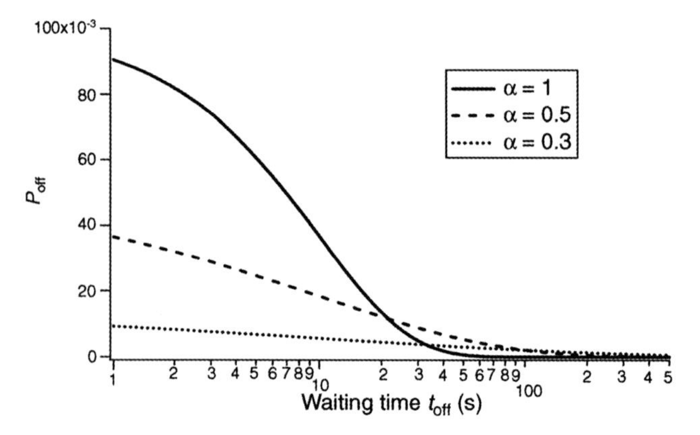

Figure 9.16 The stretched exponential pdf. When α equals 1, the pdf equals a single exponential distribution. The more α approaches zero, the more the distribution is stretched over higher orders of magnitude.

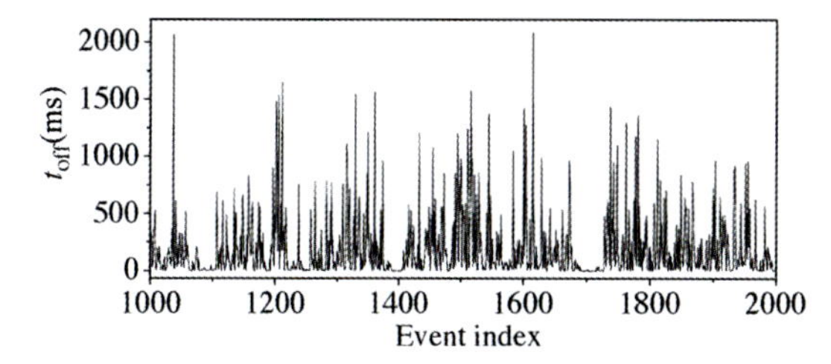

Figure 9.17 Off-times between successive turnovers as a function of their event index. Blocks of off-times of similar magnitude can be identified indicating correlation effects in the successive off-times.

9.3.2 Conformational Dynamics

As enzymatic activity fluctuations are attributed to conformational dynamics, extensive research on the exact mechanisms of these dynamics is required to unambiguously relate the two phenomena to each other.

Spontaneous protein backbone movements and oscillations have been proven through theoretical molecular dynamics studies, but also by experimental results [41]. For instance, ^{15}N-spin relaxation in proteins is influenced by the backbone dynamics and can be studied by NMR [42]. Other commonly used techniques for studying dynamic processes concerning protein conformations are (quasi-)inelastic neutron scattering, which records the energy-resolved thermal diffuse scattering of biomolecules and Mössbauer spectroscopy, in which Doppler effects due to backbone movements cause inelastic scattering of gamma radiations.

The abovementioned techniques all have one large drawback: they are based on bulk samples of the protein and therefore only contain information on the average

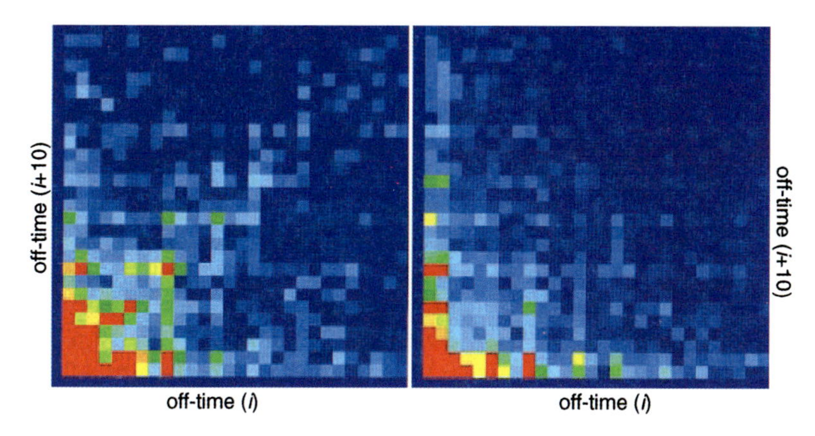

Figure 9.18 2D-correlation plot of the waiting times between successive turnovers (left) and of the waiting times separated by ten turnovers (right). For successive turnovers a clear correlation is observed, while this correlation is lost over ten or more turnovers. (Copyright Lu (1998) *Science*, **282** (5395), 1877–1882 [14].)

mobility of the protein chain. More relevant information can be acquired by studying the conformational dynamics of one single protein. Fluorescence resonance energy transfer (FRET) (which is the Förster-mechanism of energy transfer, and is proportional to the inverse of the sixth power of the distance between donor and acceptor) and electron transfer studies of a fluorophore–quencher system are very useful for this purpose, as distances ranging from a few Ångström to a few nanometer can be probed by tracking the quenching efficiency. Note that the electron-transfer efficiency exhibits an exponential dependence on distance between the fluorophore and quencher of $k_{\mathrm{ET}} = k^{0}_{\mathrm{ET}} \cdot \exp(-\beta \cdot R)$.

The group working with Xie studied Ångström-scale structural changes in single flavin reductase (Fre) enzymes [43]. The fluorescent FAD-cofactor is quenched due to electron transfer with a nearby tyrosin residue. The multidimensional decay of the fluorescence lifetime of the FAD–Fre complex indeed indicated conformational fluctuations. As the lifetimes are strongly affected by quenching, electron transfer can be monitored by lifetime analysis in a single-photon timing experiment.

For probing distances at the nanometer-scale, FRET studies are more convenient than electron transfer. Ha *et al.* probed millisecond fluctuations of the conformation of individual *Staphylococcal nuclease* using FRET, by labeling the enzyme specifically with TMR and Cy5 as the FRET pair [29]. An overview of important studies on the conformational dynamics of individual protein molecules can be found in the reviews by Blank *et al.* [5] and by Smiley and Hammes [44] (and references therein).

Single-molecule conformational studies proved the existence of the same type of dynamic disorder in the enzymatic conformation as in the enzymatic activity, indicating that these conformational fluctuations are the origin of the dynamic disorder in the activity. However, the quest for the Holy Grail in this field is still open: the *in situ* simultaneous observation of turnovers and their fluctuations together with conformational dynamics. Although the concept might seem fairly easy, there are lots of practical difficulties to overcome. Multiple labeling of the enzyme with powerful and photostable fluorophores and multiple color excitation and detection will be required.

9.3.3
Single-Molecule DNA Sequencing

The single-turnover sensitivity in detecting enzymatic activity has recently been applied for DNA sequencing. One of the first successful attempts was reported by Werner and coworkers [45]. In their work, DNA-bases were labeled by tetramethyl rhodamine (TMR) and the sequence was anchored on a microsphere in a flow upstream of the ultrasensitive detector region of a flow cytometer, and a dilute concentration of the exonuclease enzyme. When an enzyme binds to one of the DNA strands at the microsphere, the progressive digestion of the strand starts, and the labeled nucleotides flow towards the detector region where they are detected with single-molecule sensitivity (Figure 9.19). In this way, not only information on the exonuclease is obtained, but also the nucleotide sequence of the substrate can be investigated.

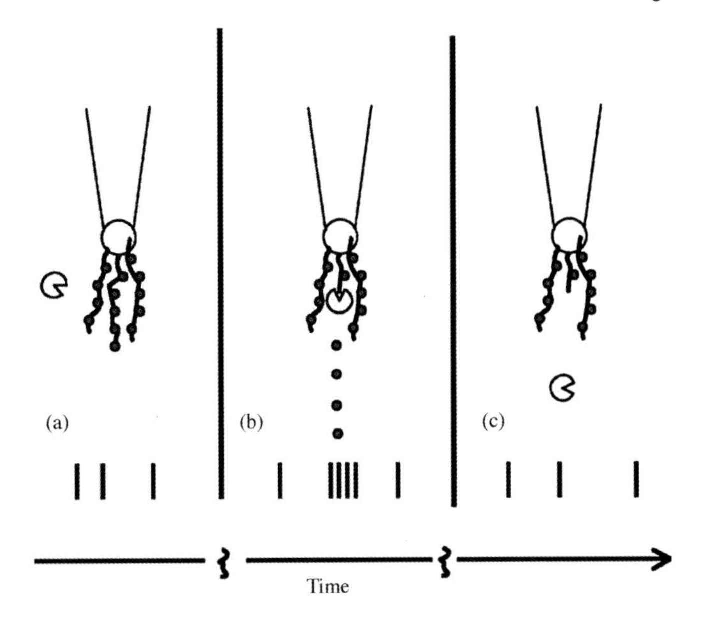

Figure 9.19 Scheme of single-molecule DNA sequencing. The fluorescently labeled nucleotides released by a progressive nuclease enzyme are lead by a solvent flow towards a fluorescence detector where they are identified one by one. (Copyright Werner *et al.* (2003) *J. Biotechnol.*, **102** (1), 1–14 [45].)

Since this pioneering approach, the research on single molecule DNA sequencing has taken a tremendous leap forward. Nowadays two strategies are operational. One is the so-called true Single Molecule Sequencing (tSMS) from Helicos Biosciences [46], while another strategy, Single Molecule Real Time Sequencing (SMRT) has been launched by Pacific Biosciences [47]. The former method (tSMS) relies on tethering a piece of ss-DNA (in their case from a M13 viral genome) on a substrate, after which the sample is incubated with one type of fluorescently labeled nucleotide and polymerase, followed by rinsing and imaging [46, 48]. A distinct fluorescent spot in the sample indicates that the first “free” nucleotide of the DNA strand is the complementary one for the fluorescently labeled nucleotide that was used for incubation. After imaging, the label is removed by chemical cleavage. Many of these cycles, each with incubation of a specific type of fluorescently labeled nucleotide, finally leads to the sequence of the complete DNA strand. Of course, because of the many necessary cycles of incubation and imaging, this technique is practically limited to rather short DNA-pieces. Typical read-lengths are in the order of 30–35 bases.

The SMRT approach is applicable on much larger DNA strands and is less time- and labor consuming, but has the major drawback of being extremely challenging from an experimental point of view. In this approach, individual polymerase enzymes are incorporated into an array of zero-mode waveguides (ZMWs) [47]. These ZMWs consist of cylindrical nano-sized holes in an aluminum film. Inside these waveguides the effective detection volume for fluorescence is several orders of magnitude smaller (approximately 20 zeptoliters) than in conventional confocal fluorescence microscopy, and therefore only fluorophores present at the active site of the polymerase will be visualized. Aside from these ZMWs, SMRT relies on the use of nucleotides with a fluorophore linked to the terminal phosphate. This minimizes the influence of the fluorophore on incorporation of the nucleotide into the DNA strand and assures that, as a natural step in the polymerase reaction, the fluorophore is released upon incorporation, by cleavage of the phosphate chain. For each type of nucleotide a fluorophore with different spectral characteristics is used. Thus, by recording fluorescence spectra from the position of a ZMW-bound polymerase enzyme, the sequence of an attached ss-DNA molecule can be sequenced by measuring the time-dependent spectral changes (see Figure 9.20). It was demon-

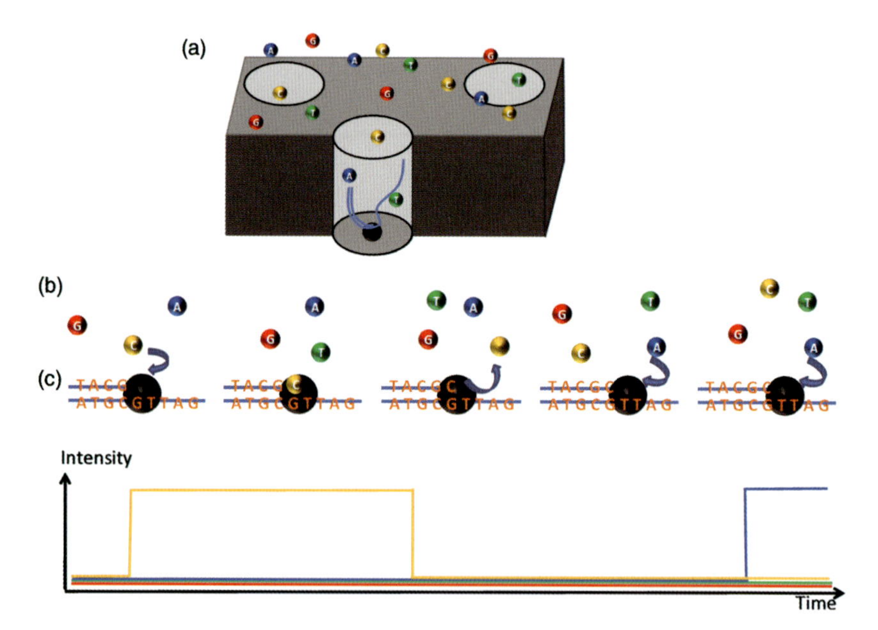

Figure 9.20 Principle of the Single Molecule Real Time (SMRT) sequencing approach. (a) Experimental geometry. A single polymerase enzyme, with a bound DNA template is immobilized at the bottom of a zero-mode waveguide (ZMW). The ZMW nanostructure provides an excitation volume in the zeptoliter range, enabling detection of individual phospholinked nucleotide substrates as they are incorporated into the DNA strand. (b) Overview of the successive events during nucleotide incorporation, together with the expected time trace of fluorescence intensity for the different colors of the four nucleotides. Firstly, a labeled nucleotide associates (1–2) with the template DNA at the active site of the polymerase, yielding an increased fluorescence in the corresponding detection channel. Upon incorporation, the terminal phosphate chain is cleaved, resulting in diffusing of the label out of the detection volume (3). The intensity decreases to its background level. Next, the DNA strand moves through the polymerase enzyme (4) so that the next position is ready for association with its complementary nucleotide (5). (c) Example of the recorded fluorescence spectra as a function of time for an ss DNA template designed for incorporation of alternating blocks of dCTP and dGTP.

strated that read-lengths of up to 1500 bases could be achieved with extremely high accuracy. More information on single molecule DNA Sequencing can be found in the recent review by Xu *et al.* [49].

9.3.4
Shedding Light on Single-Enzyme Mechanisms

Aside from unraveling dynamic fluctuations in enzyme kinetics, thereby adding a new dimension to the traditional MM-model, single-molecule enzyme studies have also proved their value by revealing some of the mechanisms of catalytic action. To illustrate how single-molecule fluorescence microscopy can contribute to a deeper understanding of enzyme mechanisms, two examples are highlighted below.

9.3.4.1 Movement of Molecular Motor Enzymes on Actin Filaments

Molecular motor enzymes convert the chemical energy released from ATP-hydrolysis into mechanical movement on actin filaments, thereby playing an important role in the functioning of the cytoskeleton of a cell. Recently, the mechanism of the moving of myosin and kinesin, two examples of molecular motors, over an actin filament has been extensively studied [50–54]. Kinesin and myosin consist of heads, held together by a coiled stalk. Each head contains a catalytic domain at the end and a light chain that connects the catalytic domain with the stalk. Such a conformation generally allows for two distinct mechanisms for movement along the actin filaments: the hand-over-hand and the inchworm movement (Figure 9.21). By labeling one of the light chains of a single myosin enzyme with a fluorescent dye, Yildiz and coworkers could assign the hand-over-hand movement as being the appropriate mechanism for myosin V "walking" along the actin filaments [52]. As depicted in Figure 9.21 a hand-over-hand movement implies a movement of the fluorescent label by alternating steps equal to a stalk step size $\pm 2x$, where x is the distance along the movement direction between the center of mass of the myosin and the label. An inchworm mechanism, on the other hand, will give rise to steps of equal distance a. By fitting the point-spread-function (psf) of a single dye, as obtained in the wide-filed image by diffraction-limited spots, with a 2D Gaussian, the dye can be localized with nanometer accuracy. This approach is termed 'Fluorescence Imaging with One-Nanometer Accuracy' (FIONA) and allows for the tracking of small displacements of the enzymes. These workers indeed found alternating steps of from 37 ± 15 to 37 ± 5 nm, depending on the location of the label on the light chain. These results strongly support a hand-over-hand movement. Subsequently, this approach was elaborated further by using defocused imaging in order to map the three-dimensional orientation of the dye while walking over the actin substrate [55].

9.3.4.2 Lipase-Catalyzed Hydrolysis of Phospholipid Bilayers

Traditionally, enzyme kinetics are described by the MM-model, where substrate and enzyme are assumed to be solubilized, although the model still holds for immobilized enzymes. However, for an immobilized substrate on which a solubilized enzyme has to act, other models have to be derived. Because enzymes concentrations

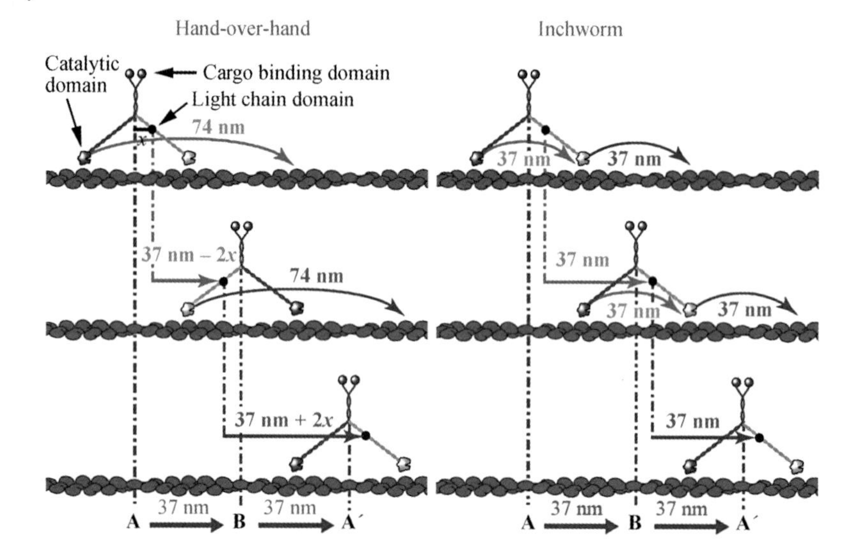

Figure 9.21 Two mechanisms for walking of myosin enzymes on actin filaments. By labeling one of the light chains and tracking the label's position with nanometer accuracy and with sub-turnover time resolution, one can distinguish between different walking mechanisms (inchworm or hand-over-hand). In the hand-over-hand mechanisms the step size is alternatively 37 nm $\pm$ 2*x*. For an inchworm movement the step size is always 37 nm. (Reproduced from Yildiz *et al.*, (2003) *Science*, **300** (5628), 2061–2065 [52].)

in their natural environments are typically very low, diffusion limitations can occur, resulting in complex kinetics.

Very recently, the action of lipolytic enzymes on immobilized monolayers of the substrate (phospholipids) has been investigated using wide-field microscopy [56–58]. Upon phospholipase action, the phospholipids are hydrolyzed, resulting in desorption of the lipids from the bilayer. By incorporating amphiphilic carbocyanin dyes in the layer, the phospholipase activity can be visualized in wide-field fluorescence microscopy by the appearance of dark areas (Figure 9.22a). For the phospholipase A1 (PLA1), three morphological appearances of enzymatic digestion of a POPC (1-palmitoyl-2-oleoyl-sn-glycero-3-phosphocholine) bilayer could be distinguished, depending on the amount of enzyme present on the bilayer. Although interesting kinetic phenomena can be examined in this way [56], these data are based on the action of multiple lipase molecules at the same time.

By using labeled enzymes, one can follow one single enzyme on a monolayer [58]. Moreover, it is possible to image individual phospholipase molecules acting on the bilayers while visualizing local structural changes to the substrate layer (Figure 9.22b). Analysis of the trajectories from single phospholipases allows correlating the mobility of the enzyme with its catalytic activity. The study of related enzymes with different activities and binding preferences provided insight into the various diffusive motions of the phospholipase at different stages in its catalytic cycle. By employing novel, high resolution image reconstruction methods, "hotspots" of

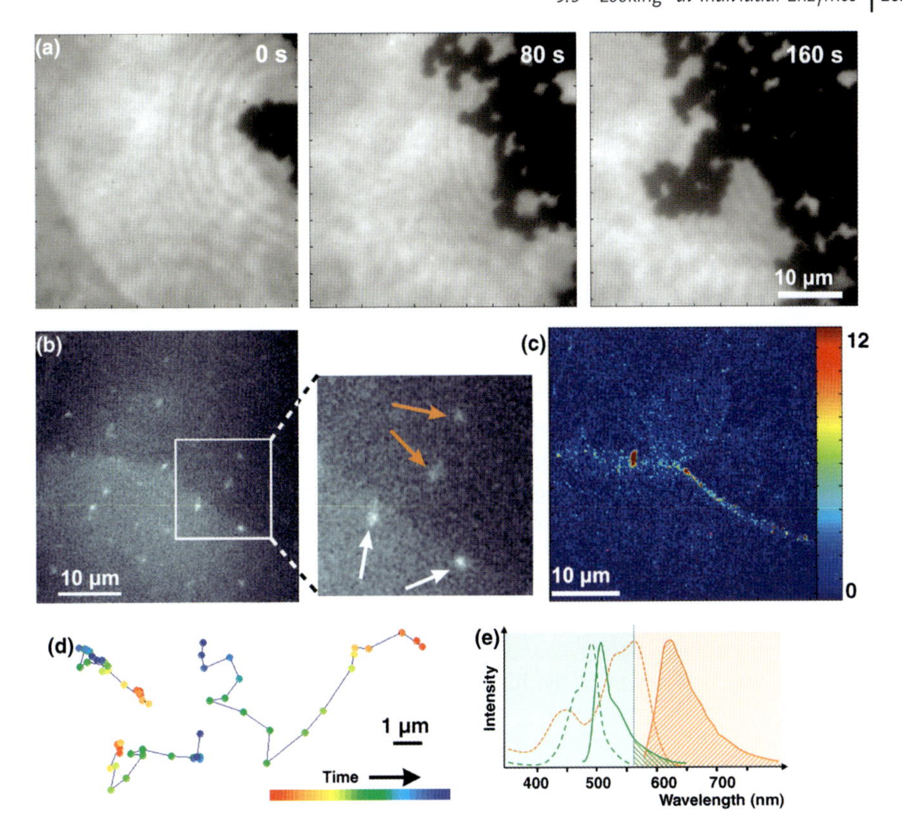

Figure 9.22 Hydrolysis of phospholipid layers by PLA1 followed by fluorescence wide-field microscopy. (a) Time-resolved fluorescence images of POPC layers labeled with DiI at progressive stages of hydrolysis. A collapse of the first bilayer in contact with the support is observed. $[PLA1] = 1.9 \times 10^{-7}$M. (b) PDI-labeled enzyme on phospholipid multilayers labeled with DiO. Discrimination of single enzymes is possible (indicated by arrows in the magnified part of the image). The magnification shows enzymes slowly diffusing on the edge (white arrows) and enzymes diffusing faster on the layer (orange arrows). (c) Histogram-based image reconstruction showing the histogram of the spatial distribution of enzyme molecules on the substrate. The probability of enzyme localization is indicated by the color bar. (d) Typical trajectories of individual aPLA1 molecules showing a heterogeneous diffusion behavior. (e) Steady-state absorption (dashed line) and emission (solid line) spectra for the DiO (green) and PDI (red). The excitation wavelengths used for each dye are indicated by arrows at the bottom. The best ratio of detected emission from the layers and from the enzyme molecules is accomplished by the use of an appropriate long pass cut-off filter (cut-off wavelength indicated by the blue dashed line).

single enzyme activity could be visualized and related to the local substrate structure (Figure 9.22c). A similar approach can easily be applied to essentially all types of interfacial enzymes and also other processes occurring at membranes, for example, active transport, signaling processes and so on.

9.4
Data Analysis of Fluorescence Intensity Time Traces of Single-Turnover Experiments

9.4.1
Threshold Method

Analyzing fluorescence intensity time traces of single turnover experiments is more than just counting peaks. After all, the key question is: when can we consider an intensity burst as a peak resulting from a catalytic turnover? Because signal-to-noise ratios (SNR) are often not that high when measuring in aqueous solutions, the determination of the threshold for distinguishing the signal from the noise is crucial for a kinetic analysis. To illustrate this method a trace with a very high SNR is chosen as an example (Figure 9.23). The trace clearly consists of very intense peaks superimposed on a low background level. In this case it can even be seen by eye that it is very easy to assign a threshold level. In a histogram of the counts, two peaks will be visible: one originating from noise counts and at higher counts a peak completely separated from the signal. The threshold simply has to be positioned in between both peaks. However, in most instances the SNR is much lower, resulting in two overlapping peaks in the intensity histogram. In this case one has to search for a good compromise between counting some of the noise and losing the signal. The simplest way is to try to fit the histogram with a superposition of two Gaussian

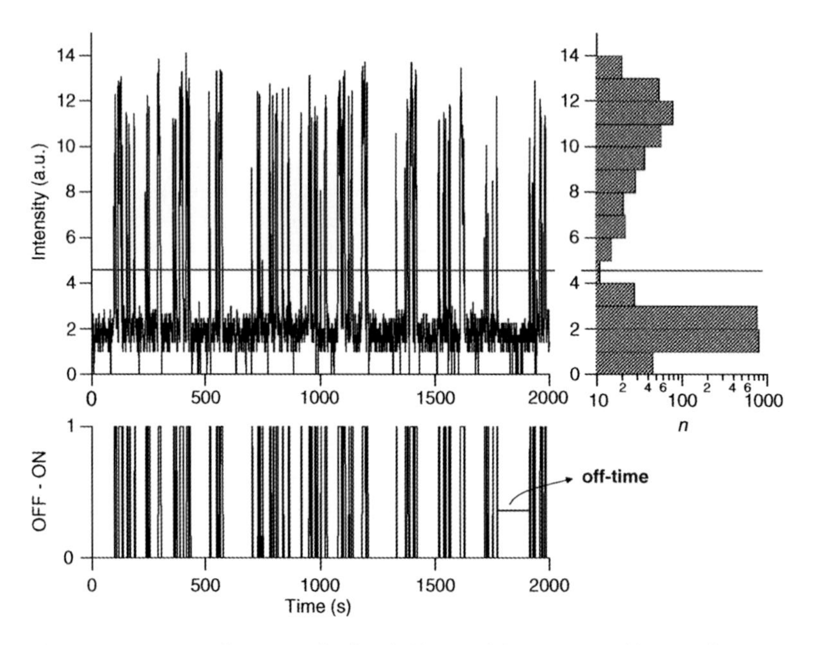

Figure 9.23 Trace analysis using the threshold method. By positioning the threshold between the noise and the signal level, based on the intensity histogram, a digitized time trace can easily be constructed by considering intensities higher or lower than the threshold as one and zero, respectively.

distributions, and to assign the threshold at the intersection of the two distributions. This is the most intuitive way of analyzing data, but often introduces artifacts due to inaccurate discrimination between the signal and noise. Recently, more advanced methods have been proposed, which are beyond the scope of this chapter [19, 61–65].

After having determined an appropriate threshold level, the intensity time trace can be easily converted into an on–off trace, as is shown in Figure 9.23. Everything above the threshold is considered to be "on", while the rest is "off". The "off-times" then correspond to the waiting times between the successive turnovers, and a profound analysis of these waiting times will reveal the detailed kinetic properties of the enzyme.

9.4.2 Autocorrelation Analysis

As is explained in the previous section, determination of the threshold level is not always very straightforward. To circumvent this problem, one can analyze the intensity trace using autocorrelation. The common equation for the autocorrelation function is:

$$C(\tau) = \frac{\langle I(t) \cdot I(t+\tau) \rangle}{\langle I(t) \rangle^2} \tag{9.11}$$

This function describes how much the intensity at a certain time is correlated with the intensity after a lag time τ, as a function of this parameter τ. In fact, when the detected photons originate from enzymatic turnovers, $C(\tau)$ is related to the probability that a turnover will occur at $t = \tau$, given that a turnover occurred at $t = 0$. This means that the autocorrelation function immediately returns the pdf of the waiting times, without the need for assigning a threshold level to distinguish between noise and signal.

Thus, autocorrelation seems to be advantageous compared with the threshold method, although the technique also has its limitations. In reality, other processes also influence the autocorrelation function. For instance, the small amount of free dye in the bulk solution can cause small intensity bursts, as every now and then a single dye molecule will diffuse through the focus volume. Instead of distinguishing between signal and noise by assigning a threshold level in the first analysis method, in autocorrelation a distinction has to be made between the turnover events and side-effects, such as diffusion, by trying to identify the time scale over which these processes occur. When there is a sufficient difference in the characteristic time scales of the various processes, a clear separation can be made.

9.5 Conclusions

Single-molecule enzymology has given scientists many new insights into the functioning of these miraculous biocatalysts. Studies on conformational dynamics and single-turnover resolved activity have illustrated the complex behavior of single enzymes. Nowadays, a lot of attention is paid to the careful statistical analysis of the recorded traces, in order to extract the maximum amount of information as possible on the enzyme's energy landscape. Some issues on the mechanism of enzyme action have also been successfully addressed thanks to SMFS. Moreover the example of single DNA sequencing proves that this technique is extremely versatile and can be applied in many ways in biochemical research. However, single-molecule spectroscopy is in no way restricted to biochemical research topics. Very recently, inspired by the single-enzyme studies, SMFS has been successfully applied for *in situ* investigation of heterogeneous inorganic catalysts [59, 60]. As the scope of applications in catalysis rapidly expands, it can be expected that SMFS will very soon become an important analytical tool in catalytic research in general.

References

1 Engelkamp, H., Hatzakis, N.S., Hofkens, J., De Schryver, F.C., Nolte, R.J.M., and Rowan, A.E. (2006) Do enzymes sleep and work? *Chem. Commun.* (9), 935–940.

2 Rotman, B. (1961) Measurement of activity of single molecules of β-D-galactosidase. *Proc. Natl. Acad. Sci. USA*, **47**, 1981–1991.

3 Funatsu, T., Harada, Y., Tokunaga, M., Saito, K., and Yanagida, T. (1995) Imaging of single fluorescent molecules and individual ATP turnovers by single myosin molecules in aqueous-solution. *Nature*, **374** (6522), 555–559.

4 Berg, J.M., Tymockzo, J.L., and Stryer, L. (2002) *Biochemistry Internatial Edition*, 5th edn, Freemann, San Fransisco.

5 Blank, K., De Cremer, G., and Hofkens, J. (2009) Fluorescence-based analysis of enzymes at the single-molecule level. *Biotechnol. J.*, **4**, 465–479.

6 Chen, Q., Groote, R., Schonherr, H., and Vancso, G.J. (2009) Probing single enzyme kinetics in real-time, *Chem. Soc. Rev.*, **38** (9), 2671–2683.

7 Lee, A.I. and Brody, J.P. (2005) Single-molecule enzymology of chymotrypsin using water-in-oil emulsion. *Biophys. J.*, **88** (6), 4303–4311.

8 Xue, Q.F. and Yeung, E.S. (1995) Differences in the chemical-reactivity of individual molecules of an enzyme. *Nature*, **373** (6516), 681–683.

9 Craig, D.B., Arriaga, E.A., Wong, J.C.Y., Lu, H., and Dovichi, N.J. (1996) Studies on single alkaline phosphatase molecules: Reaction rate and activation energy of a reaction catalyzed by a single molecule and the effect of thermal denaturation - The death of an enzyme. *J. Am. Chem. Soc.*, **118** (22), 5245–5253.

10 Rondelez, Y., Tresset, G., Tabata, K.V., Arata, H., Fujita, H., Takeuchi, S., and Noji, H. (2005) Microfabricated arrays of femtoliter chambers allow single molecule enzymology. *Nat. Biotechnol.*, **23** (3), 361–365.

11 Gorris, H.H., Rissin, D.M., and Walt, D.R. (2007) Stochastic inhibitor release and binding from single-enzyme molecules. *Proc. Natl. Acad. Sci. USA*, **104** (45), 17680–17685.

12 Shoemaker, G.K., Juers, D.H., Coombs, J.M.L., Matthews, B.W., and Craig, D.B. (2003) Crystallization of beta-galactosidase does not reduce the range of activity of individual molecules. *Biochemistry*, **42** (6), 1707–1710.

13 Tan, W.H. and Yeung, E.S. (1997) Monitoring the reactions of single enzyme molecules and single metal ions. *Anal. Chem.*, **69** (20), 4242–4248.

14 Lu, H.P., Xun, L.Y., and Xie, X.S. (1998) Single-molecule enzymatic dynamics. *Science*, **282** (5395), 1877–1882.

15 Antikainen, N.M., Smiley, R.D., Benkovic, S.J., and Hammes, G.G. (2005)

Conformation coupled enzyme catalysis: Single-molecule and transient kinetics investigation of dihydrofolate reductase. *Biochemistry*, **44** (51), 16835–16843.

16 Brender, J.R., Dertouzos, J., Ballou, D.P., Massey, V., Palfey, B.A., Entsch, B., Steel, D.G., and Gafni, A. (2005) Conformational dynamics of the isoalloxazine in substrate-free p-hydroxybenzoate hydroxylase: Single-molecule studies. *J. Am. Chem. Soc.*, **127** (51), 18171–18178.

17 Edman, L., Foldes-Papp, Z., Wennmalm, S., and Rigler, R. (1999) The fluctuating enzyme: a single molecule approach. *Chem. Phys.*, **247** (1), 11–22.

18 Velonia, K., Flomenbom, O., Loos, D., Masuo, S., Cotlet, M., Engelborghs, Y., Hofkens, J., Rowan, A.E., Klafter, J., Nolte, R.J.M., and de Schryver, F.C. (2005) Single-enzyme kinetics of CALB-catalyzed hydrolysis. *Angew. Chem. Int. Ed.*, **44** (4), 560–564.

19 Flomenbom, O., Velonia, K., Loos, D., Masuo, S., Cotlet, M., Engelborghs, Y., Hofkens, J., Rowan, A.E., Nolte, R.J.M., Van der Auweraer, M., de Schryver, F.C., and Klafter, J. (2005) Stretched exponential decay and correlations in the catalytic activity of fluctuating single lipase molecules. *Proc. Natl. Acad. Sci. USA*, **102** (7), 2368–2372.

20 English, B.P., Min, W., van Oijen, A.M., Lee, K.T., Luo, G.B., Sun, H.Y., Cherayil, B.J., Kou, S.C., and Xie, X.S. (2006) Ever-fluctuating single enzyme molecules: Michaelis-Menten equation revisited. *Nat. Chem. Biol.*, **2** (2), 87–94.

21 De Cremer, G., Roeffaers, M.B.J., Baruah, M., Sliwa, M., Sels, B.F., Hofkens, J., and De Vos, D.E. (2007) Dynamic disorder and stepwise deactivation in a chymotrypsin catalyzed hydrolysis reaction. *J. Am. Chem. Soc.*, **129** (50), 15458–15459.

22 Martinez, V.M., De Cremer, G., Roeffaers, M.B.J., Sliwa, M., Baruah, M., De Vos, D.E., Hofkens, J., and Sels, B.F. (2008) Exploration of single molecule events in a haloperoxidase and its biomimic: Localization of halogenation activity. *J. Am. Chem. Soc.*, **130** (40), 13192–13193.

23 Carter-Franklin, J.N. and Butler, A. (2004) Vanadium bromoperoxidase-catalyzed biosynthesis of halogenated marine natural products. *J. Am. Chem. Soc.*, **126** (46), 15060–15066.

24 Gribble, G.W. (1998) Naturally occurring organohalogen compounds. *Acc. Chem. Res.*, **31** (3), 141–152.

25 Yarnell, A. (2006) Nature's X-factors. *Chem. Eng. News*, **84** (21), 12–18.

26 Rajagopalan, P.T.R., Zhang, Z.Q., McCourt, L., Dwyer, M., Benkovic, S.J., and Hammes, G.G. (2002) Interaction of dihydrofolate reductase with methotrexate: Ensemble and single-molecule kinetics. *Proc. Natl. Acad. Sci. USA*, **99** (21), 13481–13486.

27 Zhang, Z.Q., Rajagopalan, P.T.R., Selzer, T., Benkovic, S.J., and Hammes, G.G. (2004) Single-molecule and transient kinetics investigation of the interaction of dihydrofolate reductase with NADPH and dihydrofolate. *Proc. Natl. Acad. Sci. USA*, **101** (9), 2764–2769.

28 Chen, Y., Hu, D.H., Vorpagel, E.R., and Lu, H.P. (2003) Probing single-molecule T4 lysozyme conformational dynamics by intramolecular fluorescence energy transfer. *J. Phys. Chem. B*, **107** (31), 7947–7956.

29 Ha, T.J., Ting, A.Y., Liang, J., Caldwell, W.B., Deniz, A.A., Chemla, D.S., Schultz, P.G., and Weiss, S. (1999) Single-molecule fluorescence spectroscopy of enzyme conformational dynamics and cleavage mechanism. *Proc. Natl. Acad. Sci. USA*, **96** (3), 893–898.

30 Kuznetsova, S., Zauner, G., Aartsma, T.J., Engelkamp, H., Hatzakis, N., Rowan, A.E., Nolte, R.J.M., Christianen, P.C.M., and Canters, G.W. (2008) The enzyme mechanism of nitrite reductase studied at single-molecule level. *Proc. Natl. Acad. Sci. USA*, **105** (9), 3250–3255.

31 Shi, J., Palfey, B.A., Dertouzos, J., Jensen, K.F., Gafni, A., and Steel, D. (2004) Multiple states of the Tyr318Leu mutant of dihydroorotate dehydrogenase revealed by single-molecule kinetics. *J. Am. Chem. Soc.*, **126** (22), 6914–6922.

32 Shi, J., Dertouzos, J., Gafni, A., Steel, D., and Palfey, B.A. (2006) Single-molecule kinetics reveals signatures of half-sites

reactivity in dihydroorotate dehydrogenase A catalysis. *Proc. Natl. Acad. Sci. USA*, **103** (15), 5775–5780.

33 Comellas-Aragones, M., Engelkamp, H., Claessen, V.I., Sommerdijk, N.A., Rowan, A.E., Christianen, P.C., Maan, J.C., Verduin, B.J., Cornelissen, J.J., and Nolte, R.J. (2007) A virus-based single-enzyme nanoreactor. *Nat. Nanotechnol.*, **2** (10), 635–639.

34 Edman, L. and Rigler, R. (2000) Memory landscapes of single-enzyme molecules. *Proc. Natl. Acad. Sci. USA*, **97** (15), 8266–8271.

35 Hassler, K., Rigler, P., Blom, H., Rigler, R., Widengren, J., and Lasser, T. (2007) Dynamic disorder in horseradish peroxidase observed with total internal reflection fluorescence correlation spectroscopy. *Opt. Express*, **15** (9), 5366–5375.

36 Hatzakis, N.S., Engelkamp, H., Velonia, K., Hofkens, J., Christianen, P.C.M., Svendsen, A., Patkar, S.A., Vind, J., Maan, J.C., Rowan, A.E., and Nolte, R.J.M. (2006) Synthesis and single enzyme activity of a clicked lipase-BSA hetero-dimer. *Chem. Commun.*, 19, 2012–2014.

37 van Oijen, A.M., Blainey, P.C., Crampton, D.J., Richardson, C.C., Ellenberger, T., and Xie, X.S. (2003) Single-molecule kinetics of lambda exonuclease reveal base dependence and dynamic disorder. *Science*, **301** (5637), 1235–1238.

38 van Oijen, A.M. (2007) Honey, I shrunk the DNA: DNA length as a probe for nucleic-acid enzyme activity. *Biopolymers*, **85** (2), 144–153.

39 Kim, S., Blainey, P.C., Schroeder, C., and Xie, X.S. (2007) Multiplexed single-molecule assay for enzymatic activity on flow-stretched DNA. *Nat. Methods*, **4**, 397–399.

40 Kou, S.C., Cherayil, B.J., Min, W., English, B.P., and Xie, X.S. (2005) Single-molecule Michaelis-Menten equations. *J. Phys. Chem. B*, **109** (41), 19068–19081.

41 Karplus, M. and McCammon, J.A. (2002) Molecular dynamics simulations of biomolecules. *Nat. Struct. Biol.*, **9** (9), 646–652.

42 Kay, L.E. (1998) Protein dynamics from NMR. *Nat. Struct. Biol.*, **5**, 513–517.

43 Yang, H., Luo, G.B., Karnchanaphanurach, P., Louie, T.M., Rech, I., Cova, S., Xun, L.Y., and Xie, X.S. (2003) Protein conformational dynamics probed by single-molecule electron transfer. *Science*, **302** (5643), 262–266.

44 Smiley, R.D. and Hammes, G.G. (2006) Single molecule studies of enzyme mechanisms. *Chem. Rev.*, **106** (8), 3080–3094.

45 Werner, J.H., Cai, H., Jett, J.H., Reha-Krantz, L., Keller, R.A., and Goodwin, P.M. (2003) Progress towards single-molecule DNA sequencing: a one color demonstration. *J. Biotechnol.*, **102** (1), 1–14.

46 Harris, T.D., Buzby, P.R., Babcock, H., Beer, E., Bowers, J., Braslavsky, I., Causey, M., Colonell, J., Dimeo, J., Efcavitch, J.W., Giladi, E., Gill, J., Healy, J., Jarosz, M., Lapen, D., Moulton, K., Quake, S.R., Steinmann, K., Thayer, E., Tyurina, A., Ward, R., Weiss, H., and Xie, Z. (2008) Single-molecule DNA sequencing of a viral genome. *Science*, **320** (5872), 106–109.

47 Eid, J., Fehr, A., Gray, J., Luong, K., Lyle, J., Otto, G., Peluso, P., Rank, D., Baybayan, P., Bettman, B., Bibillo, A., Bjornson, K., Chaudhuri, B., Christians, F., Cicero, R., Clark, S., Dalal, R., Dewinter, A., Dixon, J., Foquet, M., Gaertner, A., Hardenbol, P., Heiner, C., Hester, K., Holden, D., Kearns, G., Kong, X.X., Kuse, R., Lacroix, Y., Lin, S., Lundquist, P., Ma, C.C., Marks, P., Maxham, M., Murphy, D., Park, I., Pham, T., Phillips, M., Roy, J., Sebra, R., Shen, G., Sorenson, J., Tomaney, A., Travers, K., Trulson, M., Vieceli, J., Wegener, J., Wu, D., Yang, A., Zaccarin, D., Zhao, P., Zhong, F., Korlach, J., and Turner, S. (2009) Real-time DNA sequencing from single polymerase molecules. *Science*, **323** (5910), 133–138.

48 Braslavsky, I., Hebert, B., Kartalov, E., and Quake, S.R. (2003) Sequence information can be obtained from single DNA molecules. *Proc. Natl. Acad. Sci. USA*, **100** (7), 3960–3964.

49 Xu, M., Fujita, D., and Hanagata, N. (2009) Perspectives and challenges of emerging single-molecule DNA sequencing technologies. *Small*, **5** (23), 2638–2649.

50 Nishiyama, M., Muto, E., Inoue, Y., Yanagida, T., and Higuchi, H. (2001) Substeps within the 8-nm step of the ATPase cycle of single kinesin molecules. *Nat. Cell Biol.*, **3** (4), 425–428.

51 Hua, W., Chung, J., and Gelles, J. (2002) Distinguishing inchworm and hand-over-hand processive kinesin movement by neck rotation measurements. *Science*, **295** (5556), 844–848.

52 Yildiz, A., Forkey, J.N., McKinney, S.A., Ha, T., Goldman, Y.E., and Selvin, P.R. (2003) Myosin V walks hand-over-hand: Single fluorophore imaging with 1.5-nm localization. *Science*, **300** (5628), 2061–2065.

53 Asbury, C.L., Fehr, A.N., and Block, S.M. (2003) Kinesin moves by an asymmetric hand-over-hand mechanism. *Science*, **302** (5653), 2130–2134.

54 Yildiz, A., Tomishige, M., Vale, R.D., and Selvin, P.R. (2004) Kinesin walks hand-over-hand. *Science*, **303** (5658), 676–678.

55 Toprak, E., Enderlein, J., Syed, S., McKinney, S.A., Petschek, R.G., Ha, T., Goldman, Y.E., and Selvin, P.R. (2006) Defocused orientation and position imaging (DOPI) of myosin V. *Proc. Natl. Acad. Sci. USA*, **103** (17), 6495–6499.

56 Rocha, S., Verheijen, W., Braeckmans, K., Svenson, A., Skjot, M., De Schryver, F.C., Uji, H., and Hofkens, J. (2007) Imaging of enzyme catalysis by wide field microscopy. Paper presented at Proceedings of the 3rd International Nanophotonics Symposium, Nano Biophotonics - Science and Technology Handai. July 6–8 2006, Osaka, Japan, Vol. 4, pp. 133–141.

57 Peneva, K., Mihov, G., Nolde, F., Rocha, S., Hotta, J., Braeckmans, K., Hofkens, J., Uji-I, H., Herrmann, A., and Mullen, K. (2008) Water-soluble monofunctional perylene and terrylene dyes: Powerful labels for single-enzyme tracking. *Angew. Chem. Int. Ed.*, **47** (18), 3372–3375.

58 Rocha, S., Hutchison, J.A., Peneva, K., Herrmann, A., Muellen, K., Skjot, M., Jorgensen, C.I., Svendsen, A., De Schryver, F.C., Hofkens, J., and Uji-I, H. (2009) Linking phospholipase mobility to activity by single-molecule wide-field microscopy. *ChemPhysChem*, **10** (1), 151–161.

59 Roeffaers, M.B.J., Sels, B.F., Uji-i, H., De Schryver, F.C., Jacobs, P.A., De Vos, D.E., and Hofkens, J. (2006) Spatially resolved observation of crystal-face-dependent catalysis by single turnover counting. *Nature*, **439** (7076), 572–575.

60 Roeffaers, M.B.J., Sels, B.F., Uji-i, H., Blanpain, B., L'hoest, P., Jacobs, P.A., De Schryver, F.C., Hofkens, J., and De Vos, D.E. (2007) Space- and time-resolved visualization of acid catalysis in ZSM-5 crystals by fluorescence microscopy. *Angew. Chem. Int. Ed.*, **46** (10), 1706–1709.

61 Flomenbom, O. and Silbey, R.J. (2006) Utilizing the information content in two-state trajectories. *Proc. Natl. Acad. Sci. USA*, **103** (29), 10907–10910.

62 Baba, A. and Komatsuzaki, T. (2007) Construction of effective free energy landscape from single-molecule time series. *Proc. Natl. Acad. Sci. USA*, **104** (49), 19297–19302.

63 Li, C.B., Yang, H., and Kornatsuzaki, T. (2008) Multiscale complex network of protein conformational fluctuations in single-molecule time series. *Proc. Natl. Acad. Sci. USA*, **105** (2), 536–541.

64 Talaga, D.S. (2007) Markov processes in single molecule fluorescence. *Curr. Opin. Colloid Interface Sci.*, **12** (6), 285–296.

65 Watkins, L.P. and Yang, H. (2005) Detection of intensity change points in time-resolved single-molecule measurements. *J. Phys. Chem. B*, **109** (1), 617–628.

Index

Handbook of Fluorescence Spectroscopy and Imaging. M. Sauer, J. Hofkens, and J. Enderlein

ISBN: 978-3-527-31669-4

t